Praise for Joel Garreau and

Radical Evolution

"Captivating . . . brilliant . . . enthralling . . . dazzling."
—*The New York Times Book Review*

"Deserves serious attention for its potential to turn the world upside down and inside out in the relatively near future. . . . Great news: It's extremely readable!"
—Tom Peters, author of *In Search of Excellence: Lessons from America's Best-Run Companies*, on tompeters.com

"Wonderful." —*The Washington Times*

"So cutting edge it seems mind-boggling." —*National Geographic* Online

"Provocative . . . intriguing . . . compelling and important and could change our world forever. There's no one better than Joel Garreau to explain this . . . Garreau is an underappreciated national treasure." —*BookReporter*

"Mr. Garreau is a zesty storyteller, a gonzo futurist who builds alternative universes from solid science, and he's neither a technology booster nor a Luddite. The questions his moral quandaries raise are among the deepest questions we know how to ask: What kind of creatures are we—the apelike animals from which we evolved, or the angels we imagine we can become?"
—A *Strategy+Business* Best Business Book of the Year

"It isn't often an author gets to herald the biggest news in the last 10,000 years. But you'll get the full, uncensored, mind-blowing report here in this entertaining and surprisingly deep book. Meet soldiers who don't sleep, animals controlled with joy sticks, computers controlled by merely thinking, the blind driving cars, and parents designing their kids—and that is just what is happening right now. Veteran scout Joel Garreau prepares ordinary readers for the ultimate question of this century: Who do you think we should be? He makes it clear that as of today, human nature is now under the control of humans, and we *are* doing something about it—but we aren't aware of it. To guide you through this boggle Garreau offers astonishments, conundrums, and sanity."

> —Kevin Kelly, author of *Out of Control: The New Biology of Machines, Social Systems and the Economic World* and *New Rules for the New Economy: 10 Radical Strategies for a Connected World* and *Wired* editor-at-large

"One of the most provocative, entertaining and, yes, frightening science books in years. . . . [Garreau] is a solid researcher with a fine sense of story-telling."
> —*San Jose Mercury News*

"One of my favorite cutting-edge thinkers."
> —John Tierney, *The New York Times*

"Clarifying and companionable . . . the technoscenarios Garreau explicates are riveting, and of acute importance, as is his reminder that there is much more to life than technology, no matter how amazing it gets."
> —*Booklist*

"Courageous."
> —Barnes and Noble

"[Garreau] has a knack for asking questions that deliver unexpected answers. I have known some of the people in this book for over two decades, and yet I caught myself surprised again and again. Alone, these details would not amount to much, but collectively they are 'ground truth' in a map that Joel builds of this emergent landscape of techno-human co-evolution."
> — Paul Saffo, Global Business Network

"*Radical Evolution* is truly outstanding. I was hooked. It reads like a thriller, informs like a textbook, engages like a love note. Garreau is, truly, an artist."
> —Eamonn Kelly, author, *Powerful Times: Rising to the Challenge of Our Uncertain World*

"The truly outstanding feature of Garreau's highly literate work is its scope and complexity. . . . *Radical Evolution* doesn't just present the Big Ideas of pundits as diverse as transhumanist Raymond Kurzweil or rogue techno-luddite Theodore Kaczynski, but puts them in context with thoughts from Blaise Pascal, Henry David Thoreau, and Mark Twain."

—NuSapiens.com: Biology, Technology, Philosophy; Human Evolution in Progress

"Fascinating and disturbing." —*New York Sun*

"A magnificent job. . . . We'd better start deciding, now, what kind of world it will be." —Straight.com: The Georgia Straight, Vancouver

"Self-consciously fair, and very readable." —GreenBiz.com

"How weird, how soon? That's the question that dominates the debates about the coming of 'post-humanity.' With his customary journalistic acumen and wry humor, Garreau has the answer: much weirder than you imagine, much sooner than you expect."

—Stewart Brand, author of *The Clock of the Long Now: Time and Responsibility* and *The Media Lab: Inventing the Future at M.I.T.*

"Joel Garreau lives well ahead of the curve—even the really big Curve he describes in these pages. One of our foremost chroniclers of change and historians of the future, he's done it again."

—Bill McKibben, author of *Enough: Staying Human in an Engineered Age* and *The End of Nature*

"Joel Garreau has hit upon something critical here, something most of us see daily and struggle to make sense of: that human technology may be advancing faster than our ability to adapt, leaving us ill-equipped to measure and manage the consequences. This is a timely, important book, and a fascinating read."

—Nathan McCall, author of *Makes Me Wanna Holler: A Young Black Man in America*

OTHER BOOKS BY JOEL GARREAU

Edge City: Life on the New Frontier
The Nine Nations of North America

Radical Evolution

The Promise and Peril of Enhancing

Our Minds, Our Bodies—and

What It Means to Be Human

Joel Garreau

BROADWAY BOOKS

New York

PRINTED IN THE UNITED STATES OF AMERICA

BROADWAY BOOKS and its logo, a letter B bisected on the diagonal, are trademarks of
Random House, Inc.

Visit our Web site at www.broadwaybooks.com

First Broadway Books trade paperback edition published 2006.

Library of Congress Cataloging-in-Publication Data is on file with the Library of Congress.

ISBN-13: 978-0-7679-1503-8
ISBN-10: 0-7679-1503-8

To Roland and Gloria Garreau, who got me through the past,
Simone and Evangeline, who are guiding me into the future,
and Adrienne, who helps me survive the present.

"I was wondering when you'd notice there's lots more steps."

Contents

Radical Evolution

Prologue

The Future of Human Nature

Confusion is a word we have invented for an order which is not understood.

—Henry Miller, *Tropic of Capricorn*

THIS BOOK CAN'T BEGIN with the tale of the telekinetic monkey.

That certainly comes as a surprise. After all, how often does someone writing nonfiction get to lead with a monkey who can move objects with her thoughts?

If you lunge at this opportunity, however, the story comes out all wrong. It sounds like science fiction, for one thing, even though the monkey—a cute little critter named Belle—is completely real and scampering at Duke University.

This gulf between what engineers are actually creating today and what ordinary readers might find believable is significant. It is the first challenge to making sense of this world unfolding before us, in which we face the biggest change in tens of thousands of years in what it means to be human.

This book aims at letting a general audience in on the vast changes that right now are reshaping our selves, our children and our relationships. Helping people recognize new patterns in their lives, however, is no small trick, as I've discovered over time.

For example, there's the problem you encounter when asking people what they'd do if offered the chance to live for a very long time—150 years or more. Nine out of 10 boggle at this thought. Many actually recoil. You press on. Engineers are working on ways to allow you to spend all that time with great physical vitality—perhaps even comparable to that of today's 35-

year-olds. How would you react if that opportunity came to market? There's a question that gets people thinking, but you can tell it is still quite a stretch.

We live in remarkable times. Who could have imagined at the end of the 20th century that a human augmentation substance that does what Viagra does would sponsor the *NBC Nightly News*?

Discussing this sort of change, however, can be hard. Take the United States Department of Defense program to create the "metabolically dominant soldier." In one small part of that agenda, researchers hope to allow warriors to run at Olympic sprint speeds for 15 minutes on one breath of air. It might be indisputably true that human bodies process oxygen with great inefficiency, and this may be a solvable problem, and your taxpayer dollars unquestionably are being spent trying to remedy this oversight on the part of evolution. Nonetheless, it takes effort to hold some readers with this report. It just sounds too weird.

One fine spring evening, I found myself at a little table outside a San Francisco laundry, pondering how to bridge this divide between the real and the credible. The laundry, called Star Wash, is on a lovely but quite ordinary street. In the window there is an American flag and a sign that tastefully spells out "God Bless America" in red, white and blue lights. It is run by a woman named Olga, from Guatemala City. I was traveling, interviewing the people who are creating the vastly enhanced human abilities that *Radical Evolution* discusses, and was waiting for my shirts to be finished.

Most of the prospective readers of this book, it occurred to me, are probably like Olga. They don't care about gee-whiz technology. Why should they? Neither do I, truth be told.

What they care about is what it means to be human, what it means to have relationships, what it means to live life, to have loves, or to tell lies. If you want to engage such people, you have to tell a story about culture and values—who we are, how we got that way, where we're headed and what makes us tick. That's what has always interested me; it's what my reporting has always been about. The gee-whiz technology is just a window through which to gaze upon human nature.

———∾∾∾———

FOUR INTERRELATED, intertwining technologies are cranking up to modify human nature. Call them the GRIN technologies—the genetic, robotic, information and nano processes. These four advances are intermingling and feeding on one another, and they are collectively creating a curve of change unlike anything we humans have ever seen.

Already, enhanced people walk among us. You can see it most clearly wherever you find the keenest competition. Sport is a good example. "The current doping agony," says John Hoberman, a University of Texas authority on performance drugs, "is a kind of very confused referendum on the future of human enhancement." Extreme pharmacological sport did not begin or end with East Germany. Some athletes today look grotesque. Curt Schilling, the All-Star pitcher, in 2002 talked to *Sports Illustrated* about the major leagues. "Guys out there look like Mr. Potato Head, with a head and arms and six or seven body parts that just don't look right." Competitive bodybuilding is already divided into tested shows (i.e., drug free) versus untested shows (anything goes). That's merely the beginning. Scientists at the University of Pennsylvania who created genetically modified "mighty mice" have been deluged by calls from athletes and coaches who want to try this technology themselves. These mice are shockingly large and muscular. They are built like steers, with massive haunches and necks wider than their heads. Could such gene doping work in humans—assuming it isn't already? "Oh yeah, it's easy," H. Lee Sweeney, chairman of Penn's Department of Physiology, told *The New York Times*. "Anyone who can clone a gene and work with cells could do it. It's not a mystery. . . . You could change the endurance of the muscle or modulate the speed—all the performance characteristics. All the biology is there. If someone said, 'Here's $10 million—I want you to do everything you can think of in terms of sports,' you could get pretty imaginative."

Then there's the military. Remember the comic-book superheroes of the 1930s and 1940s, from Superman to Wonder Woman? Most of their superpowers right now either exist or are in engineering. If you can watch a car chase in Afghanistan with a Predator, you've effectively got telescopic vision. If you can figure out what's inside a cave by peering into the earth with a seismic ground pinger, you've got X-ray vision. Want super strength? At the University of California at Berkeley, the U.S. Army has got a functioning prototype exoskeleton suit that allows a soldier to carry 180 pounds as if it were only 4.4 pounds. At Natick Labs in Massachusetts, the U.S. Army imagines that such an exoskeleton suit may ultimately allow soldiers to leap tall buildings with a single bound.

"My thesis is that in just 20 years the boundary between fantasy and reality will be rent asunder," writes Rodney Brooks, director of the Artificial Intelligence Laboratory at the Massachusetts Institute of Technology. "Just five years from now that boundary will be breached in ways that are as unimaginable to most people today as daily use of the World Wide Web was 10 years ago."

We are at an inflection point in history.

For all previous millennia, our technologies have been aimed outward, to control our environment. Starting with fire and clothes, we looked for ways to ward off the elements. With the development of agriculture we controlled our food supply. In cities we sought safety. Telephones and airplanes collapsed distance. Antibiotics kept death-dealing microbes at bay.

Now, however, we have started a wholesale process of aiming our technologies inward. Now our technologies have started to merge with our minds, our memories, our metabolisms, our personalities, our progeny and perhaps our souls. Serious people have embarked on changing humans so much that they call it a new kind of engineered evolution—one that we direct for ourselves. "The next frontier," says Gregory Stock, director of the Program on Medicine, Technology and Society at the UCLA School of Medicine, "is our own selves."

The people you will meet in *Radical Evolution* are testing these fundamental hypotheses:

- We are riding a curve of exponential change.
- This change is unprecedented in human history.
- It is transforming no less than human nature.

This isn't fiction. You can see the outlines of this reality in the headlines now. You're going to see a lot more of it in just the next few years—certainly within your prospective lifetime. We have been attempting to transcend the limits of human nature for a very long time. We've tried Socratic reasoning and Buddhist enlightenment and Christian sanctification and Cartesian logic and the New Soviet Man. Our successes have ranged from mixed to limited, at best. Nonetheless, we are pressing forward, attempting once again to improve not just our world but our very selves. Who knows? Maybe this time we'll get it right.

In 1913, U.S. government officials prosecuted Lee De Forest for telling investors that his company, RCA, would soon be able to transmit the human voice across the Atlantic Ocean. This claim was so preposterous, prosecutors asserted, that he was obviously swindling potential investors. He was ultimately released, but not before being lectured by the judge to stop making any more fraudulent claims.

With this legal reasoning in mind, flash forward a decade and a half from today. Look at the girl who today is your second-grade daughter. Fifteen years from now, she is just home for the holidays. You were so proud of her when she not only put herself through Ohio State but graduated summa cum laude. Now she has taken on her most formidable challenge yet, competing with her generation's elite in her fancy new law school. Of course you want to hear all about it. It is her first time home in months. But the difference between this touching tableau and similar ones in the past is that in this scenario—factually grounded in technologies already in development in the early years of the 21st century—changes in human nature are readily available in the marketplace. She is competing with those with the will and wherewithal to adopt them.

"What are your classmates like, honey?" you ask innocently.

"They're all really, really smart," she says. But then she thinks of some of the students in contracts class—the challenging stuff of *One L* fame. And she stops.

How does she explain what the enhanced kids are like? she wonders. She knows her dear old parents have read in their newsmagazines about some of what's available. But actually dealing with some of her new classmates is decidedly strange.*

- They have amazing thinking abilities. They're not only faster and more creative than anybody she's ever met, but faster and more creative than anybody she's ever imagined.

- They have photographic memories and total recall. They can devour books in minutes.

- They're beautiful, physically. Although they don't put much of a premium on exercise, their bodies are remarkably ripped.

- They talk casually about living a very long time, perhaps being immortal. They're always discussing their "next lives." One fellow mentions how, after he makes his pile as a lawyer, he plans to be a glassblower, after which he wants to become a nanosurgeon.

- One of her new friends fell while jogging, opening up a nasty gash on her knee. Your daughter freaked, ready to rush her to the hospital. But her friend just stared at the gaping wound, focusing her mind on it. Within minutes, it simply stopped bleeding.

- This same friend has been vaccinated against pain. She never feels acute pain for long.

---

* For the sources on these and other statements in this book, please check The Notes.

- These new friends are always connected to each other, sharing their thoughts no matter how far apart, with no apparent gear. They call it "silent messaging." It almost seems like telepathy.

- They have this odd habit of cocking their head in a certain way whenever they want to access information they don't yet have in their own skulls—as if waiting for a delivery to arrive wirelessly. Which it does.

- For a week or more at a time, they don't sleep. They joke about getting rid of the beds in their cramped dorm rooms, since they use them so rarely.

Her new friends are polite when she can't keep up with their conversations, as if she were handicapped. They can't help but condescend to her, however, when she protests that embedded technology is not natural for humans.

That's what they call her—"Natural." In fact, that's what they call all those who could be like them but choose not to, the way vegetarians choose to abstain from meat.

They call themselves "Enhanced." And those who have neither the education nor the money to even consider keeping up with enhancement technology? These they dismiss as simply "The Rest." The poor dears—they just keep falling farther and farther behind.

Everyone in your daughter's law school takes it as a matter of course that the law they are studying is changing to match the new realities. The law will be upgraded, The Enhanced believe, just as *they* have new physical and mental upgrades installed every time they go home. The technology is moving that fast.

In fact, the paper your daughter is working on over the holidays concerns whether a Natural can really enter into an informed-consent relationship with an Enhanced—even for something like a date. How would a Natural understand what makes an Enhanced tick if she doesn't understand how he is augmented?

The law is based on the Enlightenment principle that we hold a human nature in common.

Increasingly, the question is whether this still exists.

—⁂—

I CALL THE SCENARIO above "The Law of Unintended Consequences." It is not a prediction—I have no crystal ball, alas. But this scenario is a faithful rendition of what our world could well be like if some of the engineering currently being funded turns out to work. "Forget fiction, read the newspaper," notes Bill Joy, the former chief scientist at Sun Microsystems. Scenario planning is intended to prod people to think more broadly and view events with a new perspective. How did I arrive at this scenario? Let me give you some background.

In the late 1990s, when this book started, the rules of cause and effect seemed to have become unhinged. The problem was that the world was going through astounding change. First came the Internet, and then the World Wide Web. Cell phones the size of candy bars, palm computers the size of a deck of cards, and music players not much bigger than credit cards proliferated and merged in a primordial evolutionary silicon stew. A walk through a dark house in the middle of the night became an easy navigation. All the tiny lights marked the way in festive red or green, winking and shining from microwaves and clocks and phones and televisions and music players and video players and fax machines and laptops and printers and smoke detectors and docking stations and recharging stations and game players. Each signaled the presence of yet another microprocessor— part of that march in which the average American inexorably is becoming surrounded by more computers than she has lightbulbs, as is already the case in as utilitarian a vehicle as a Honda Accord.

The raging argument back then was whether this Cambrian explosion of intelligence marked the biggest thing since the printing press or the biggest thing since fire. And yet socially, the decade was a snooze. From my perch as an editor and reporter at *The Washington Post,* it seemed like the headlines, such as they were, involved little except peace, prosperity and Monica.

How could this be? I asked myself. *Where is the social impact of all this change? Where is the Reformation? Who are the new Marxists?* After all, human organization is always influenced by the technology of the time. "We shape our buildings, and afterwards our buildings shape us," as Churchill put it.

During the Agrarian Age, for example, the family was the fundamental economic and social unit. Commercial enterprises were basically family-run, even the big ones in Renaissance Venice. Governments descended through family in the case of kingdoms. The French army or the Spanish navy was quite literally a band of blood brothers. Nations were defined by people of genetic kinship.

All this changed, however, with the rise of the telegraph and the railroad in the mid-1800s. Suddenly vast swaths of time and distance had to be managed. Entire continents and oceans had to be spanned. To handle the challenge, new kinds of organizations were forced to emerge. The Ford Motor Company, for example, ripped the planet's very dirt for its iron ore at one end of its operations. At the other end it sold finished Model T's. Such a globally complex enterprise was impossible to run as a mere family enterprise. How could you produce enough trusted cousins? Thus the Industrial Revolution created fertile ground for steeply hierarchical corporations to blossom. It changed us. By the 1950s an employee of one of those corporations thought of himself as "The Organization Man" and "The Man in the Gray Flannel Suit"—cogs in the machine. The Industrial Age's contradictions also created a reaction to it—Marxism. Indeed, the entire 20th century can be described as an era of ideological, economic and military warfare over how to handle the great social upheavals created by this shift in technology and social affairs.

Ahem. So, okay? This technological change in the nineties is supposed to be the biggest thing since fire, and the best we can do for headlines is a tawdry Oval Office sex scandal? You can see the reason for my confusion. Where was this social upheaval that history taught us to expect?

As it happens, this is not the first time I've found myself covering worlds that do not seem to add up. In fact, I've come to welcome such assignments. They allow us to examine what's going on really. My previous books on what makes our world tick—*The Nine Nations of North America* and *Edge City: Life on the New Frontier,* which identified realities already in play that were not yet obvious to everyone—were similarly preceded by bouts of unsettling perplexity.

This time, two "aha" moments occurred in the course of my reporting. The first was the reminder that innovation arrives more rapidly than does change in culture and values. Perhaps, it occurred to me, the nineties were like the fifties. The fifties were a period of great technological upheaval— missiles with nuclear warheads, mass-produced suburban housing, mainframe computers. From television to Sputnik, the list was endless. And yet the fifties were the boring Eisenhower decade. The cultural upheaval of sex, drugs and rock and roll—enabled by The Pill, synthetic psychedelics and the transistor—did not occur until the sixties. You see similar upheaval in the earlier half of the century with the dawning of the age of automobiles, refrigeration, radio and telephones. The twenties, too, were a frivolous decade, promptly followed by the social upheaval of the thirties.

Perhaps that is the way history works. Perhaps because culture and values lag technology, when upheaval occurs, it is often of seismic proportions. If that is so now, then the cultural revolution for which we are due is just beginning to emerge. That's how tracing the outlines of that transformation became my beat during the early years of the 21st century.

The second "aha" moment was more formidable. I remember it as being like the scene in *Jaws* where the captain finally glimpses the shark. He responds, famously: "We're gonna need a bigger boat."

Such a moment came as I realized that this story was not about computers. This cultural revolution in which we are immersed is no more a tale of bits and bytes than the story of Galileo is about paired lenses. In the Renaissance, the big deal was not telescopes. It was about realizing that the Earth is a minor planet revolving around an unexceptional star in an unfashionable part of the universe. Today, the story is no less attitude-adjusting. It is about the defining cultural, social and political issue of our age. It is about human transformation.

The inflection point at which we have arrived is one in which we are increasingly seizing the keys to all creation, as astounding as that might seem. It's about what parents will do when offered ways to increase their child's SAT score by 200 points. It's about what athletes will do when encouraged by big-buck leagues to put together medical pit crews. What fat people will do when offered a gadget that will monitor and alter their metabolisms. What the aging will do when offered memory enhancers. What fading baby boomers will do when it becomes obvious that Viagra and Botox are just the beginning of the sex-appeal industry. Imagine that technology allows us to transcend seemingly impossible physical and mental barriers, not only for ourselves but, exponentially, for our children. What happens as we muck around with the most fundamental aspects of our identity? What if the only thing that is truly inevitable is taxes? This is the transcendence of human nature we're talking about here. What wisdom does transhuman power demand?

It's been a long time since the earth has seen more than one kind of human walking around at the same time. About 25,000 years if you believe that Cro-Magnons were critters significantly different from "behaviorally modern" *Homo sapiens*. About 50,000 years if your reading of the fossil evidence suggests you have to go back to the Neanderthals with their beetle brows and big teeth to discover an upright ape really different from us. The challenge of this book is that we may be heading into such a period again, in which we will start seeing creatures walk the Earth who are enhanced beyond recognition as traditional members of our species. We are beginning to see the outlines of such a divergence now. In 2003,

President Bush signed a $3.7 billion bill to fund research that could lead to medical robots traveling the human bloodstream to gobble up cancer or fat cells, for those who can afford the procedures. At the University of Pennsylvania, male mouse cells are being transformed into egg cells. If this science works in humans, it opens the way for two gay males to make a baby, each contributing 50 percent of his genetic material—and blurring the standard model of parenthood.

As you get further into these pages, you will meet real people with real names and faces working today toward just such modifications of what it means to be human. The powerful driver of this roller coaster is the continuing curve of exponential change. Evolution is accelerating so fast, some claim, that the last twenty years are not a guide to the next twenty years, but at best a guide to the next eight. By the same arithmetic, the last fifty years perhaps are not a guide to the next fifty years. They are, some guess, a guide to the next fourteen. As I type this, the evening news is airing yet another report describing some advance as "science fiction coming close to reality." Remember that phrase. You're going to be hearing it a lot in the coming years. When that occurs, I would like you to remember this book.

At least three alternative futures flow from this accelerated change, according to knowledgeable people who have thought about all this, as you will see in ensuing chapters. The first scenario is one in which, in the next two generations, humanity is rapidly replaced by something far more grand than its motley self. Call that The Heaven Scenario. The second is the one in which in the next 25 years or so, humanity meets a catastrophic end. Call it The Hell Scenario. You will find chapters on each, because both scenarios are plausible, and either would lead to the end of human history as we know it, and soon. The third scenario is more complex. It is the one we might call The Prevail Scenario. In this scenario, the future is not predetermined. It is full of hiccups and reverses and loops, all of which are the product of human beings coming to grips with their own destinies. In this world, our values can and do shape our future. We do have choices; we are not at the mercy of large forces. We can prevail.

I approach these three scenarios with an open mind, but critically. I try not to advocate any of them—I report them. Nor am I aiming this book at the 90-percent-male alpha-geek population who devours *Wired* magazine, that talisman of the digitally hip. If they find merit in my work, I am honored. But I hope for a broader audience. I try to speak to some very bright people I know—my mother, my daughters—who care far more about humans than they do machines. Me too.

If my interest in that third scenario—Prevail—marks me as an optimist, so be it. Heaven and Hell each might make a good summer blockbuster movie, featuring amazing special effects. But they tend toward the same story line: We are in for revolutionary change; there's not much we can do about it; hang on tight; the end. The Prevail Scenario, if nothing else, has better literary qualities. It is a story of struggle and action and decision. In that way, it is also more faithful to history, which can be read as a remarkably effective paean to the power of humans to muddle through extraordinary circumstances.

Scenario work shows that the future is usually a combination of all the stories you can construct to anticipate it. So I have done my best to present entertaining but accurate depictions of people who hold wildly different views. These are important thinkers and pioneers who deserve to be taken seriously. Most of them. Some are in there because I just couldn't resist telling their tales.

I hope this book serves as a road map and a guide to what we'll all be living through, pointing out significant landmarks along the way, as well as the turns and forks we can expect in the road. At the very least, however, I hope *Radical Evolution* ends up saying something about the present. George Orwell's most renowned work was entitled *1984* because he was really writing about 1948. Scenarios are always about the present, really. The fact that they exist today teaches us something about who we are, how we got that way, what makes us tick and, most of all, where we're headed.

There's one thing that I've already learned writing this book.

If you have a choice between starting your story with a telekinetic monkey or an attractive teenager in a wheelchair whose life might be changed by the technology the monkey represents, you have to lead with the bright young woman every time. For that's what people care about. And that's why the focus of this book is not on engineering—it is on the future of human nature.

CHAPTER TWO

Be All You Can Be

The future is already here;
it's just not evenly distributed.

—William Gibson

A T A CERTAIN ANGLE, seated behind the dining room table in her ponytail, khaki slacks and pinstriped shirt, Gina Marie Goldblatt does not appear in any way remarkable.

This particular January, she is a college sophomore home for the holidays from the University of Arizona. In Gina's serious moments, she wants to go on for a master's degree in business administration and a law degree and someday run her own company. But this week she's focused on going skiing with her friends for the first time. So finding a good time to visit with her is an experience in teenage time management. "We're the most last-minute people you'll ever meet," she says of her posse's complicated lives. To find pattern in the way her crowd swarms, it helps to remind yourself that college kids, like the proteins that underlie much of human nature, really are much more organized than a tangle of spaghetti. There is logic in the complexity. Events do work out.

Nonetheless, the idea of Gina going skiing is as astounding as is her ambition to someday go skydiving. Because of what Gina casually refers to as "medical malpractice," she's in a wheelchair for what is projected to be the rest of her life. "I have cerebral palsy. It's a brain injury. It can be caused by a trauma or by a lack of oxygen to the brain," she says of her birth. "I think I was pretty much a combination of both. Most of the kids with cerebral palsy are also born premature. I was, yes—by, like, two months. It's just something that you're born with."

It's not like a spinal cord injury, she explains, which paralyzed the late *Superman* actor Christopher Reeve after he was thrown from his horse, landing on his head. It's not like diabetes, which can appear when you're older. Nor is it like muscular dystrophy, which can kill the young. Gina can live long, but with lasting difficulties. "Part of the reason why cerebral palsy patients have development difficulty is because I didn't even start crawling until I was like maybe one or two, which meant that I didn't start, like, picking up things until that age," she says. "Doctors say whether you know it or not, like when you're picking up stuff, your brain is learning how to count little by little. So since you don't start that until you're older, you can't really catch up.

"It's something that I've learned to compensate for. Like, I read at the speed of a fifth-grader. I can understand everything that a college sophomore should understand, but I can't read it quickly. It just takes me longer." She speaks and types fluidly, although sometimes her writing needs editing because she can put sentences in the wrong order without recognizing it. Her handshake is soft, revealing low muscle tone. She needs help putting a clip on her shirt.

Gina's father, Michael Goldblatt, did not want her ever to think that she should be conquered by her limitations. So she became a pioneer. He enrolled her in public schools in affluent Oak Brook, outside Chicago, where they lived. She was the first seriously handicapped person to be fully mainstreamed at any of these schools. Her father was so determined that she be treated like everyone else that he even ran for the local school board and won. Nonetheless, Gina remembers the experience as "horrible."

"I had a teacher who would tell me that I didn't deserve to be in her class. And when I asked her why, the only reason she could come up with was 'Because you're in a wheelchair and have a disability.' " This was a Spanish honors class her freshman year in high school. Gina is fluent in Spanish, as is her patrician Mexican-born mother, Marta. "I was just, like, 'Okay, fine, whatever. You're not the first person to tell me that I can't do something.' "

At college, she has hired pre-med students to help her with the nitty-gritty details of life—getting bathed, getting into her chair, getting her backpack strapped onto the chair, taking out the trash. She's got a companion dog, Jinx, a yellow Lab, who can pick up books and take her socks off—he can even hold open doors with a harness on his back that includes suction cups and hooks. She also has what she describes as a "really cool" wheelchair that features two internal computers so that she can lie down in it, and put her feet up and sit back, at her command.

So now she's pretty dauntless. "We come up with these crazy ideas," she says of her crew. "Like, I decided I wanted to go skiing. Yes. Because, like, I want to go skiing. They have sit skis. Yes. They're like skis but you sit on them. It will be my first time." In the past she has talked about sky-diving. "My mom will just laugh—she'll sit there and laugh."

Gina is impressive, but she is not yet a world changer. She simply hasn't been around long enough.

The telekinetic monkey may change all that, though.

Gina's father, Michael, is awfully proud of that monkey. He likes to talk about how the work being done with it someday may change Gina's life. She may no longer need her wheelchair. Someday, because of that monkey, she may be able to control machines with her thoughts. Those machines may be embedded in her body. They might allow her to walk.

Gina is getting a little tired of hearing about the telekinetic monkey. "In fact, I heard about the monkey over dinner last week while we were at a restaurant. We had family friends over and they wanted to know, like, what he was doing, and so he mentioned the monkey."

What her father is doing at the particular moment of the early 21st century captured here is running the Defense Sciences Office of the United States' Defense Advanced Research Projects Agency. DARPA is one of the world's foremost drivers of human enhancement. Goldblatt readily acknowledges that his daughter is his inspiration. What he is doing is spending untold millions of dollars to create what might well be the next step in human evolution. And yes, it has occurred to him that the technology he is helping create might someday allow his daughter not just to walk but to transcend.

The first telekinetic monkey that DARPA funded is named Belle.

Belle is a cute monkey—an owl monkey, tiny, with huge brown globular eyes framed in white ovals two-thirds the size of her head. Her fur is russet and gray. Belle is astonishingly quick. One of her accomplishments is her prowess at an electronic game. She intently watches a horizontal series of lights in her lab at Duke University in Durham, North Carolina. She knows that if a light suddenly shines and she moves her joystick left or right to correspond to its position, she gets a drop of fruit juice. Treats may not matter now, though. She's gotten way into the game.

Belle is not really telepathic, strictly speaking. That would mean that she could communicate from her mind directly to another mind. DARPA's researchers haven't gotten that far—yet. Although Michael Goldblatt can clearly see how they might.

Belle is telekinetic. That means that simply by thinking, she can get a mechanical arm far away—in Massachusetts, in fact—instantly to move exactly the way her mind commands. Her Duke researchers line up probes thinner than the finest sewing thread right next to individual neurons in different regions of Belle's motor cortex—the part of the brain that plans movements. These are linked to two computers, one in the next room and another 600 miles north, at MIT, via the Internet. The computers each control a robotic arm. Then the researchers disconnect her joystick and start Belle's game. Sure enough, not only is she able to play it splendidly using just her thoughts, but the two robotic arms instantly mimic the motions that Belle's arm would make to control the joystick, "like dancers choreographed by the electrical impulses sparking in Belle's mind," her researchers report. The first time she did it, the two labs, in North Carolina and New England, erupted into loud celebration.

Needless to say, there's quite a story behind this. Especially since the reason you create a telekinetic monkey is ultimately to create a connection between any intelligence, silicon or human—any mind and any machine—anywhere. It is meant to lead to the day when a human might, for example, with her very thoughts control a robot orbiting Jupiter, causing its sensors to zoom in this way and that.

The next step is to rig a distant machine such that it can pipe what it is sensing directly into the brain of its human host. The goal is to seamlessly merge mind and machine, engineering human evolution so as to directly project and amplify the power of our thoughts throughout the universe.

If this sounds like superpowers, that is not far-fetched. In the 1930s and 1940s, in the hopes and dreams for society that we record in our comic books, we began to imagine what it would be like for people to transcend the mortal bonds of everyday humanity.

Take young Billy Batson, for example. He was a Depression-era orphan who sold newspapers on the street and slept in the subway. One night, he was led to a subterranean cavern and introduced to an ancient Egyptian wizard. For 3,000 years this mage had battled evil with the wisdom of Solomon, the strength of Hercules, the stamina of Atlas, the power of Zeus, the courage of Achilles and the speed of Mercury. Hence his name, S-H-A-Z-A-M. Knowing Billy to be virtuous, and realizing it was his time to pass, Shazam anointed the youth with his abilities. By uttering the sorcerer's name, Billy could become the grown-up Captain Marvel, with powers that included super strength. He could leap great distances and repel bullets with his body.

In today's terms, Billy Batson is no fantasy. He's somebody who's got hold of the nanotech Future Warrior exoskeleton—think of it as a wearable robot suit with superhuman strength—now in development as part of a $50 million program for the U.S. Army at Natick Labs in collaboration with MIT.

Or take the story of the sickly Steven Rogers, who lived in Depression poverty with his widowed mother, Sarah. She died overworking herself to provide for her son, leaving him to survive as a delivery boy. Alarmed by the rise of Nazism, Rogers decided to join the military but was deemed "too frail." After begging to be accepted, Rogers was tapped for Operation Rebirth, given a "secret serum" and subjected to a rain of "vita-rays." The weakling was reborn as Captain America, who could lift over a quarter of a ton and run 30 miles per hour, with reflexes 10 times as fast as normal.

Nowadays, his treatment would be called gene doping, a biotechnology already successful in lab animals and one that Olympic committees fear will make its human debut well before the 2008 Olympic games in China.

Throughout the cohort of yesterday's superheroes—from Spider-Man to the Shadow, who knows what evil lurks in the hearts of men—one sees the outlines of technologies that today either exist or are now in engineering. The Green Lantern has a ring that can create any physical object out of little but his imagination and an energy source. (He has a nanotech assembler—imagine a computer printer that can create any object from its constituent atoms.) Superman has telescopic and X-ray vision. (This is current military technology, from reconnaissance robots to cave pingers.)

In the middle of the 20th century, the powers of these superheroes were dreams. Today, we are entering a world in which such abilities are either yesterday's news or tomorrow's headlines. What's more, the ability to create this magic is accelerating. In 1985, the human genome was thought to be a code that would resist being cracked until 2010 or 2020. When the feat was accomplished in 2001 at a fraction of the estimated price, it was no more surprising than was the cascade of cloned mice, cats, rabbits, pigs and cattle that followed the first cloned sheep. Who is not braced for the first renegade human clone?

What will this mean? Will human nature itself change? Will we soon pass some point where we are so altered by our imaginations and inventions as to be unrecognizable to Shakespeare or the writers of the ancient Greek plays?

Many are trying to envision such a world. They describe our children and children's children as no longer really being like us. They call them transhuman or posthuman. They see our lives changing more dramatically in the next few decades than in all of recorded history. Who knows? They may be right.

After all, how many in the early 21st century expected an American soldier in Asia to display supernatural powers by shining a little red light on a target, confident that soon that laser would cause missiles precisely to vaporize the tank he had illuminated?

Shazam!

THERE ARE VERY FEW organizations in the world that routinely look as far forward as the Defense Advanced Research Projects Agency. It regularly thinks—and funds—20 and 40 years out. It's already changed your life. In the early sixties, there was no field of computer science. There were no computer science departments in universities and certainly no computer networks, much less personal computers. That's when J.C.R. Licklider—director of command and control research for the Pentagon's Advanced Research Projects Agency, DARPA's ancestral organization—envisioned something he called the Intergalactic Computer Network. He imagined it as an electronic commons open to all, "the main and essential medium of informational interaction for governments, institutions, corporations, and individuals." On this Intergalactic Computer Network, people using computers at home would be able to make purchases, do banking, search libraries, get investment and tax advice, and participate in cultural, sport and entertainment activities, he believed.

By the late 1960s, such prescience would inspire those who followed Licklider at ARPA to fire up a halting, primitive version of his Intergalactic Computer Network. They called it the Arpanet. This was a decade before the first commercial personal computer. In the 1970s, they expanded it into a network of networks.

You now know it, of course, as the Internet.

Today, DARPA is in the business of creating better humans.

"Soldiers having no physical, physiological, or cognitive limitations will be key to survival and operational dominance in the future," Goldblatt once told a gathering of prospective researchers at an event called DARPATech. "Indeed, imagine if soldiers could communicate by thought alone. . . . Imagine the threat of biological attack being inconsequential. And contemplate, for a moment, a world in which learning is as easy as eating, and the replacement of damaged body parts as convenient as a fast-food drive-through. As impossible as these visions sound or as difficult you might think the task would be, these visions are the everyday work of the Defense Sciences Office. The Defense Sciences Office is about making dreams into reality. . . . These bold visions and amazing achievements . . . have the potential to profoundly alter our world. . . . It is important to remember we are talking about science action, not science fiction."

DARPA is by no means the only or even the largest organization in the business of creating the next humans. DARPA's publicly acknowledged $3 billion annual budget is less than that of the National Science Foundation and is dwarfed by that of the National Institutes of Health, just to name two near the nation's capital. For that matter, its "bio-revolution" program represents only a fraction of DARPA's overall agenda.

The significance of DARPA trying to improve human beings, however, is that few if any institutions in the world are so intentionally devoted to high-risk, high-return, explicitly world-changing research. The cast at DARPA does not have kind words for incremental research. DARPA's "*only* charter is radical innovation," its strategic plan says. The swagger at DARPA is that of players who always go for the long ball, even at the risk of frequently striking out. Its program managers actively seek out problems they call "DARPA-esque" or "DARPA-hard." These are challenges verging on the impossible. "We try not to violate any of the laws of physics," says DSO's deputy director, Steve Wax. "Or at least not knowingly," adds Goldblatt. "Or at least not more than one per program."

The reason they reach that far is because they believe that's where they might find earthshaking results. That's why it becomes common to hear, wherever areas of astounding human transformation are discussed, "Oh, DARPA is working on that." That's why DARPA is at the forefront of the engineered evolution of mankind.

DARPA has a track record. Not only did it pioneer the Internet and e-mail, but DARPA helped fund the computer mouse, the computer graphics industry, very-large-scale integrated circuits, computers that recognize human speech and translate languages, the computer workstation, reduced-instruction-set computing, the Berkeley Unix operating system, massively parallel processing and head-mounted displays. It was a key player in the global positioning system, the cell phone, "own-the-night" night-vision sensors, weather satellites, spy satellites and the Saturn V rocket, which got humans to the moon. It also helped to create super-capacitors, advanced fuel cells leading to the next generation of cars and telesurgery. All of the military's airplanes, missiles, ships and vehicles, including the materials and processes and armor that went into them, and especially everything with the word *stealth* as part of its name, has "DARPA inside." Various ray guns, including laser, particle-beam and electromagnetic-pulse weapons, started with DARPA. So did the M16 rifle. Then there are the legions of air, land and sea robots, including the Predator, which, when it successfully fired a Hellfire missile at an al-Qaeda leader's SUV in Yemen in 2002, had the distinction of becoming arguably the first robot known to incinerate a human being.

The whole point of DARPA is to "accelerate the future into being," its strategic plan says—to identify discoveries now on the far side of useful-ness and bring them to the near side as quickly as possible. One program manager, in his DARPA job interview, was asked to describe where he thought science would be in 20 years. Then he was asked whether he would like to try to make it happen in three.

Particularly significant, DARPA creates institutions to support the fu-ture it desires. DARPA invests 90 percent of its budget outside the federal government, mainly in universities and industry. Academic centers at MIT, Stanford and Carnegie Mellon that made fundamental contribu-tions to information technology coalesced because of DARPA. If it feels companies need to exist, DARPA helps foster those, including Sun Microsystems, Silicon Graphics and Cisco Systems. If standards need to exist, DARPA sometimes steps in, too, promoting, for example, Unix, and the TCP/IP protocol that is the foundation of the Internet.

President Eisenhower created DARPA after the shock of Sputnik. Americans believed the United States' Cold War adversary had seized "the ultimate high ground." The military wanted to seize back the lead. But most of all it wanted never again to be surprised by the technological advances of potential adversaries.

As a result, DARPA's brag list starts with space. (NASA was spun off from DARPA.) Today's list is heavily loaded toward the information in-dustries, because that's where the payoff has been in the past few decades. But just as DARPA in the mid-eighties began to invest heavily in biolog-ically inspired robots, since the late nineties it has increasingly focused on human biology through the Defense Sciences Office. Goldblatt describes human enhancement as "our future historical strength"—what DSO and DARPA will be known for.

The denizens of the Defense Sciences Office treasure shirts with the legend "DSO: DARPA's DARPA." The notion is that if DARPA is at the cutting edge, DSO is the cutting edge of the cutting edge. In enhancing human performance, the program managers of DSO see a "golden age" of opportunity for radical, high-risk, high-reward change. As Goldblatt puts it, the old Army slogan " 'Be All You Can Be' takes on a new dimension."

DARPA is headquartered across the Potomac from the District of Columbia in Arlington, Virginia, convenient to the Pentagon. Its neighborhood should be considered impressive. Within blocks are the vast digs of the National Science Foundation, the campus of the Federal Deposit Insurance Corporation and the Arlington grounds of George Mason University. Nonetheless, the area comes off as cheesy. It remains punctuated by low-rent medical centers, a funeral home, a storefront where you can learn ballroom dancing and an International House of Pancakes with a spectacularly garish blue roof. These are the remnants of not long ago, when Arlington was a shabby inner suburb. Today it is an increasingly trendy and thriving collection of edge cities. Nonetheless, it will take more bulldozing before a transition to physically distinguished is anywhere near complete.

DARPA is housed in a substantial 10-story building with a sort of male, burgundy marble façade, smoked glass windows and an outdoor plaza that sounds hollow when your heel hits it because of a parking cavern below. Headquarters, nonetheless, is easy to miss. There are no signs advertising the tenant. It blends so thoroughly with the other blocky office buildings in the area that it is possible to miss the turn even on the fourth visit. The landmark to watch for is the uncommon number of police cars, marked and unmarked, guarding the place.

In the visitor control center, three guards in blazers process guests, while two more stand behind them, beneath a fashionably designed lighting array with "DARPA" backlit in aqua. On the coffee table phone is a red sign. It reads: "Do not discuss classified information. This telephone is subject to monitoring at all times. Use of this telephone constitutes consent to monitoring." One day, a high-ranking, uniformed aide accompanied retired admiral John Poindexter into visitor control. The aide asked the guards if he could stash the admiral's bag in the visitor center for a short time. The look of fierce incredulity on the part of the guards was so sufficient an answer that the aide hurriedly gathered up the bag.

At the elevators, the guards are conspicuously armed. Outside, if you seem headed in a direction the guards don't like, unmarked cars head you off. Visitors are escorted even to the men's room. Actually, security there has been relaxed—the guides no longer have to accompany you into the can. Now, they are relieved to report, they can just wait for you outside.

This buttoned-up environment contrasts sharply with the spirit on the fifth floor. Goldblatt, the leader of the Defense Sciences Office, is a quick, curly-haired elf with strikingly long blond eyelashes. He compares his program managers to Jason and the Argonauts. It's an interesting choice.

Jason and the crew of the *Argo* were among the first legendary explorers in human myth. Tales have been told about them now for 3,300 years. They were the greatest pioneers ever to light out for the Territory. They included Amphiaraus, the seer; Atalanta of Calydon, the virgin huntress and only woman; Caeneus the Lapith, who had once been a woman; Calais, the winged son of Boreas, the north wind; Heracles of Tiryns (Hercules), the strongest man who ever lived and the only human to be granted immortality among the gods; Periclymenus, the shape-shifting son of Poseidon who could take any form in battle; and 44 more. These ancient Greeks set sail when most of the eastern Mediterranean was an unknown realm full of inexplicable gods and monsters and witches. They met every challenge and faced every unknown. They performed impossible feats. Jason yoked two fire-breathing bulls to plow the field of Ares, sowed it with dragon's teeth from which armed men immediately sprouted, defeated that army single-handedly, and then got past a loathsome and immortal dragon of a thousand coils, larger than the *Argo* itself, to snatch the Golden Fleece of a magic ram.

We still celebrate what their story says about human nature. Lewis and Clark's Corps of Discovery, the first Americans to traverse what would become the United States, echo the same myth. Take a close look at the bridge of the *Enterprise* in *Star Trek*. That's Jason and the Argonauts rendered in modern terms.

Heavy company. Nonetheless, of his crew, Goldblatt proclaims, "We do not fear the unknown, and we relish exploring the unknowable." And who knows—history may not view his comparison as preposterous. For his program managers have been handed the keys to all creation and asked if they would like to take it out for a spin.

DARPA, for example, is very interested in creating human beings who are unstoppable. Three things that slow humans down in combat are pain, wounds and bleeding. So Navy commander Kurt Henry, a tall, dark, muscular, mustached and affable physician who radiates the cool of a movie leading man, is directing researchers who are working on those. He is the manager of a program called Persistence in Combat (PIC).

In California, there is a biotech company in Silicon Valley called Rinat Neuroscience. Henry is funding its "pain vaccine." What the substance does is block intense pain in less than 10 seconds. Its effects last for 30 days. It doesn't stifle your reactions. If you touch a hot stove, you still have the initial shock; your hand will still automatically jerk away. But after that, the torment is gone. The product works on the inflammatory response that is responsible for the majority of subacute pain. If you get shot, you feel the bullet, but after that, the inflammation and swelling that trigger agony are substantially reduced. The company has already hit its first milestones in animal testing and is preparing reports for scientific conferences. The commercial implications are formidable. If you were to get $400 per dose for a quarter million troops, there's your first $100 million. Rinat is a spin-off from Genentech, the world's first biotech firm. It has attracted venture capital funding; an initial public offering is expected soon. This product could revolutionize pain management. Think what it could do for cancer patients.

Blinded rats are being made to see by Harry Whelan, a professor of neurology at the Medical College of Wisconsin. In a battlefield, a laser powerful enough to burn is a very lethal thing if it is aimed at pilots' eyes. Using light in the near-infrared spectrum, however, in a process called photo-biomodulation, wound healing is accelerated. Vision in rats is being largely restored in anywhere from 5 to 24 hours—not yet quick enough to help pilots, but this is a work in progress. The research is sufficiently advanced that it is about to be tried on monkeys. The hope is that it will also mend wounds to skin, bone, neurons, cartilage, ligaments and tendons within four days. Whelan is also exploring what the process might do for spinal cord injuries, Parkinson's disease and brain tumors, as well as tissue and organ regeneration. If it works, he will have created something akin to the "physiostimulator" of the original *Star Trek,* the curative device Bones waves over injuries to heal them. The Navy SEALs are deeply interested in that.

Henry is also directing a gaggle of researchers who have discovered that the natural chemical cascades in the body that stop bleeding can be triggered by signals from the brain. The implication of this is that you might be able to train people to stop hemorrhaging within minutes, simply by concentrating their mind on their wound. Henry is directing another group of researchers who have discovered that if you inject millions of microscopic magnets into a creature and then wave a wand over them to get them all to point in the same direction, that can stop bleeding.

Those are not the most challenging of Henry's programs, however. That one would be Regenesis. Regenesis starts with the observation that if you cut off the tail of a tadpole, the tail will regrow. If you cut off an appendage of an adult frog, however, it won't. This raises the question of what mechanism has been shut off in the adult frog. If you could answer that question, you might be able to figure out what mechanism in humans has been shut off that prevents us from regrowing a blown-off hand or a breast removed in a mastectomy. "We had it; we lost it; we need to find it again" is Henry's slogan. As one of his principal investigators, Robert Fitzsimmons, points out, it is possible to grow an entire human from only a few cells. Every human ever conceived demonstrates that. So why can't you regrow an arm? What are the rules? And if you think the answer to that question will be available in a thousand years, the next question is, why not now?

You ask Henry if he is modeling his program after the lines in *Macbeth*:

> *In the cauldron boil and bake;*
> *Eye of newt, and toe of frog,*
> *Wool of bat, and tongue of dog.*
> *Double, double toil and trouble;*
> *Fire burn and cauldron bubble.*

His response is—no, not particularly, why do you ask?

Did you know that dolphins and whales never sleep? At least not the way we do. They can't. They're mammals. If they slept, they'd drown. What they have evolved instead is an ability to allow only one portion of their brain to sleep at a time. While the right lobe sleeps, the left lobe is on guard. Then they switch brains. What would happen if humans could control which portion of their brain is working while another portion recharges? The goal of the Continuous Assisted Performance (CAP) program, managed by John Carney, is to find out.

"As combat systems become more sophisticated and reliable, the major limiting factor for operational dominance in a conflict is the warfighter," the CAP mission statement says. "Eliminating the need for sleep during an operation, while maintaining the high level of both cognitive and physical performance of the individual, will create a fundamental change in warfighting. . . . The capability to resist the mental and physiological effects of sleep deprivation will fundamentally change current military concepts of 'operational tempo.' "

The plan is to create a "24/7" soldier—one who can easily navigate, communicate and make good decisions for a week without sleep. Any enemy who does have to sleep would be at a profound disadvantage. Small groups of sleep-free warriors could run rings around much larger forces. Logistics would fundamentally change. This is no small deal. Military savants like to say, "Amateurs talk strategy; professionals talk logistics." The Marines call the supply of "beans, bullets, and Band-Aids"—food, ammunition and medical supplies—a major limit to battle. If they were provided on a true round-the-clock basis, especially when the military is flying them from North Carolina to wherever the battle is, it changes a lot of equations. (Think of the equations that might be changed for civilians working for Federal Express. Think what this will do for college students and medical residents pulling all-nighters for a week.)

"In short, the capability to operate effectively, without sleep, is no less than a 21st century revolution in military affairs that results in operational dominance," the mission statement says.

Carney is pink-skinned and soft, with gold wire-framed glasses. He somehow manages to look like a plumber, which he is, in a neuropharmacological sort of way. He holds over 150 patents. The Silicon Valley company he founded, Centaur Pharmaceuticals, commercializes his research on stroke medication. It now has several drugs in phase two and three clinical trials, meaning they may soon come to market.

"Through evolution certain species have already solved the problem of how not to sleep; they actually don't care about sleep," he notes. Yet what happens in humans is that "after 24 hours you start getting a little bit irritable, by 48 hours you're frankly irritable and not fun to be around, and you're making bad judgment calls. But if you happen to have stars on your shoulder"—if you're a general—"nobody's going to challenge you. You're still going to be out there at the command center making bad decisions and nobody's going to come up to you and tell you that you're making bad decisions because you'll bark at them. And then by 72 hours you're frankly not useful for anybody. Even though you're still standing."

CAP's major research efforts include preventing or reversing changes in the brain caused by sleep deprivation; expanding available memory space within the brain, especially short-term memory; and developing problem-solving circuits within the brain that are sleep-resistant. As it happens, finding out how to redirect function from one pathway in the brain to another also has enormous potential for civilians with Alzheimer's, stroke and brain damage.

Another program of Carney's is Unconventional Pathogen Countermeasures. The point is, for example, to "take anthrax off the table" as a threat, as Carney puts it. Also smallpox. What's unconventional about it is that "despite the fact that you're in the middle of nowhere and you have no way of getting medical help to diagnose what you've got, the drug will work."

What's more, as a side benefit, it apparently could cure malaria, and probably the common cold. "Yes. Anything that can infect you," says Carney. "It's not going to cure Alzheimer's disease or arthritis. But anything that came from the living world that can cause disease in you."

We're talking about Pestilence as in the Four Horsemen of the Apocalypse?

"Right," says Carney.

And you're going to knock that baby right out of its—?

"Yes," Carney continues.

The object of the game is to discover the essential part of life common to many of these pathogens—no matter how they might be genetically reengineered—and interrupt them. An example would be finding an enzyme that appears only in bacteria but not in us. It might exist only for a very brief time in the bacteria, but without it, that life form cannot exist. Then you attack it. Another is "genomic glue"—something that sticks onto the genome of the pathogen so tightly that it prevents the genome from being read, translated and in any way replicated. It's like laying logs on a train track. Nothing in the cell gets through. The nice part, so far, is that the bugs have not been able to develop resistance to the treatment no matter how hard the researchers have tried to induce it. There are half a dozen approaches to viruses and bacteria in the works, but one antigenomic drug is at the last stages of testing in mice. This one seems to work on smallpox, malaria, anthrax and tularemia. It stops the Black Death—the plague—in its tracks. And yes, it also works on the flu. Researchers are ready to go to the FDA for human-safety trials. They hope the substance will be stockpiled against biological warfare, for which no clinical trials are ethically possible.

Will these approaches throw out some side effect that makes them unusable? If we were to tamper with the ecology of bugs in our system, killing off whole classes of them, might a potentially explosive Darwinian niche be opened into which all kinds of fearful unknowns might pour?

"It's like wildcatting for oil," says Carney. "It's a high-reward, high-risk environment."

"This is a paradigm shift, yes," Carney says. He's interested in making your immune system invulnerable. "One of the things that DARPA does historically is get into an area, give it the kind of credibility and experience that it needs to become accepted, and then we move on. I would say that in the world of immune modulators, we've done that."

Just to make things clear, "DARPA has no laboratory space," Goldblatt says. "DARPA does no work which we would consider execution. The actual work products—the milestones, the goals and objectives—are all done by independent investigators. They have the common tie—that they applied for—of funding coming from the Unconventional Pathogen Countermeasures program."

"These hands never get dirty," Carney jokes. Program managers like him—almost all PhD's or MD's or the equivalent, with years of experience in their fields—are compared to horse-racing pros, like jockeys. They pick and choose and encourage those who actually do the work. In fact, DARPA has been described as 140 decision makers united by a common travel department. (Usually, 23 of these are in the Defense Sciences Office.) They build communities of principal investigators who otherwise might not know much about each other. "The people who do our work are the smart people outside this building," says another program manager, Alan Rudolph. "You can have all the visions you want, but if there isn't a horse to ride, you're not going anywhere in the race. Now, sometimes you've got to go out and convince the horse to run your race. Of course, the way we do that is to incentivize them with big money."

"People around here get desensitized to what a million dollars means," Goldblatt says. "People around here get desensitized to what *ten* million dollars mean."

What would happen if these DARPA program managers weren't around?

"Probably a lot of these projects would never be done," Carney replies. "In fundamentals, we are tolerant to risk. Others are less tolerant to risk. So it might take a lot longer for somebody to get support. Or you don't know if they would ever pass muster to be able to get money from other agencies."

What else is on Carney's mind?

He wonders if it might be possible to recognize "a genetic personality."

"Is there a way to identify people who have a particular behavioral vulnerability that makes them more likely to be involved in lying and deceiving people from the standpoint of the terrorism issue?" he asks.

Has it been established that such a gene exists?

"No. Only in the figment of my imagination. But Michael can tell you that I'm a little different from the normal cut of people. I like to dream about things."

Hunger, exhaustion and despondency also slow humans down. Dealing with that is the province of the Metabolically Dominant Soldier program, managed by Joe Bielitzki. Bielitzki is a proud son of St. Sylvester's parish on Chicago's Near Northwest Side, near Logan Square. He has the broad shoulders and chest of a triathlete, which is the event for people who think marathons are for sissies. It is an endurance race combining three long-haul events—swimming, bicycling and running. He still competes, even though he is in his fifties.

Bielitzki jokes that "Metabolically Dominant Soldier" sounds like he's trying to create Spider-Man. No, not exactly, he explains. But his aim is high. He is tinkering with the internal machinery of human cells—controlling cellular metabolism and other activity within the cells—with the aim of tuning up every soldier's metabolism to the level of Olympic endurance athletes. "We want every war fighter to look like Lance Armstrong as far as metabolic profile," he says, referring to the American cycling champion. "A metabolically dominant soldier has strength and endurance that doesn't quit. The Energizer Bunny in fatigues kind of does it. Keeps going and going."

He claims he is not talking about creating superhuman strength. But he is interested in improving human cells from the extremely small parts up. Take mitochondria, for example. They produce the energy to power the cell. He is interested in modifying the number of mitochondria in muscle cells and their efficiency at creating energy. He is confident that he can take an individual now formidably trained to perform 80 pull-ups before exhaustion and render him capable of 300. Not to mention being able to walk forever with a 150-pound pack.

He likes the slogan "Be all that you can be and a lot more."

One of the ways Bielitzki would like to do this is by eliminating the need for food. "One of the things we know about war fighters is that we can't get enough calories in them to maintain high levels of strength and endurance over time," he says. "And so metabolic dominance is really focused on—if you can't get enough into 'em, why not just do away with food for three to five days completely?

"Special Forces guys working a 14-hour day are going to burn 6,000 to 7,000 calories a day. If we increase it to 24 hours a day"—that would be if Carney's program works and these guys don't sleep—"they're going to need 12,000 calories a day. You can't eat that much. Well, you can, but you're not going to feel good about it. It boils down to one Meal, Ready to Eat, and 46 PowerBars. You can't eat 46 PowerBars in a day. You can't even carry 'em. And so the question is, if we can only get 15 to 20 percent of your calories into you in a rational way, why put any into you at all? Why not, say, live off what you've got? We've all got stored calories—we just don't have access to them right now. So this is about improving the muscle and mitochondria so they can utilize the energy that's available. Maybe instead of deploying you lean and mean, we deploy you mean and plump.

"And the other issue is how do you deploy the soldiers at peak and keep them at peak the whole time they're out there so there's no degradation in their performance level, either cognitively or physically and maybe most importantly emotionally?

"When I say 'emotionally'—I don't know if you've ever done endurance sports?"

Ah, no.

"The one thing we know is that when you get hypoglycemic"—when you run out of carbohydrate energy—"you start to get depressed. You become despondent. You lose focus, mental acuity, response time, but mostly you just don't care. And that's a bad thing to have happen to you on the battlefield.

"We know that happens with Special Forces guys." Twenty-four hours after they go into action, "their physical levels are 40 percent below where they were when they started. And we want to get rid of that degradation in performance."

Bielitzki insists he is not talking about building supermen.

The Department of Defense "says this is about winning. This is not about losing," Bielitzki says. "It's not about having the war fighters sent out on the field to die. When you look at reasons for failure and you look at reasons why people die, they are getting weak, getting hungry, making bad decisions, being unable to continue. Those are the reasons you get killed. We want to remove that if we're going to have to put people in harm's way. This is not about Arnold Schwarzenegger in *The Terminator*."

And this would make you metabolically dominant relative to whom?

"Everybody else. My nutrition and my ability to utilize stored energy supplies for three to five days is such that I don't need to eat calories, but I always have calories available for energy."

When you start asking questions at DARPA, one reply comes up a lot: "The civilian implications of this technology have not escaped us."

Take the moment, for example, when it finally sinks in that Bielitzki is talking about fixing your cells so that you could live off your fat. A man who has worked out for years in an unsuccessful attempt to control his potbelly quickly raises his hand. "Me, me," he croaks. "Give some to me."

Bielitzki acknowledges the potential for spin-off technologies. "Forty billion dollars a year goes into the weight loss industry in this country," he muses. "This will change it."

A science and technology policy wonk, deeply worried about engineered human evolution in all its forms, stops dead when told about the potential for cell enhancement to conquer fat. "It does what?" she asks. "Okay, so I burn in hell for this. Sign me up."

"Will it have significant dual use?" Bielitzki asks. "Probably. Will the International Olympic Committee ban it? Absolutely. My measure of success for this is that the IOC bans everything that we do. We know that Lance Armstrong is different than everybody else. Can we safely induce it in anybody in a short period of time? That's really what metabolic dominance is about. Will there be a commercial market for it? Probably. Somebody has to make it. Is this a classified project at this point? No. This is all open."

Does this change human nature?

"I don't think human nature changes very much. Cognitive carrying capacity to hold information hasn't changed," Bielitzki says.

Ah yes. This brings us to Alan Rudolph.

Alan Rudolph is the godfather of the telekinetic monkey.

Rudolph has a goofy, boyish grin, stylish rimless glasses, a PhD in cell biology from the University of California at Davis and an MBA from The George Washington University. He is the program manager for an extraordinarily broad portfolio of DSO's projects. He jockeys hundreds of principal investigators. But he has also gotten his hands dirty. He has 15 patents in biological self-assembly, biomaterials, tissue engineering and neurosciences. He makes a distinction between DARPA and think tanks such as RAND, Brookings and the Highlands Forum.

"There are a lot of people who think about the future. This is one of those places where you can put money behind those fantasies. You get a vision, and then you start throwing money at it and trying to roll the ball down the road. It makes it an interesting place, no doubt."

He likes to describe himself as a "combat zoologist."

"Let me give you a little bit of my background so you understand my perspective," he says. "I'm a zoologist, and I think there's maybe three of us in the whole DOD. I come from systems taxonomy, physiology, the thinking about populations, ecologies, communities and organisms, how they adapt and evolve. Then I went off and got an MBA because I wanted to figure out how to make these things happen. A lot of things happen because somebody's got to make some money. Bad or good."

So now he's working on everything from multilegged robots to computerized human eye implants to brain-machine interfaces—the famous telekinetic monkey.

"The culture here allows you to say what if, and I'm willing to cross the boundary. What's born here is a fundamental philosophy that says what if we can just increase the number of interconnects between living systems and the nonliving world—hardware or software—what could happen?"

One thrust ends up making machines more lifelike. "Neurotechnology for biomimetic robots," he says. "Getting robots to jump, run, crawl, do things that nature does well. We're evolving our machines to be more like animals."

The other thrust connects life more directly with machines. "Let's create higher-density interconnects with living systems. And let's do so with the brain and with neural tissue." What does this mean? Rudolph describes it as "listening to the orchestra of neurons."

"Much of what you and I do that's different than a cockroach is based on our central nervous system and our brain," he observes. "It was just a leap of faith that if we could create interfaces with living systems"—devices that can "listen to the plethora of signals, that good things would happen." If you listen to the "orchestra" of the brain, you might be able to detect patterns in the signals "and try to make a song or a piece of music that had meaning."

The result is massive connections between individual living neurons inside the skulls of humans and wires that lead to computers. The first commercial step was cochlear implants—tiny machines that allow the profoundly deaf to hear by wiring tiny computers directly to the nervous system. "You talk about transforming humans. That was one of the earliest examples of a successful brain-machine interface. Thirty-five thousand people now have cochlear implants and are doing pretty well," Rudolph observes. The next step is retinal implants—computer eyes—wired to the brain of the blind.

But that's not the hard part. The DARPA-hard part is hearing the symphony for motion. "That's a really hard defense problem," Rudolph says.

"One of the fun things is you go play combat zoologist," he says. "You take a bunch of scientists and you go stick 'em in an infantry unit and you run around and play the war game. You go down to Combat Town in Quantico. You go on a carrier and get a ride in an F-16. It's a real fantasy camp. It's a great way to get a sense of the problems. We can be future thinkers, but when you have a customer you can't lose sight of that. And as long as we have a D in front of DARPA, we have to be cognizant of the customer's interest.

"I'm sure people raise their eyes and say, 'What the hell is a brain-machine interface going to do for the Defense Department?' Well, you draw them back to *Foxfire* with Clint Eastwood. The cockpit of a jet fighter is a complex place to be, filled with challenges for human performance. If the brain can somehow play a role in command and control of those systems, that would be a good thing."

Here's where Gina Goldblatt comes in.

"We've got a program in exoskeletons here. A guy's going to put on a suit and run and jump high or whatever." This is the Superman program in which the suit picks up on its wearer's muscle movements and greatly amplifies them so, for example, you can carry a 500-pound pack on your back for a very long time. But suppose the exoskeleton was responding not to your muscle movements but to your brain commands. "The reality is if you had a brain-machine interface, that's going to be an integral part of how he does that well. We're asking a fundamental question: Can the brain accommodate control and command of new devices?"

That was the point of Belle controlling the mechanical arm with her thoughts. Or of Ivy and Aurora, the monkeys that came after Belle. Eventually you might fly an F-22 with your thoughts. Or Gina might control artificial muscles with her mind.

Right now the way you make it happen with monkeys is not great. You drill a hole in the head of the monkey and implant a device that looks like a very tiny hairbrush bristling with hundreds of wires, each of which lines up next to individual neurons. "We are thinking now of ways to get these interfaces in without opening your head, drilling the thing in," says Rudolph.

"We've got our team of 70 crazed academics charging this dream. There are commercialization efforts sort of working now in parallel. A new company started called Cyberkinetics. They're going to sell a device to quadriplegics, paraplegics and locked-in patients who can't move a damn thing. Christopher Reeve had intact central nervous system motor function. So you can implant a chip in those people."

Then comes the question of how you run a device that you're not used to running. "I've got a little robotic dog and a little robotic cockroach. The cool thing is can I figure out how to control and experience that from a [human] central nervous system point of view?"

Isn't the hard part of that getting the information from the robotic cockroach and porting it back into your skull?

"Yes, the same chip will close the loop and will allow you to experience—whatever—controlling a robot, maybe flying a plane. Not from the standpoint of feeling it as a joystick, but as other sensory input. Visual, mechanical, force dynamics."

In other words, you should be able to pipe any sense from any sensor, anywhere, into your brain. You might directly sense the images from a remote camera, for example, allowing you to feel as if you had eyes in the back of your head. For that matter, you might feel a color or taste a sound.

"That's the powerful vision that I think this could enable. There's no reason why one can't input with a chip other types of experiences."

Closing the loop—allowing the human brain to receive signals directly, not just send them—opens the door to genuine telepathy. Rudolph's researchers are working on creating telepathic marmosets. Marmosets are very small South American monkeys with thick, soft, richly variegated fur. He's trying to get them to conduct brain-to-brain communication.

"Marmosets have distinctive calls associated with fear and threat, food and familial identification. We're going to use that device—that hairbrush. Both of them are going to have hairbrushes" embedded in their brains. "We're going to send the pattern with that call to a second one. If he hears the right call, we're going to look for a response. Then the question is, well, what if the monkey says, 'Fuck you'—you know what I mean? There are a number of issues and challenges that we're going to face with this. A National Academy member is involved with this one, at Vanderbilt—Jon Kaas, the guy who mapped the motor cortex. I mean, I couldn't have been more pleased to get a senior guy willing to take that kind of chance. It is amazing what real critical thinkers are willing to do."

Why stop at transmitting speech directly into the brain? Why not pictures?

"Can I alter what you see, change what you see, or put something in that you see what I see or you see what my camera sees?" Rudolph says. For example, "I want to see over the hill and I send a micro air vehicle or a robot over there and now I'm experiencing the visual image of the robot. I see what it sees."

There's no reason, by the way, that these images have to be in the spectrum of ordinary light. If you want to see in infrared or ultraviolet or whatever else the machine can sense, patch those night-vision puppies right into your visual cortex. Why should owls have all the fun?

"The third project," says Rudolph, "has profound implications for neuro degeneracies as well as augmented humans. I think it's the closest in terms of thinking about the evolution of man in the context of cognition. You're going to put the hairbrush in with this chip that mimics the circuitry of the brain." You replace a damaged portion of the brain with a chip that works like the brain. "For the first time you have a chip that now participates with brain function. Such a chip could be used to augment brain function. You start to increase processing speed. You will enhance memory. We'll know in the next year or two whether we can replace a lost circuit."

How are you going to change the batteries?

"It's a good point. It's a good point. Power is a big issue. Our battery technology sucks. Our power problems are huge. I think all these implants will be run off the energy in the body, ATP. There's low-temperature fuel in the body. The body is amazing in terms of its chemical conversions of energy. So we have a whole program that we launched."

Is that yours, too?

"Biomotors, yes. I share it with another program manager, Anantha Krishnan. We have some implantable batteries that work off of the natural body constituents. Tissue engineering is going to give us muscle. Building robots with living muscle."

Why would you want to do that?

"Right now we can keep it alive longer than we can get a battery to work. Yes. Outside the body. Yes. We've got a thing called the 'lox bot.' " It's a little biorobotic device that resembles a piece of smoked salmon. "It uses skeletal muscle from a frog, and the damn thing swims using skeletal muscle. It swims through its energy source. It's in a bath of glucose and ATP and the thing swims for like 20 hours. That's the University of Michigan and MIT."

The challenge, I tell Rudolph, is going to be convincing my mother that this is not science fiction.

"Me too." He laughs. "You're just trying to write about it. Try telling them I'm spending your tax dollars doing this.

"And we haven't talked about bees. I do a lot of stuff with bees. That's 10,000 flying dogs. They can be trained to sniff for things. They are little electrostatic dust mops so they collect things on their body and bring back spores and all kinds of information. A honeybee hive of 10,000 or so makes 100,000 trips per day over a five-kilometer radius. So the amount of information coming back to the hive is huge. So we place technology at the hive and just monitor the hive. We looked at one where you pull in a truck with concealed explosives and train honeybees to smell explosives. Put the honeybee hives on either side and with a camera just look at the truck and count bees swarming around the truck."

Does former UN weapons inspector Hans Blix know about this?

"Oh, you know, the military is way too conservative to use a beehive. We told this to the Israeli national police, and they grilled me for, you know, 45 minutes on what happens if somebody gets stung. So I said, 'Look, at the end of the day you've got a choice. You can get stung by a bee or blown up by a terrorist. I'll take the bee.' "

Have you read Michael Crichton's book, *Prey*?

"I haven't read his book yet. Is it good?"

Well, the bad guys are funded by DARPA.

"Oh, Jeez," he replies.

The list of DARPA-inspired human enhancements goes on and on. General Dynamics has a development and production contract potentially worth $3 billion intended to transform muddy-boot soldiers into nodes on a network. This involves manufacturing what the Pentagon calls "uniforms" for use by soldiers in the field by 2010. But this Objective Force Warrior Ensemble is far more than clothing. The soldier wears an undershirt fitted with body sensors that keep track of and broadcast his vital signs. His helmet receives video from robots. It also holds a camera, night-vision amplifiers, infrared sensors, laser finders, a global positioning system and a skull-mounted transmitter and receiver. A retractable eyepiece is useful for reading text messages or to view images sent from command centers or drones. Body armor is lighter, contributing to an equipment weight reduction of 50 percent, to 50 pounds. In this configuration, unmanned vehicles, known as mules, carry supplies. "Can you imagine traversing the mountains of Afghanistan with 100 pounds on your back?" asks program engineer Jean-Louis "Dutch" DeGay.

The original vision of the Engineered Tissue Constructs (ETC) program is based on the idea of rebuilding customized organs and body parts on demand, with the construction going on inside your body, not transplanted.

One of the goals of the Metabolic Engineering program is to allow badly injured soldiers to go into suspended animation or hibernation. It would allow them to survive even without oxygen for short periods of time, until the area is safe enough for help to arrive. This is also the program interested in allowing soldiers to run Olympic-quality sprints for 15 minutes on one breath of air. Turns out humans are very inefficient in the way we process resources. There's a whole lot of oxygen in one breath, and we waste most of it.

The Bioinspired Dynamic Robotics program is trying to replicate the foot of the gecko, the tropical lizard with amazing feet that perform Spider-Man–like feats. It would be handy for robots also to be able to climb straight up walls and hang from ceilings.

The Mesoscopic Integrated Conformal Electronics (MICE) program has already succeeded in printing electronic circuits on the frames of eyeglasses and helmets, weaving them into clothes, even putting them on insects. These include electronics, antennas, fuel cells, batteries and solar cells.

The Biological Input/Output Systems program is designed to enable plants, microbes and small animals to serve as "remote sentinels for reporting the presence of chemical or biological" particles. They'd do this by changing color, lighting up fluorescently, dropping their leaves or changing the color of their flowers.

The Brain-Machine Interface program is investigating how you would put wireless modems into people's skulls.

And that's just the Defense Sciences Office, the department of DARPA most directly involved with human enhancement. Meanwhile, on the floor where the Information Processing Technology Office (IPTO) resides, its director, Ron Brachman, former research vice president at AT&T Labs and previously at Bell Labs, and president of the American Association for Artificial Intelligence, wants to complete DARPA's vision from the sixties. When the original IPTO was created in 1962, its director, J.C.R. Licklider, focused the office on his novel conception of computers and humans working in symbiosis. That idea resulted in the Internet. Now the new IPTO "wants to realize this vision by giving computing systems unprecedented abilities to reason, to learn, to explain, to accept advice, and to reflect, in order to finally create systems able to cope robustly with unforeseen circumstances," according to Brachman. The object of the game is to produce machines—and the italics are his—"*that truly know what they're doing.*"

Some of this is so far-out-sounding that it beggars description. Don't even ask about the "special focus area" called Time Reversal Methods, for example.

Devotees of the film *Men in Black* may recall the scene near the end when the two protagonists sit wiping their faces of intestinal slime from the interstellar cockroach they have just vanquished. Will Smith turns to Tommy Lee Jones and says, "This definitely rates about a 9.0 on my weird-shit-o-meter."

Many is the time, cruising DARPA, that it is easy to recall that scene.

Readers with eclectic historical memory may by now be asking, "Aren't these the same guys who, during the Cold War, poured our taxpayer dollars into crackpot schemes like extrasensory perception and remote viewing?" Yes, and the guys at DARPA still don't apologize for it. In fact, it's a perverse badge of honor. "If those had worked, wouldn't you like to have known about it?" asks Goldblatt. "As long as it can be investigated rigorously and systematically and step by step, very little is too far out for us."

Readers with an interest in civil liberties, meanwhile, may by now be asking "Aren't these the same guys who created that Total Information Awareness project with the crazy logo of the Egyptian pyramid and this all-seeing eye, the implication of which was that in the name of anti-terrorism, we would never again have the slightest shred of privacy?" Wasn't that the one run by former admiral John Poindexter of Iran-contra fame, who later was canned from DARPA in a flap over a proposed futures market in terrorism? Actually, Poindexter worked not in the Defense Sciences Office but in a different department. Whole other floor. And they've canned the logo and changed the name. And they're feeling hurt and misunderstood. The press jumped to all sorts of inaccurate conclusions, they say. Not surprising given that DARPA, as usual, displayed its maddening reflex of not wanting to discuss what's up, even when the outlines of their projects are on the public record. But yes, that's DARPA.

Will all these projects work? Unquestionably, no. Not all. DARPA's attitude is that if an idea looks like a sure thing, let somebody else fund it. The National Science Foundation. Or venture capitalists, more to the point. A project is regarded as "DARPA-esque" only if few others would tackle it, but it would be earth-jolting if it did work. If you don't have failures, you're not far enough out. DARPA managers view themselves as instigators. By the time something new is mainstream enough to attract academic conferences attended by several hundred researchers, DSO usually sees its midwife work as done and moves on to new challenges. At the same time, DSO ruthlessly cuts off money to projects that fail to achieve their milestones, goals and objectives. An effective program manager knows when to cut bait.

Will all these projects bear fruit soon? Some more than others, and for the same reasons. The bulk of DARPA's projects operate in the 5- to 10-year time frame. But especially in the Defense Sciences Office, by the time a project is sufficiently mature that manufacturers are asking what color seats you'd like in it, program managers, like the Lone Rangers they are, have disappeared in a cloud of dust and a mighty "Hi-yo Silver." At the same time, history is moving fast enough that portions of this list may very well become part of your life between the time this is written and the time you read it.

Do defense dollars cause weird bounces in the research efforts? No question. The program to grow replacement organs from scratch, for example, as exciting and promising as it seemed, didn't initially do much for the Pentagon. "That's Veterans Affairs," they said.

What this recitation does demonstrate, however, is that engineered evolution of humanity is in the works right now. This inventory is hardly theory, much less fiction. This is about real flesh-and-blood human beings doing substantive, material things. It is a snapshot of what one small portion of one organization is working on in the first decade of the 21st century. It does not begin to include everything else coming out of labs and institutes around the world. Yet it shows that researchers are hardly waiting for some vague future. They are not, for example, waiting for some technology such as gene modification to mature in the next decade. They are working on enhancing people in important ways right now.

As Richard Satava, a DSO program manager whose portfolio includes cyborg moths and robot surgeons, puts it, we are entering the "biointelligence age." If we master this revolution, he believes, "we will be the first species to control our own evolution."

––––⁊⁊⁊––––

A MONG THOSE AT DARPA who are working on changing what it means to be human, the word you most commonly hear is *fun*.

Fun comes up all the time. Program managers view what they're doing as the greatest fun of their lives. Whatever day of the weekend or time of the night you e-mail them, it is common to get a quick response. They seem always to be on. Their tours of duty are usually only three or four years, and they clearly view it as the most intense experience of their lives. They know they will never see its like again. They describe what they're doing in terms that make it sound like the greatest adventure since Tom and Huck.

What you don't get is much of a sense of introspection. The program managers at DARPA can clearly see the individual steps that it would take to achieve telepathy, for example. But they don't talk much about what the impact of telepathy would be. Or what a world full of telepaths might be like.

If you point out that technology has a history of biting back, delivering unintended consequences, and ask whether DARPA worries about that, Goldblatt replies, "Yes, of course. It's your job. We even have a bioethicist on staff. But you can't let the fear of the future inhibit exploring the future."

Are there no limits on what we should try based on potential for evil?

"I don't think you should stop yourself because you can dream up scenarios where things didn't go the way you wanted them to go. We probably wouldn't be flying people into space if we really understood the risks, and now that we understand the risks more clearly, I guess there's a question of whether we will put more people in space."

If you ask Joe Bielitzki, the self-proclaimed pacifist who's creating the metabolically dominant soldier, about the implications of creating supermen, he sounds tortured. He replies, "There's potential for contradictions in all of science, but the intent is not to create a superman. The intent is to send the war fighter out there best equipped to come back alive. And those are big differences. I mean, the results may look similar, but the intent is not to create a superhuman. There's no reason to have a superhuman. But get somebody who can carry a little more, go on a little longer, drag their butt off the battlefield even if they're injured—keeping people alive is really what it's about. And this is coming from probably the ultimate pacifist. War is not a good thing to be in. But if people are going to fight you might as well give them every chance to come home to the people who love 'em."

If you ask Kurt Henry, who's trying to regrow arms that are blown off, about the meaning of what he's doing, he replies with a grin, "That's above my pay grade. That's not my department."

Gina Goldblatt is not at all phobic about technology. She's accustomed to relying on it. "Technology is assisting a disabled person to reach her full potential," she says. "That means that I started using computers in third grade. A lot of people don't think of their computer as their pen and paper, where I did. So therefore it allowed me to remain in mainstream classes."

So what about brain implants like Belle's? I ask her. Are you looking forward to getting one of those? Cyberkinetics, that Massachusetts company funded by DARPA, has received the Food and Drug Administration's permission to test just such a device on humans.

"Like, people are asking me that, too," she said. "My friends will ask me, 'do you ever look at the future as being able to find a cure for cerebral palsy?' But I don't know. I know my cerebral palsy is—whether or not I want to admit it—part of me. It always has been and it always will be."

Gina Goldblatt sees her cerebral palsy as part of her human nature.

———

WHEN MICHAEL GOLDBLATT and I first met, we ended up at a nearby restaurant called Tara Thai. There he started to open up about the importance of the work DARPA was doing, creating bolder, better, stronger, faster, smarter human beings.

He mentioned the impact DARPA's work would have on us all. For example, he said, he had a daughter with cerebral palsy. She had spent her whole life in a wheelchair. While her accomplishments were many and remarkable, he was actually spending many millions of taxpayer dollars to save his daughter, and mentioned the work with Belle, the North Carolina monkey. Thus I heard the story for the first time.

So, I said, in order to save your daughter, you're willing to fundamentally alter human nature?

There was a four-beat pause.

"Fundamentally altering human nature," Michael Goldblatt finally said, "would be an unintended consequence."

CHAPTER THREE

The Curve

Ch-ch-ch-changes, turn and face the strange.

—"Changes," by David Bowie

O NCE UPON A TIME, a peasant rescued from death the daughter of a very rich king. The king, overcome with gratitude, offered to grant the peasant any wish, whether it be gold, jewels or even a tenth of all his lands.

The peasant, however—who was not as simple as he seemed—asked only for a chessboard and some corn. "Tomorrow," he told the king, "I would like you to put a single kernel of corn on the first square. The next day, I would like you to put two kernels of corn on the second square. Then each additional day, I would like you to double the number of kernels you place on each of the succeeding squares."

The king, who in his youth had spent more time learning to joust than to cipher, was baffled by such a humble-appearing request. But a promise is a promise, so he agreed.

You probably know how this story ends. On the 3rd day, the peasant got 4 kernels. On the 4th day, 8 kernels. On the 5th day, 16. Even onto the 10th day, the peasant had barely received enough corn to make a decent porridge. But by then The Curve of accelerating returns was taking off. By the 20th day, the king owed the peasant 524,288 kernels of corn that day alone. All the king's horses and all the king's men were consumed with bringing the peasant his corn. By the 30th day the peasant was owed half a billion kernels of corn just that day, and the entire king's navy was devoted to importing corn from far and near to add

to the peasant's vast store. The king finally realized that there were 64
squares on a chessboard, and 34 more doublings to go. On the 40th day
alone, he would have to deliver 549 billion kernels of corn. He became
appropriately distraught and summoned the peasant. "What can I do to
end this?" he asked.

"Tell you what," the savvy peasant said. "I'll take your crown, and your
scepter, and your throne. In fact, I'll take the entire kingdom. And by the
way, what did you say was the name of your daughter?"

In such a fashion are vast upheavals in society and its values caused by
agreeing to ride such a curve.

It is just such a period in which we now find ourselves.

—⁓⁓—

D own the peninsula from San Francisco lies Santa Clara County. As
late as the sixties it was still most famous for its apricots and prunes.
When its orchards erupted into bloom in the spring, they attracted
tourists like the leaves in New England's autumn. Since the 1920s, this
area just south of Palo Alto and Stanford University had been known as
the Valley of Heart's Delight.

By the mid-1960s, however, the Santa Clara Valley was changing. In
1938, two young Stanford grads, Bill Hewlett and Dave Packard, started
the area's first technology company in their now-famous garage in Palo
Alto. Their first big customer was Walt Disney, who bought eight of their
"audio oscillators" for use in his new animated film, *Fantasia*. In 1959,
Robert Noyce of Fairchild Semiconductors figured out how to etch
thousands of transistors on one piece of silicon and mass-produce these,
thereby sharing credit with Jack Kilby for inventing the computer chip as
we know it. Not for a dozen more years, nonetheless, would the area acquire the name that would make it legendary. Only in 1971 was the Valley
of Heart's Delight first referred to in print as Silicon Valley.

Well before then, in 1965, the 36-year-old director of Fairchild's Research and Development Laboratories, Gordon E. Moore, made an interesting discovery.

Moore defied many of the stereotypes we have today of nerds. He was
an outdoorsman who loved to camp and fish. He had an athletic build
from his days as a 5-foot-11 football player and gymnast. He was a California native, having grown up just over the Santa Cruz Mountains on
the Pacific coast side of the San Andreas Fault, in Pescadero. He was
thoughtful and cautious, hardly an egomaniacal blowhard. Nor was he

born to geekdom. When he went to Caltech, he was the first of his family to go to college. He never dreamed of coming to work in jeans and a T-shirt. He always wore a tie and a dress shirt with the top button buttoned, although his white shirts were sometimes known to have short sleeves and looked suspiciously like they might feature polyester. The only reason he didn't view a slide rule as a routine fashion accessory (remember, this was before the pocket calculator) was that he was a physical chemist, not a constantly enumerating engineer.

"You know the movie *Apollo 13*? You know those guys in NASA mission control?" says Howard I. High, a longtime associate. "He looked just like that. He would have fit right in. He would have made a good spy. He just looked so *normal*."

Moore helped move the earth on July 18, 1968, when he and Noyce left Fairchild to found a company called Intel. (Moore's assistant director of R&D at Fairchild, Andy Grove, was their first employee.) They would usher in a new age. The desks of the whole world would wind up featuring strange new appliances, these dun-colored boxes with "Intel inside." Moore and his early compatriots would become billionaires many times over.

But it was back in 1965 that Moore made the observation that may truly secure his place in history, for it may have the most consequence for the future of the human race. What he noted, in an article for the 35th anniversary issue of *Electronics* magazine, was that the complexity of "minimum cost semiconductor components" had been doubling once a year, every year, since the first prototype microchip had been produced six years before. Then came the breathtaking prediction. He claimed this doubling would continue every year for the next 10 years. Carver Mead, a professor at Caltech, would come to christen this claim "Moore's Law."

Moore's Law has sprouted many variations over time. As the core faith of the entire global computer industry, however, it has come to be stated this way: The power of information technology will double every 18 months, for as far as the eye can see.

Sure enough, in 2002, the 27th doubling occurred right on schedule with a billion-transistor chip. A doubling is an amazing thing. It means the next step is as tall as all the previous steps put together. Twenty-seven consecutive doublings of anything man-made, an increase of well over 100 million times—especially in so short a period—is unprecedented in human history. To put this in the context of our peasant and his corn, the

27th doubling just precedes the moment when even the king begins to realize the magnitude of his situation.

Doublings of this extent have never before happened in the real world. This is exponential change. It's a curve that goes straight up. The closest most people had come to the idea of such doublings was when, back in grammar school, they first tried to wrap their minds around the "miracle of compound interest." In that version of exponential growth, if you put a dollar in your savings account and you, the tax man and catastrophe don't mess with it, in several lifetimes it will wonderfully turn into a million dollars. It really will. The curve of accelerated returns is the principle that underlies saving early for your retirement. Time produces astonishing and transformative results when you can count on a doubling and redoubling curve.

But such continuity usually doesn't happen. Take the time before the American Civil War. The number of miles of railroad track doubled nearly seven times in the 10 years between 1830 and 1840, from 23 miles to 2,808. That was impressive. It was a curve of exponential change that would be as steep and world-altering as was the chip in the 1960s. Nonetheless, in the beginning, most people still viewed railroads as a curiosity for the elite. They still traveled by water, on horseback or on foot. Most people didn't use computers in the 1960s, either.

Here's the difference. For railroads, the pace of growth was not sustainable. You needed more and more land and steel and coal to expand the system. Those are finite resources. There's only so much of any of those. So The Curve began to level off.

It took 40 more years, until 1880, to get the next five doublings, to almost 100,000 miles. By then, The Curve was really losing steam. It took another 36 years, to 1916, for U.S. railroads to make their final doubling plus a bit, reaching their peak mileage of over 254,037 miles.

Make no mistake. The railroads changed whatever they touched. America was transformed. A struggling, backward, rural civilization mostly hugging the East Coast was converted into a continent-spanning, world-challenging, urban behemoth. New York went from a collection of villages to a world capital. Chicago went from a frontier outpost to a brawny goliath. The trip to San Francisco went from four months to six days, and that Spanish-mission gold-rush town became a sophisticated anchor of the Pacific. Not for nothing do historians still celebrate the driving of the Golden Spike in Promontory, Utah, on May 10, 1869, uniting the continent. The West became a huge vacuum, sucking record numbers of immigrants across the Atlantic. The frontier was settled. Distance was marked in minutes. Suddenly, every farm boy needed a pocket watch. For many of them, catching the train meant riding the crest of a new era that was mobile and national. A voyage to a new life cost 25 cents.

Of course, these railroad doublings, like most transformative curves, soon ran out of critical fuel—including money and demand for the services. At this point, things leveled off, and society tried to adjust to the astounding changes it had seen during the rise of The Curve. Historically, adapting to this sort of upheaval has been like shooting the rapids. We start in the calm waters to which we are accustomed, bump and scream and flail through the unprecedented, then emerge around the bend into a very different patch of calm water, where we catch our breath and assess what we've done.

This process is represented by an S-curve. At the flat bottom of the S, you have a period of stability such as the early 1800s. You leave that for the rapid change represented by the steep middle of the S. That's when The Curve rises exponentially, as in the mid-1800s. Then things level out at the top of the S. The last transcontinental railroad completed in the United States was the Milwaukee Road in 1909. After that, the market for transcontinental rail was saturated. In part, that was because of the rise of a new transformative technology: The one millionth Model T rolled off the assembly line in 1915.

Moore's Law would have been revolutionary enough if chip power had leveled out in the 1980s at the top of an S-curve of 14½ doublings— comparable to that of the railroads over 85 years. Our world today, marked by ubiquitous personal computers, would have ensured that.

But The Curve did not stop. In 1975, Gordon Moore revised his Law to predict doublings "only" every two years. But he turned out to be too conservative. The computer industry regularly beats its clockworklike 18-month schedule for price-performance doubling.

Another way of expressing Moore's Law is far more recognizable to many people. The price of any given piece of silicon can be expected to drop by half every 18 months. Who hasn't eyed a whiz-bang $2,000 computer as a Christmas present, only to see an equivalent machine drop in price to $1,300 by the next holiday? Before 10 Christmases pass, the gift becomes a ghost. It has been cast aside. Not because it doesn't work; it chugs along just fine. But we have changed. It now seems so clunky. The power that could have only been bought with $2,000 10 years before can be expected to be available for $31.25, according to Moore's Law. By then the power is so unremarkable that you can get it for free with a subscription to *Newsweek*. Of course it no longer sits on a desktop. It has disappeared into watches, cell phones, jewelry and even refrigerator magnets with more power than was available to the entire North American Air Defense Command when Moore first prophesied in 1965. In some cases that power seems to dissolve into pocket lint—so unremarkable you don't even register that it's there. It essentially disappears. Take smart cards. You may have some and not even know it. They frequently look like credit cards. But they have chips in them, so they have significant powers. They allow you to enter especially secure buildings or store your medical records or pay for your subway fare. Passports come equipped with them. Full-blown versions are tiny computers without a keyboard or a screen. By 2002 those smart cards matched the processing power of a 1980 Apple II computer. By the middle of the decade they matched the power of a 386-class PC, circa 1990. Before 2010 they will have Pentium-class power. All for under $5 apiece. Think about that—a $4 Pentium. Retail items such as disposable razors increasingly come with radio identification chips, smaller than a grain of rice, that deter shoplifting. Those chips have the power of the state-of-the-art commercial computers of the 1970s. They cost pennies. They are designed to be thrown away.

This astonishing power has become almost free because, unlike the railroads, its expansion does not have the material limits of, say, Grand Central Station. The cost of shipping a ton of grain was halved perhaps three times during the railroads' heyday. The cost of computing had halved almost 30 times by the early 21st century. There are only four limits to computer evolution: quantum physics, human ingenuity, the market and our will. Actually, it's not at all clear that there are any practical limits represented by quantum physics, human ingenuity and the market, at least not in our lifetimes. Whether our will can shape limits is the core issue of the rest of this book.

You can see the effects all around you. In the same April 1965 issue of *Electronics* in which Gordon Moore laid out Moore's Law, Daniel E. Noble of Motorola also made some stunning predictions. In less than 50 years, he boldly prophesied, not only would computers become common in the home, but "the housewife will sit at home and shop by dialing the selected store for information about the merchandise wanted."

Okay, it only took 35 years. And he missed the possibility of how few "housewives" there'd be by then. But you get the point of the power of

The Curve. The practical outcome of this juggernaut is that IBM is expecting to fire up a machine around the time this book is published. Called Blue Gene, it is 1,000 times more powerful than Deep Blue, the machine that beat world chess champion Garry Kasparov in 1997. It is designed to handle 1,000,000,000,000,000 instructions per second. "If this computer unlocks the mystery of how proteins fold, it will be an important milestone in the future of medicine and healthcare," said Paul M. Horn, senior vice president of IBM Research, when it was announced.

Probably by the time you read this, Blue Gene will be probing the deepest underpinnings of human biology. Proteins control all cellular processes in the body. They fold into highly complex, three-dimensional shapes that determine their function. Any change in shape dramatically alters a protein's activity. Even the slightest change in the folding process can turn a desirable protein into an agent of disease.

That means that breakthroughs in computers now are creating breakthroughs in biology. "One day, you're going to be able to walk into a doctor's office and have a computer analyze a tissue sample, identify the pathogen that ails you, and then instantly prescribe a treatment best suited to your specific illness and individual genetic makeup," Horn said.

What's remarkable, then, is not this computer's speed but our ability to use it to open new vistas in entirely different fields—in this case, the ability to change how our bodies work at the most basic level. We will be able to do so because at a thousand trillion operations per second, this computer might have something approaching the raw processing power of the human brain itself, depending on whose measurements you trust of the abilities of that organ between your ears. Nathan Myhrvold, the former technology chief of Microsoft, points out that it cost $12 billion to sequence the genome of the first human. He expects it soon to cost $10 for anyone who wants theirs done.

Other vistas that are opening up because of The Curve of information technology include genetics, robotics and nanotechnology. The ability to tinker with our genes offers the astounding promise—and peril—of immortality, which mythically has been the defining difference between gods and mortals. It also offers the possibility of an even greater variety of breeds of humans than there is of dogs. Robotics allows machines increasingly to behave like living things, and living things increasingly to be enhanced by machines—blurring the line between the made and the born. Nanotechnology is the means of creating objects by working at the scale of individual atoms and molecules, allowing the creation of materials with astonishing properties, such as clothing that not only is bullet-proof but also stores electricity, making batteries obsolete.

And still The Curve rises. The limits to making chips on flat pieces of silicon are widely expected to be reached somewhere around 2015 as the lines etched on them approach the width of molecules. Does that mean Moore's Law will top out? Or will it simply shift to another means of computing? Historically, the upward arc of computer power has actually been a cascade of S-curves, with a new technology appearing just as the old one begins to peter out. As a result, although one can find a plethora of predictions throughout the 20th century claiming that computing was about to reach one insurmountable barrier or another, they all turned out to be wrong. As the limits of mechanical calculating machines were hit around World War II, they were succeeded by an entirely different kind of machine, the first electronic computers, filled with vacuum tubes, so big that buildings often were built specially to contain them. Just as the practical limits of vacuum-tube-filled machines were being hit in the 1950s, machines featuring individual transistors replaced them. Just as the transistor machines were reaching their peak, they were replaced by the silicon integrated circuit. What will flat silicon be replaced by, and when? Quite a few alternatives are in the works. You didn't expect Intel to just give up, did you?

The next generation may include machines that have circuits in three dimensions, like the brain. Also machines that harness the power of the vastly spooky realm of quantum mechanics. There is even work being done on "meat machines"—machines that use nature's own DNA to compute. More to the point, rapidly evolving machines allow us un-precedented opportunities to see how we might create the next genera-tion of rapidly evolving machines. Even the bootstrapping process is accelerated. For example, the information technology that enables the manipulation of atoms in nanotechnology—a nanometer is a billionth of a meter—allows the creation of new materials that may be used to keep The Curve rising well after flat silicon is obsolete. Whether Moore's Law continues, however, matters. How fast these successor technologies will prove to be feasible is a critical uncertainty. It determines whether we will have enough time to handle the way the world will change.

The driver of this incessant change is the need to compete or die. The classic example is Wal-Mart. The way it got to be the largest employer in 21 states, with more people in uniform than the U.S. Army, with its daily sales of $1.42 billion exceeding the gross domestic product of 36 countries, is its intimidating speed. Seventy percent of its merchandise is rung up at the register before the company has paid for it, *Fortune* magazine noted. Speed is why Wal-Mart does not route all of its ships from China the obvious way, across the Pacific. Instead, many take the long trip through the Suez Canal, into the Mediterranean and across the Atlantic. As a result, exactly half of Wal-Mart's imports end up on each North American coast. More expensive, but it ultimately gets the merchandise to your hands faster. The interior of a Wal-Mart distribution center may look like the labyrinthine warehouse in the final scene of *Raiders of the Lost Ark*—think 42-foot-high corridors of toilet paper stretching toward the horizon—but much of the stuff never touches the depot's floor. It moves from one truck to another truck along 24 miles of conveyor belts. All of this is accomplished with Wal-Mart's ever-improving technology. Now Wal-Mart is pioneering the idea of putting those radio frequency identification flecks onto every object it sells. These tiny midges are replacing bar codes. They call out "Hi, I'm here," to anyone with a transceiver. By 2006 Wal-Mart expects to keep track of inventory from factory to consumer by having one such chip hidden on each of the more than 5 billion crates of stuff it handles in a year. It is also moving to tag individual items on the shelves, starting with the most frequently shoplifted. (Did you know one of the most boosted items is Preparation H? Go figure.) Wal-Mart loses billions a year to theft. If these chips dramatically reduce this shrinkage, all other retailers will have to compete to lower their costs or die.

Retailing is hardly the only arena of competition, although by the middle of the first decade of the new century, eBay was poised to be one of the nation's 15 largest retailers, with Amazon.com joining the top 40. Small, casually run antique stores are closing because they can't take the competition from eBay. This echoes the way small, casually run bookstores closed because of Amazon.

Take manufacturing. "The choice facing Dell's rivals, from Gateway Inc. to Hewlett-Packard Co., is simple: adopt many of Dell's Net-efficient methods or exit the business," *BusinessWeek* noted.

Think of services. Expedia became the biggest leisure-travel agency in America, with profit margins higher even than American Express, starting from nowhere when the Web was created. Unable to keep up, 13 percent of traditional travel agency locations closed in one year.

How about finance? In the first few years of the 21st century, online mortgage service LendingTree was growing by 70 percent per year.

Or consider the pharmaceutical industry. There are three groups of people who will ultimately be attracted to any new enhancement. In order, they are the sick, the otherwise healthy with a critical need and the rest of us.

This became immediately obvious when a new drug called modafinil entered the market in the early 21st century. What it does is shut off your urge to sleep. It works without the jitter, buzz, euphoria, crash, addictive characteristics or potential for paranoid delusion of stimulants such as amphetamines, cocaine or even caffeine, researchers say. The FDA has approved modafinil for the sick—narcoleptics who fall asleep frequently and uncontrollably. But this widely available prescription drug with the trade name Provigil immediately was tested on the needy well—healthy young U.S. Army helicopter pilots. It allowed them to stay up safely for almost two days while remaining practically as focused, alert and capable of dealing with complex problems as the well rested. Then, after a good eight hours' sleep, it turned out they could get up and do it again for another 40 hours, before finally catching up on their sleep.

It's the future of the third group—the millions who, in the immortal words of Kiss, "wanna rock and roll all nite and party every day"—that deeply concerns the sleep industry. "It's a standing joke among sleep doctors that nobody sleeps in New York or Washington," says Helene Emsellem, director of the Center for Sleep and Wake Disorders in Chevy Chase, Maryland. "Except in New York they do it for pleasure, while in Washington they do it to work."

Modafinil and its follow-on improvements hold the potential for changing society. "This could replace caffeine," says Joyce Walsleben, director of New York University's Sleep Disorders Center. Caffeine is as old as coffee in Arabia, tea in China and chocolate in the Americas. It is the globe's most widely used drug—a bigger food additive in dollar terms than salt. Will people feel that they need to routinely control their sleep in order to be competitive? Will unenhanced people suffer fewer promotions and raises than their modified colleagues? Will this start an arms race over human consciousness?

Similarly, at the turn of this century, a little boy was born. His doctors immediately noticed he had unusually large muscles bulging from his tiny arms and legs. By the time he was four and a half, it was clear that he was extraordinarily strong. Most children his age can lift about 1 pound with each arm. He can hold a 6.6-pound dumbbell aloft with each outstretched hand. Otherwise, the boy appears normal—at least so far. He is the first human confirmed to have a genetic variation that scientists believe could lead to new approaches for building extraordinary muscles in people.

Wyeth Pharmaceuticals has begun preliminary testing of a drug designed to mimic the effects of his mutant gene as a possible treatment for the most common form of muscular dystrophy. At the same time, the discovery is raising concerns that athletes will try to exploit the discovery to enhance their abilities. "Athletes find a way of using just about anything," says Elizabeth M. McNally of the University of Chicago, who wrote an article accompanying the findings in *The New England Journal of Medicine*. "This, unfortunately, is no exception." What happens when such a drug moves from the sick to the healthy with an urgent need to the rest of us who work out only sporadically and with mixed results? Will abdominal six-packs be just a pill away? Similarly, what happens when brain-enhancement procedures are developed to fight Alzheimer's? Will they also be eagerly embraced by the ambitious?

Consider the effects The Curve has had on the arts. The traditional music industry is being gutted by tens of billions of online downloads. Sales were down 20 percent in one year alone. Next up: the same for the movie industry as films become increasingly available online.

Warriors in all of these realms soon learn that planning to compete in the age of The Curve is a lot like shooting skeet. You can't aim at where the clay pigeon is at this moment. You have to aim at where the clay pigeon is going to be, especially if you're trying to draw a bead on sophisticated global players such as China and India, who might not be in the game were it not for the way The Curve drops prices and shrinks communication distances. You have to figure out where the competition is headed and plan to beat it three, five, ten years out. Thus is acceleration of The Curve constantly fueled.

This compete-or-die imperative, of course, is ancient. That's why you can see echoes of The Curve over the millennia. For most of civilization, economic growth was so slow as to be essentially invisible. Increased wealth was not part of most people's life expectations. Annual per capita income in Western Europe at the time of Christ was $450 in today's dollars. It took more than 18 centuries to see fewer than one and a half doublings, to $1,269 in 1820, according to the economist Angus Maddison. Yet less than 200 years later, it was $17,456, more than six doublings since the Romans, almost all of it in modern times, and The Curve is continuing up. In the 20th century alone, the U.S. gross domestic prod-

uct marched inexorably up a curve, through booms and depressions, doubling five times—from a few hundred billion dollars right after World
War I (in year 2000 dollars) to more than $10,000 billion by the end of
the century. Just in the last half of the 20th century, the world's gross
domestic product doubled almost three times in constant dollars. The
world's exports doubled six and a half times in constant dollars during
that period. Remember, this means an increase of almost a hundred
times. The fastest-growing category was "miscellaneous," according to
Don Kash, the distinguished professor of innovation at the School of
Public Policy at George Mason University. This means, he says, that innovation was occurring so fast that people were increasingly incapable of
categorizing what it is they were inventing.

You even can see The Curve in human evolution. To get from the formation of the Earth to the first multicellular organisms took perhaps 4
billion years. Getting from tiny organisms to the first mammals took 400
million years. Getting from mammals to the first primitive monkeys took
150 million years. Getting from monkeys to hominid species such as
chimpanzees took something like 30 million years. Notice how the pace
accelerates? Getting from hominids to walking erect took 16 million
years. Getting from walking erect to humans painting on cave walls at
Altamira, Spain, took 4 million years. Getting from cave painting to the
first permanent settlements took some 10,000 years. Getting from settlements to the invention of writing in Sumeria took about 4,000 years. At
that point, biological evolution was trumped by cultural evolution. We
could now store, recall and widely share our thoughts and insights. Intelligence became less the property of isolated bands and more the sum of
civilization. As humans increasingly became capable of acting collectively,
they could make advances in the arts, sciences and economics far beyond
the capabilities of any individual, and The Curve really started to take off.
Four thousand years to the Roman Empire, 1,800 years to the Industrial
Age, 169 years to the moon and 20 more years to the Information Age.
Where we now find ourselves. Wondering if and how this Curve ever
stops and whether or not we like this game. Thinking about whether we
are about to enter another transition. Considering the likelihood that we
are engineering our own evolution.

Meanwhile, the amount of computer memory you can get for a dollar
is doubling every 15 months. The cost-performance ratio of Internet
service providers is doubling every 12 months. The modem cost-performance ratio is doubling every 12 months. Internet backbone band-

width is doubling every 12 months. The size of the Internet is doubling every 12 months. In short, the number of other curves of accelerated change unleashed by Moore's Law have themselves begun to proliferate exponentially.

Human genes mapped per year—doubling time, 18 months. Resolution of brain-scanning devices—doubling time, 12 months. Growth in personal and service robots—doubling time, 9 months.

U.S. manufacturing productivity is increasing on a curve. U.S. patents granted have been rising on a very steep curve. The number of scientific journals has doubled every 15 years since 1750. The number of "important discoveries" has doubled every 20 years. The number of U.S. engineers doubles every 10 years. Even dollars spent on U.S. education are rising on a curve.

In 2003, 35 years after its founding, Intel shipped its one billionth chip. It expects to ship its second billionth chip in 2007. Compete or get out of the game.

Actual physical mechanical devices are dropping exponentially in size. Entire motors smaller than a human cell exist in the lab. The size of computers is also dropping exponentially. In late 2002, Elizabeth Mullikin took a cream-colored oval object from her doctor. Rob Stein of *The Washington Post* described it as looking "like a big multivitamin, except one end was a clear dome. And a white light was flashing from the tip of it like a lighthouse beacon. She popped the blinking object into her mouth and washed it down with a drink of water. 'Honey, it was not any worse than taking any old pill,' said Mullikin, 77, of Columbia, Maryland. 'You just take a sip of water and down it goes.' "

What she had swallowed was an M2A disposable diagnostic capsule, also known as the "gut cam." It's a self-contained, wireless color video system designed to travel through the digestive tract, continuously taking pictures of any tumors, internal bleeding, and lesions that might show up—just like in the 1966 movie *Fantastic Voyage*.

"People want their cell phones small. They want their garage door openers to be small," noted Martin Schmidt, director of the Microsystems Technology Laboratories at the Massachusetts Institute of Technology. "Some of the fruit of that effort is what you're seeing in medical devices." With the gut cam, "it was the push toward digital cameras and circuitry from cell phones. So the question is, 'What is the next thing like that?' "

All of these exponential curves are adding up to a world profoundly different from the one humans are used to living in. We have crossed some line. "My son today wakes up in the morning certain of one thing," says Kash, the professor of innovation. "And that's that the world will be different by nightfall. He expects it.

"Humans didn't used to live that way."

———~~~———

IN THE NORTH AMERICAN Great Plains in 1928, the greatest thing since radio was hybrid corn. It caused farmers' yield per acre to soar. It was one of the most astonishing new agricultural technologies since the plow. But it was costly, which meant that to pay for it, you needed more land per farmer, more tractors per farm and more bank loans per tractor. It pushed so many small-scale, change-averse farmers off the land that some states' populations actually shrank. It reshaped the heartland for generations.

In 1941, two sociologists at Iowa State University, Bryce Ryan and Neal Gross, began to wonder how this chain of events had occurred. So they sought to find out how Iowa farmers had made the decision to try this newfangled innovation.

Ryan and Gross were not the most obvious academics for the job, notes Everett M. Rogers in his classic work, *Diffusion of Innovations*. One farmer asked Gross for advice about horse nettles. Gross, a city slicker who would end up at Harvard, had no idea the farmer was talking about noxious weeds. My God, man, said Gross to the farmer. Call a veterinarian to look at the poor sick horse.

Nonetheless, Ryan and Gross' 1943 report is to this day the most influential study of the ways humans adapt to innovations. Turns out the diffusion of any new technology—air-conditioning, cable TV, laser eye surgery—follows the same pattern.

First there are the Innovators. These are the geeks for whom venturesomeness is an obsession. Are you old enough to remember the first Apple Newtons? These were the first palm computers that were supposed to recognize your handwriting, back in 1993. Remember the people who rushed to buy them? Gary Trudeau in his comic strip *Doonesbury* lampooned them. These are the Innovators, the kind of people who always want to be first with anything. They love to be rash, daring and risky. They usually exist on the fringe of any social group. Their opinions are not necessarily respected. They add up to 2.5 percent of the population, according to the model.

Then come the Early Adopters. They are just slightly ahead of the crowd. But they are much more connected to the social fabric. They are hip. They frequently populate newspaper trend stories. Think of the first crowd of singles to bring their palm computers to nightclubs to beam their phone numbers rather than shout over the din. Because they value their reputations as trendsetters, these people are judicious about which innovations they adopt. (Remember former presidential candidate Bob Dole shilling for Viagra?) Early Adopters constitute a seventh of the population.

The Early Majority follows. They are numerous—a third of the population. They never lead, but they don't want to be stick-in-the-muds. Fitting into this group are the CEOs who, in the nineties, may not have known how to type, but nonetheless insisted that their company have a Web site of some sort. This group moves deliberately but does try to maximize return.

The Late Majority then kicks in. For them, change has become inevitable, usually because clinging to their old ways is killing them economically. Recall the businesses that finally realized e-mail was not a fad. This Late Majority is another third of the population. By the time a technology reaches them, it can be irritating to deal with those who are not with the program: "You have to find a pay phone because you don't have a cell phone? Are you serious?"

Finally come The Laggards. They tend to be suspicious of any change and stick to other people like themselves. They frequently don't have a lot of money, and they hate risk. There are as many of them as there are Innovators and Early Adopters combined. Laggards sometimes get treated like the Amish in sentimental newspaper accounts. Think of the heart-tugging tale of the last person in Mississippi to plow with mules, or the last dairy farmer to hand-milk cows in the Los Angeles metropolitan area.

This is the human dynamic that feeds the radical evolution of The Curve. "What people mean by the word technology," says Alan Kay, who first conceived the laptop computer, "is anything invented since they were born."

But now we've crossed a line. Patrick J. Fee of Germantown, Maryland, a consultant, once told me about the reptilian rocket moment of shock when he realized it. He found himself staring at the blue screen of death on his laptop. Gone. Vanished. Everything. He pounded on the computer, hitting the same key again and again. He cursed so dramatically that his small dog fled. It was no use. The hard drive was fried, and with it years of work, addresses, phone numbers, overdue projects—"my life," as he put it. With it, too, went the press on Fee's dress shirt, which became sweat-soaked while his heart pounded.

Fee suffered the classic anxiety attack of our new century—a fight-or-flight reaction when you lose control of the machines that have become part of you. Such a reaction is involuntary. It's not rational. It's the same alarm that goes off when you look over a cliff or somebody drops a snake in your lap. You pant and feel nausea, dizziness and a sense of impending doom.

It starts in a tiny part of your brain called the *locus ceruleus,* way down in that very dim bulb at the tip of your spinal cord, the reptilian brain, which is at least 300 million years old.

But something new is going on here. Let's see—when our computers die, we react as if attacked by a velociraptor? Our reptilian brain is recognizing something: We have bonded with these new machines. They have become part of us and we part of them. We are Borg, as they say on *Star Trek*—cyborgs, enhanced creatures. We have crossed the line.

Take Soo-Yin Jue. As she wobbled into an office north of California's Silicon Valley, she trembled, her knees buckled and she grabbed the table to keep from falling. She held her hands up in front of her face, went pale and then lost all expression. She was, quite literally, in shock. Nine years of her life she had poured into a book about China, crossing the Pacific a dozen times. All of that was embedded on her Mac. She thought she had backed it up. Only when she heard the horrible grinding sound did she discover how wrong she'd been. She felt as if she had "lost her soul."

The woman who talked Jue down from flash-frozen terror was Nikki Stange, who has a degree in psychology and knows exactly how, psychically and emotionally, we view our new machines as extensions of ourselves. A former suicide prevention counselor, she now has the title "data crisis counselor." She works for DriveSavers, a large data recovery company.

"You can hear the white knuckles," Stange says. "They are in total despair, and you have to let them know there is hope; there is a reason to live. Personally and professionally, it's like working in the emergency room of a hospital. You know how you hear that when someone is near death, your life flashes before your eyes? I can't tell you how many people tell me about having that sensation when their hard drive crashes. The intensity of emotions is certainly similar."

You could see it with Jue. In front of a reporter and a photographer from the *San Jose Mercury News,* she embraced everyone at DriveSavers after they recovered her book, as if they had saved her life.

Thomas Lewis, a psychiatrist at the University of California, San Francisco, who studies how mind and brain link to other beings, agrees. He is the co-author of *A General Theory of Love*, a groundbreaking book that looks at the most intimate emotional workings of the human brain. Lewis says that dealing with machines that are more and more a part of us is comparable to dealing with a parallel personality.

"When you look at another person, you are reacting to them and they to you. You are engaged in that kind of synchronous duet or ballet," he says. "It's a novel development to expect that from a machine. With your hand on the mouse, you do something and you expect the machine to do something back. It really bugs people if you interrupt that loop. People are only designed to make that loop with other people. In our mad frenzy to make computers more and more responsive—to our voices, to our facial expressions—we're attempting to duplicate in silicon that kind of reaction duet.

"There's two ways to think about it," says Lewis. "Emotionally, people have a general disposition to a bond of affection with their regular companions that help them out. Nowadays, they are distraught if they're separated from their computer, their helpful mechanical friend. They turn to the computer for emotional support, to be entertained by it, to encounter a social presence in the form of online communities and chat groups. Out here in Silicon Valley, I have spoken to people who say they consider regular human relationships superfluous and outdated, that they get everything they need from the computer. They say that and mean it; they're not kidding around.

"And then cognitively, it's become an auxiliary part of your mind. If you lose it, you lose part of your mind."

In 1960 two NASA scientists, Manfred Clynes and Nathan Kline, coined the term *cyborg*—cybernetic organism—to describe human bodies that had been altered and augmented with machines. Trust the National Aeronautics and Space Administration to figure that there's nothing better for those long-haul space trips than a niftier class of human.

Since then, there has been no end to the titanium and plastic stuck into our bodies—pacemakers, hips, knees, heart valves, eye lenses. But these are mechanical and primitive. They don't wire to human consciousness or brainpower, much less to emotions. Until now.

It quickly came to the point that Deanna Kosma banned her husband's new machine from the bedroom. When it vibrated in the middle of the night, skittering across the nightstand, beckoning him—and especially when he responded to it—it was too much. The apparatus in question is called a BlackBerry. At one level, it is only a primitive handheld wireless device that allows you to send and receive e-mail anytime, anywhere. But it has widely and quickly become notorious as the "CrackBerry." Everyone from colleagues to lovers use words like "junkies" to describe its users. "It can compete with the children for his attention," Deanna Kosma says.

I call Montgomery Kosma, a Washington attorney, and demand he defend himself. He talks about the two weeks he took off when his son was born. His BlackBerry was in for repairs. He felt as if a part of him had been removed. "I was at home, with no access, forced to rely on ancient technology—voice mail. It was an incredible burden to me. I felt withdrawal."

Kosma is cheerfully defensive about the time he devotes to the machine even when he could be examining the miracle of tiny baby hands. "There are times when the baby's asleep. I'm not interrupting the baby. I deal with it when it's convenient. There's often five minutes of downtime. I take it into the bathroom when the kids are washing their hands," he says.

Deanna half jokes about Montgomery's "addiction behavior." She stresses that she is "married to a very good man. My husband's still pretty human. He doesn't allow it to do his thinking for him. Or his living." Then she pauses.

"Now, you want to talk about an addict? My son will have withdrawal symptoms with his GameBoy. He will get irritable if you take it away. It's hard enough for him to be away from the Internet or cable TV. But he will get snappy if you take the Game Boy away. Almost like an addict. You can see it in his eyes."

That's why the next time you jack an MP3 player into your skull to shut out the world, or the next time you can't put down that solitaire game, or the next time you talk in the food court to noncorporeal companions rather than the person who is serving you lunch . . .

The next time you pay more attention to your e-mail than to your children, the next time you feel like throwing up when your connection to the cosmos is ruptured, the next time the innermost recesses of your brain recognize a machine as part of you when it dies, remember this:

You have crossed the line. For you, the revolution has occurred. The machines have not only changed you, they have become you. You have become Borg. Not metaphorically, but in a way as real and tangible as that keyboard you clutch.

Resistance, apparently, is futile.

———⁓———

T HE CURVE WARPS our sense of past and future. It is at the center of our feeling that we are at a hinge in history, at a time when the earth is beginning to move beneath our feet.

Human memory is a wonderfully elastic thing. Its greatest trick is to see any point on a curve as part of a straight line stretching directly and infinitely back into the past and forward endlessly into the future. *Things have always been this way.* Haven't they? *They will always be this way.* Won't they? *This is normal, no big deal.* Right?

By the arithmetic of The Curve, however, the last 20 years is not so much a guide to the next 20 years. It is a guide to the next eight. Similarly, the last 50 years is not a guide to the next 50 years; it is rather a guide to the next decade and a half.

Test your perceptions. Take a not terribly distant year, such as 1990. Below is a list of 15 events. Which of these would you say occurred in that year? Which would you say occurred much earlier or later?

The environment:

- Africanized "killer" bees first entered the United States.
- The Clean Air Act passed.
- The spotted owl was added to the threatened species list.

Statecraft:

- Boris Yeltsin was elected president of the Russian Republic in that new nation's first free elections.
- Saddam Hussein invaded Kuwait.
- Nelson Mandela was freed.

Cultural news:

- Dr. Jack Kervorkian assisted his first suicide patient.
- Johnny Carson left *The Tonight Show*.
- Smoking on U.S. domestic airplane flights was banned.

Technology:

* The Hubble space telescope was launched into orbit.

* Microsoft replaced DOS, in which users made computers do things by typing in commands, with Windows 3.0, which matched Apple's mouse-driven point-and-click system.

* In one entire year, *The New York Times* mentioned the Internet in only 27 articles, and then often thought it necessary to explain to its readers that this was "a network of business, government and military computers."

The economy:

* The Dow Jones Industrial Average was at 2,629.21.

* The federal government's estimate of the cost to bail out the savings and loan industry was doubled to $130 billion.

* The corporation with the largest market capitalization in America was International Business Machines.

Doesn't some of this recitation seem like ancient history? The answer is they all occurred in 1990. If you guessed that any of these events had occurred much longer ago, that may say something about the way you are responding to The Curve.

Confronted with such reminders, the tendency is then to say, *Okay, well that era was weird. But things are leveling out now.*

Our mental maps have changed, however. It used to be that societies preserved their traditions and transmitted their values by telling stories about their past. Americans were no different. Once, the Western was our most popular genre. The great cowboy movies—*High Noon, The Magnificent Seven, The Good, the Bad and the Ugly, The Man Who Shot Liberty Valance, The Ox-Bow Incident, The Treasure of the Sierra Madre*—told us how to live, how to act, how to be human.

At some point, however, we turned our gaze. We started exploring and explaining ourselves by telling stories of our future. Now the great blockbusters are *Star Wars, The Matrix, Men in Black*. We are awed by *2001: A Space Odyssey, Close Encounters of the Third Kind, Blade Runner*. Our emotions are tugged by the little darlings of *E.T., A.I.,* and *Lilo and Stitch*. We cheer *Independence Day, Contact, Minority Report, X-Men*. The star of *The Terminator* runs for governor of California, promising vaguely super-human power to balance the state budget, and wins! Even our romantic myths are wrapped in characters with astonishing abilities. Never before have three series of films—*Star Wars, Lord of the Rings,* and *Harry Potter*—proved so fabulously profitable and internationally popular.

In these films, check out how often we seem interested in anticipating everyday affairs suddenly interrupted by vast changes in the rules about how the world works, from the *Harry Potter* series to *Close Encounters of the Third Kind.*

The Curve implies one of the all-time changes in the rules. Those who study it call it "The Singularity."

———ᴍ———

A T DUSK, through brightly lit dust, the stadium in which the Super Bowl is to be played the next day looks like the most impressive craft extraterrestrials ever imagined landing. Strobes flash from its sides, smoke from test fireworks wafts from its interior, unexpected pulses of illumination shear through banners, blades of focused light shoot out for miles. To embrace all of San Diego's Qualcomm Stadium you have to sweep your eyes from left to right. The city's red trolley cars pull up like toys. Even the blimps overhead—flying in formation as if they were alien escort craft, radiating orange and yellow light from within—seem puny.

The figure striding across the top of the stadium is barely discernible. He's walking on the light ring. That's the thin concrete halo below which dangle the flood lamps, well above the highest seats. The light ring is no wider than the hood of an SUV. It has no handrails. It's an 80-foot drop off one side and a 300-foot drop off the other. Every 10 paces, the figure moving on it has to dance over yet another crate of high explosives that will be part of tomorrow's fireworks display. Dressed in black, with his baseball cap reversed, he prances out to the very end of the strip with the sort of abandon that curdles your innards even if you are sitting firmly on the ground watching it through binoculars.

He's out there to aim the laser cannon at his friends. Again.

His friends are just 400 yards from the stadium in an unadorned industrial garage. It is just across that deep ditch of a ravine grandly named the San Diego River. Beyond that, the I-8 roars. The garage has roll-up doors at either end and is big enough to handle a couple of tractor-trailers. But that's not what's in it. The garage is a fluorescent-lit void being filled with possibilities. A crowd of young men have pushed tables and desks to the center and covered them with screwdrivers, cables flowing down from the ceiling, monitors, keyboards, headsets, surge suppressors, garage door openers, FedEx boxes, Pizza Hut boxes, Bud Light boxes, Snapple Pink Lemonade boxes, satellite dishes, joysticks and robots made out of Legos. In the corner is a Honda 600R competition motorcycle. A movie-quality blonde occasionally swings by in a wine-colored Jaguar, bringing in more food, like Wendy tending the Lost Boys. A pink inflated bubble suit wanders through. Turns out there is a man inside. He is equipped to investigate unhappy nuclear reactors.

From this tumult, a fog of intelligence is emerging. This project is called The Shadow Bowl. This hive, hosted by San Diego State University, is nicknamed "The River." Very fat pipes connect it to the Supercomputer Center at the University of California at San Diego. What this is about is wiring the Super Bowl for human cognition.

Some of the cables lead up to the roof. There stands another laser cannon. If you precisely aim one laser cannon at another, you can create a beam of conjoined light along which you can transmit anything you can imagine and some things you might not think exist. Lasers have very narrow beams, though. So aligning them is no small trick. It seems Shania Twain's band—the one that will be playing at halftime and which is now rehearsing on the 50-yard line—is so amped that its bass notes vibrate the light ring. This throws the beams out of whack, breaking the connection. That is why the leader of this enterprise, Dave Warner, is up there dancing on top of that damn fool light ring, re-aiming that laser.

This gathering is actually a sophisticated collection of perhaps a hundred people with biological, chemical, radiological, temperature, weather, motion and video sensors who are attempting to conduct an unusual experiment. They are engaged in an exercise that resonates to that possible outcome of The Curve called The Singularity. They are trying to make an entire multi-square-mile environment intelligent.

In the parking lot outside The River, antennas bristle. Truck-mounted robot uplinks scan the skies with their dishes, looking for their satellites like baby birds searching for their mothers. This is the Super Bowl of January 2003, only 16 months after the 9/11 attacks. The worry is what happens if there is an assault on this biggest secular holiday event of the American calendar. It's no small concern. Some idiot has allowed a gasoline storage depot to operate just uphill from Qualcomm Stadium. If somebody were to fly a plane into that tank farm, flaming petroleum would head right for the stadium. The ravine of the river is heavily shielded with brush. If someone were to infiltrate it with mortars filled with biological, chemical or radiological weapons, it would be an easy lob to the 50-yard line.

The more you look at the festive stadium from the roof of the nearby garage brain center, the more uneasy you become. There are so many ways to attack those happy, innocent football lovers colorfully garbed as pirates for the game between the Raiders and the Buccaneers, teams named for outlaws of three centuries past. It's almost too perfect. It gives one a shudder. Overhead, Blackhawk helicopters and jet fighters roam. Somehow this is not comfort-inducing.

The Defense Department, of course, is funding much of this work down by The River. Warner has been a DARPA principal investigator. He carries himself with the swagger of a lifeguard (which he used to be), sporting long, rock-star-quality hair, now receding at the forehead. (The blonde carting in supplies who has a smile so big you can see her molars is Janice Robertson, his girlfriend.) He favors fashionable sunglasses and a cell phone in a hip holder that he perpetually twirls like a six-gun. He has an MD and a PhD and variously describes himself as a cultural engineer and a neuroscientist. He has a tiny but powerful light-emitting diode taped to the bill of his baseball cap, demonstrating how often he has to perform surgery on small, dark pieces of gear. He refers to his funders as "DARPA Vader Ville."

As far as Defense was concerned, what was being demonstrated here was how it might be possible to recognize a weapon of mass destruction and react to mass casualties. The practical result emerging was something quite different.

There is a uniform one comes to recognize at a gathering of those who are inventing the future. At The River, everyone wears black jeans and black sneakers, out of sheer habit. If you arrange to have little else in your wardrobe, getting up in the morning involves two fewer decisions. The ideal topper for such an ensemble is a black T-shirt. As it happens, Warner has provided those for the two dozen stalwarts at the core of this exercise. High up on the chest of these Shadow Bowl staff shirts there is a symbol of a figure that seems to have archangel wings, surrounded by a ring of palm fronds. It gives the staffers the curiously authoritative look of an intergalactic peacekeeping team. Warner gave careful psychological consideration to these symbols. The message: Don't mess with the guys in the black T-shirts.

Among this hard core, the significant marks of individuality and identity involve the weird stuff hanging around their belts. People pull out of their fanny packs the most impressive things—an entire socket wrench set, or a knife sufficient for gutting a calf, or an Iridium phone that can connect directly to a satellite. My award for the best status display, however, goes to the fellow with a sling on his belt from which hangs a flashlight with a red lens. That object states that you are so adapted to dark rooms illuminated only by computer screens that when you need to search under the desk for misconnected cables, you view it as unthinkable to ruin your batlike night vision with a beam approaching daylight spectrum. It would harsh your mellow.

None of this, however, is to be mistaken for lack of serious purpose. All manifestations of bleeding-edge technology by definition are demonstrations of the just barely possible. Thus they usually appear ragged and unprepossessing. By the time the future has all its wires carefully tucked away in a nice metal box where you can no longer see the gaffer tape, it is no longer the future. If you had been in Steve Jobs' garage in 1976, looking at the first mock-up of the Apple personal computer, you might have been forgiven for not seeing in it an agent of massive social change. You might not have looked at it and instantly seen e-mail, much less Google, in your personal future.

Just so, you have to squint a little at this ragtag collection of boys and their toys in San Diego to imagine where all this takes us. But for one weekend in January, what happened was that the boys of the Shadow Bowl for the first time in human history made several square miles of the San Diego River smart. They made the water smart, with sensors making it alert to little biological critters meant to do harm. They made the air smart, full of sensors wary of radiation, chemicals and detonations. They made the dirt smart, sensitive to the movement of would-be attackers. Most important, they imaged all this and ported the intelligence into one place. There blossomed unprecedented simultaneous views of everything that was going on in the area, from the parking lots to the drunk tank to the end zones.

They did this in part to imagine how you'd build a superorganism. How might you rebuild the connections between human and machine if you were to adapt the machines to the human nervous system, rather than the other way around? Dave Warner calls this the "last-millimeter" problem—the stubborn and persistent lack of connection between all that our machines can gather and all that our minds can know.

How would you wire all of the senses that humans come equipped with and make them a seamless part of a network in which the distinction between human and machine blurred? How might you feed information directly to your skin so that you would know whether a potential threat was coming from the left or the right? How might you feed information directly to your ears, which can make fine distinctions that eyes cannot? If something small but bad started to happen, you might instantly hear and recognize a discordance as certainly as you could tell which violin suddenly went out of tune in a philharmonic. How might you use your nose to alert you to critical incoming information by overriding all the other senses as if with a sudden burst of ammonia?

In such a world, the superb human ability to recognize patterns would be an element in a loop that roamed far beyond what is now the human ability to sense. If the human so connected were a fighter pilot or an air traffic controller or a pollution monitor, it would allow her to actually feel, hear and smell tens of thousands of cubic miles of space, alert to discord or opportunity in the music of the spheres. In this fashion, the intelligence of millions of little networked agents would enhance human thought. Augmented perception, this is called, extending our senses out past our skin, giving humans mastery of all they survey and beyond. It is meant to be a qualitative change in what it means to be human, to be enhanced in ways beyond the imagination of any previous generation.

Back at the sponsoring university, Vernor Vinge thinks about the implications of all this. Author of *True Names* and *A Fire Upon the Deep,* the novelist Vernor Vinge is a sweet, unassuming 60-something with a fuzzy fringe of grayish white hair and silver-framed aviator bifocals. His day job, before he decided to write best-sellers full time, was at San Diego State, where he still has an office, as a professor in the Department of Mathematics and Computer Science. It shows. He's the sort of methodical chap who takes notes of ideas that occur to him while listening to you. He wants to be sure to be systematic about sharing everything with you.

Vinge (rhymes with *stingy,* which he distinctly is not) in 1993 introduced the idea of The Singularity to describe huge but unpredictable social change driven by The Curve. In a seminal academic paper delivered to a NASA colloquium he wrote, "I argue in this paper that we are on the edge of change comparable to the rise of human life on Earth." He's anticipating the possibility of greater-than-human intelligence. He's talking about some form of transcendence.

As a metaphor for mind-boggling social change, The Singularity has been borrowed from math and physics. In those realms, singularities are the points where everything stops making sense. In math it is a point where you are dividing through by zero, for example. The result is so whacked out as to be meaningless. Physics has its black holes—points in space so dense that even light cannot escape their horrible gravity. If you were to approach one in a spaceship, you would find that even the laws of physics no longer seemed to function. That's what a Singularity is like. "At this singularity," writes Stephen Hawking in *A Brief History of Time,* "the laws of science and our ability to predict the future would break down." Another borrowed metaphor is "the event horizon," the point of no return as you approach a black hole. It is the place beyond which you cannot escape. It is also the point beyond which you cannot see.

Some people think we are approaching such a Singularity—a point where our everyday world stops making sense. They think that's what happens when The Curve goes almost straight up. The sheer magnitude of each doubling becomes unfathomable.

To Vinge that's actually more than a possibility. He's gone to great effort to imagine scenarios in which it might *not* occur. Even though he has a multiple Hugo Award–winning imagination, he hasn't had much luck. If The Singularity is possible, he doubts it can be prevented. He believes some sort of fundamental transcendence will happen soon. "I'll be surprised if this event occurs before 2005 or after 2030," he says.

Vinge makes an analogy to the evolution we know. By long ago learning to do what-ifs in our head, we rapidly surpassed natural evolution. We discovered we could solve many problems thousands of times faster than nature could. Now, with our exploding technology, "by creating the means to execute those simulations at much higher speeds, we are entering a regime as radically different from our human past as we humans are from the lower animals," he writes.

The critical element of his Singularity scenario is that it is fundamentally out of control. When finally we experience it, he believes, it will be like wildfire: "Developments that before were thought might only happen 'in a million years' (if ever) will likely happen in the next century."

Vinge and others see several ways that greater-than-human intelligence might occur in our prospective lifetimes:

- The Curve drives supercomputers, intentionally or unintentionally, to cross the line to greater-than-human intelligence.

- The Curve drives the Net to interconnect so much power that, intentionally or unintentionally, it wakes up as one super-organism.

- Information industry implants into biological humans produce people with greater-than-human intelligence. (As long ago as the early nineties, Vinge notes, a PhD-level human and a decent workstation—not even connected to the Net—probably could have maxed any IQ test.)

- Biological technology, probably through genetic engineering, produces humans with greater-than-human intelligence.

The Singularity would occur this way. Suppose one of the above scenarios were to happen. Suddenly we find ourselves with an ultra-intelligent critter. Making machines is what humans do real well. So what a greater-than-human-intelligence critter naturally does is start making machines vastly better and more intelligent than humans could. And faster. Much, much faster. These vastly better and more intelligent critters then create even more intelligent critters. And the spiral never ends. This would lead to what Vinge describes as "an intelligence explosion." In fact, that first ultra-intelligent critter might be the last invention humans ever need make. Or ever are allowed to make. It would be nice if "the machine is docile enough to tell us how to keep it under control," he writes.

Today all serious discussions regarding the social impact of the coming decades of The Curve start with Vinge's notion of The Singularity. Some wonder if it is in fact inevitable. For example, Marvin Minsky, MIT's grand old man of artificial intelligence, says that we are so bad at writing software—it is so laden with bugs—that he believes the first ultra-intelligent machines will be leapingly, screamingly insane. Others wonder whether we sadly underestimate how powerful the human brain is and grossly overestimate how soon The Curve will yield hardware and software that approach it. As you will see in the next chapters, some people see the approach of The Singularity as a force for good, in a scenario I call Heaven. Some people imagine this technology getting into the hands of psychopaths, opening the door to supreme evil, in a scenario I call Hell. Perhaps most intriguingly, some people are looking at a future in which we choose to alter initial conditions leading to The Singularity. They dismiss as mechanistic how we chase the number of transistors we can put on the head of a pin. What they see is humans choosing to refocus on how many connections we can make among the qualities of the human spirit. I call this scenario Prevail.

These critiques obviously take off in radically different directions. Nonetheless, they all grapple with the question of if, how and when we might transcend human nature. It says something about the technology we fear and respect in the early 21st century that when they grapple with such questions, they all see as their starting point Vinge's notion of The Singularity.

Vinge's office at San Diego State is dingy, with gray metal shelves, yellowed linoleum tiles, askew venetian blinds, glaring fluorescents and institutional-dirty light blue walls. Sitting there, I say to him this whole Singularity business is all very well and good, and perhaps even logical, but it sounds simply incredible. His eyes squeeze almost shut in his full, easy face when he smiles, which he does a lot. Yes, he knows. It's not the first time he's heard that objection. Almost by definition, disbelief accompanies any notion that all the rules that humans have known for millennia might soon blow up. "Some economists have been playing with this quite a bit," Vinge says. He has one paper from a scholar at UCLA that discusses, with equations, the consequences of The Curve going straight up. Even he has trouble buying that. "Anyone who talks about vertical asymptotes in terms of trend-line projections has some hard explaining to do," he says.

More credible, he thinks, is his version of The Singularity. He doesn't believe change has to become almost infinite for The Singularity to occur. "Just getting applications that are good enough to support superhuman intelligence" would trigger The Singularity, he believes.

I ask for a copy of the economist's paper. "Yes, that's easy for me to find; I just have to go to my extended memory here," he says, reaching for his computer keyboard, smiling at his little cyborg joke about his superhuman intelligence.

I tell him about Belle, DARPA's owl monkey with her brain connected to computers. "The first story I ever wrote that sold is about a preliminary attempt at intelligence upgrading and they did it with a chimpanzee," he replies. He sees fiction as scenarios, written vividly.

Vinge is made of stern stuff. Although the other side of a singularity is theoretically unimaginable, that hasn't prevented him from trying. He hopes for a "soft takeoff" of The Singularity, since he dreads what a "hard takeoff" might feel like, and doesn't know whether it would be safe for humans.

"The purest scary version" of the hard takeoff, he says, "is just if you had an arms race to get to The Singularity. A hard takeoff is where the whole transition from a situation where people are still talking like we're talking, about the plausibility and implausibility of it all—the transition from that to things being incontrovertibly strange is 100 hours. It's essentially like dealing with an avalanche. It's not something that you can talk about planning for. It's not really a plannable thing.

"In a nightmare scenario, it would be part of an arms race. Suppose you have two national forces that are going after this the way the Americans went after the A-bomb. Superhuman intelligence—it's the ultimate weapon. Now, they might not actually even think of it necessarily that way. What they might want, say, is something that can really monitor the Net. You talk about your intelligence problem."

Vinge digresses: "The reason it's not going to be effectively ruled illegal to do that research is there are so many reasons for going in that direction. There are military reasons for wanting it. There are economic reasons for wanting it. There are, by God, artistic reasons for wanting it."

Back to the hard takeoff. "So let's suppose it's the military, which is in an environment where it could spend a lot of money. And the other side knows that. Both sides have some idea of what's going on, on the other side. They can occasionally steal breakthroughs. Okay, at this point, you begin to take more chances. Any sort of controls that they might otherwise do or take are kind of left behind. It's sort of a single-minded thing. As you're making it better, you might hook it up with things that are effectors." He's talking about robotic arms and legs and eyes and ears and hands. "So basically you wind up with a self-fueled advancement of The Singularity. You know, like one side says, 'Our best intelligence is the other side is 100 hours away from having this.' 'Oh, okay. Well, we can do it in 48 hours.' And then the other side says, 'Well, if we throw all this overboard, we can do it in three hours.' And what comes out of that is one of the worse scenarios." The superhuman intelligence is a war fighter.

Phew. Okay, so what's your scenario for soft takeoff look like, Professor Vinge?

"Soft takeoff, I suppose, takes 50 to 150 years. It's a scenario in which the notion of corporate identity becomes not a legal metaphor." In this version, global business is seen as an ecology in which many organisms contribute to a web of survival. Some of them are human. And some of them are increasingly intelligent machines, such as the ones that know your credit card has been stolen before you do. "The back ends become gradually more and more smart. Corporate infighting is about as unpleasant as we now imagine it, only now it's being undertaken by self-aware entities."

The good news is that customer service provided by the machines "becomes much nicer, and in fact these guys really are very nice to humans. That's the nicest form of soft takeoff that I can imagine." One of the charming things about such a gradual, ecological scenario is that the machines might still need us to change their batteries and tend to their networks for a very long time.

That's the good news in this scenario? I ask.

"It would be more safe if it took a long time to happen," Vinge replies. "For all my rampant technological optimism, sometimes I think I'd be more comfortable if I were regarding these transcendental events from one thousand years remove—instead of twenty," Vinge writes.

"Okay, well, come back to today," I say to Vinge. "How would we know if this were not all fairy tales? How would we know if The Singularity was approaching?"

Vinge has given some thought to that. "We will see automation replacing higher and higher level jobs," he has written. Think of librarians displaced by Google. In such a world, "the work that is truly productive is the domain of a steadily smaller and more elite fraction of humanity." A symptom might be that throughout the world, "ideas themselves should spread ever faster, and even the most radical will quickly become commonplace." Another mark could be filmmakers and fiction writers having an increasingly hard time imagining a credible future more than a generation or so out.

As sunset approaches, I tell him about the guys I have been hanging out with at the Shadow Bowl, and what they've been doing. Someday summoning up smartness about everything in our environment around us may be routine. It may be as easy and unremarkable as is the process today of calling up music to surround us, anytime, anywhere. One wonders what our ancestors of a few centuries ago would have thought of that ability today. Would they have viewed us as godlike?

Right now, the Shadow Bowl is basically just a pretty cool hack. I remember that the most frequent word heard at DARPA is *fun*. I think of the best definition of *fun* encountered yet—"one more variable than you can handle."

But of course the really interesting part of this is figuring out what it all means, both metaphorically and literally—what the signals are trying to tell us about our future.

You know, there's another possible early indicator of The Singularity, says Vinge, smiling as the sky turns russet over the Pacific.

Your friends out at The River, he says.

Their stuff might work.

What Are Scenarios?

Scenarios are rigorous, logical, but imaginative stories about what the future might be like, designed to help people plan. Scenarios are not predictions. They are tools for preparation. Recall how pilots just returning from combat—no matter how complex the conditions they encountered—frequently say, "It wasn't as bad as the simulator." That is the value of scenarios. Simulators do not predict the future; they allow those who use them to carefully and calmly anticipate and rehearse their response to almost any sudden eventuality. Think of them as idea maps.

Scenario planning was pioneered in the early 1970s by Shell, the multinational oil company. Since then, it has been adopted by many important organizations, including IBM, Coca-Cola, Apple, Hewlett-Packard, AT&T, the California Energy Commission, Texaco, Intel, the CIA, Cementos Mexicanos, Clorox, Dow Chemical, Deutsche Bank, DARPA, Eastman Kodak, DuPont, Fannie Mae, First Union, Glaxo-SmithKline, Heineken, Kellogg, Lucent, Mattel, Morgan Stanley, Motorola, the National Education Association, Nissan, Pacific Gas & Electric, Pitney Bowes, Procter & Gamble, Scottish Enterprise, the Republic of Singapore, Sun Microsystems, UPS, Xerox and the World Council on Sustainable Development. When such organizations use the process, they create multiple scenarios out of the same set of existing facts, in order to describe as wide a variety of possible futures as imagination allows. The point is to allow these decision makers to prepare for them all. That's because a careful reading of history suggests that all past futures have turned out to be a combination of the scenarios that might have been written to anticipate them. Finally, if the future that actually appears turns out not to match any particular scenario, that is an early warning that history is taking a turn.

Scenarios have rules:

- They must conform to all known facts.
- They must identify "predetermineds." These are future events so locked in by those of the past that they can usefully be viewed as inevitable. For example, a predetermined element of a U.S. presidential election is that it will occur every four years.

- Scenarios then identify "critical uncertainties." These are possibilities that logically might occur in the future but which are both highly uncertain and highly important. For example, a critical uncertainty about any U.S. presidential election is who might win.

- Sometimes scenarios identify "wild cards." These are possible but highly improbable eventualities that would have great impact should they occur—for example, the leading presidential candidate being assassinated just before the vote.

- Scenarios reveal "embedded assumptions." These are frequently unprovable and often unexamined foundations on which our thinking about the future rests. For example, no one expects the U.S. military to so dislike the outcome of a vote as to overthrow the government.

- It is useful to identify in advance certain "early warnings" that serve as an alert that a particular scenario is coming to pass. For example, professional political operatives have known for decades that in a close presidential race, the mathematics of the Electoral College is such that if one party has California and the other has Texas, the outcome likely will be decided in Florida.

From time to time, you will find in this book boxes like these that identify the critical elements of scenarios that have been described in the text. They are meant to help you see the structure of these scenarios, and perhaps imagine alternative ones. They also might give you early warning, either if the scenarios turn out accurately to describe events as they unfold or if it turns out that history is taking a significantly different path.

The Curve Scenario

In this scenario, information technology continues to explode at a rate comparable to that from 1959 through the early 21st century. These unprecedented rapid doublings of information power and dramatically reduced costs continue to spawn new transformative technologies, such as genetics, robotics and nanotechnology. Those in turn also proceed to grow at an unprecedented rate, merging and intertwining to produce novel opportunities and challenges. Within the current human generation, these events transform society and ultimately test the meaning of human nature itself.

Predetermined elements:

- There are Curves of exponential change.

- Many of these describe the realities of technology.

- These Curves especially describe the increase in capabilities of information technology.

- The Curves of information technology increasingly enable new Curves of exponential change to emerge in other fields, especially genetics, robotics and nanotechnology.

- All of these Curves of exponential change have major impacts on society, culture and values.

Critical uncertainties:

* Are The Curves of exponential change smoothly accelerating, or will they display unexpected slowdowns, stops or reversals?

* Are these Curves of exponential change under the control of society's culture and values, or are they impervious to human intervention?

The Singularity Scenario

(Builds on The Curve Scenario)

In this scenario, The Curve of exponentially increasing technological change is unstoppable because new discoveries confer great advantage on those who adopt them—economically, militarily and even artistically. Either intentionally or accidentally, this leads, before 2030, to the creation of greater-than-human intelligence. This greater-than-human intelligence in turn proceeds to replicate and improve itself at such a rate as to exceed comprehension. This produces an inflection point in history called The Singularity, comparable to that in which humans rose from the lower animals. (Alternatively, The Singularity is triggered simply by the rate of change accelerating so greatly as to be beyond understanding, with or without the creation of greater-than-human intelligence.) The impact on everyday life is profound, as if we are being swept up by an avalanche. Succeeding scenarios in this book do not depend on The Singularity coming about. It would, however, dramatically influence the speed and scope of their outcome.

Predetermined elements:

* There are Curves of exponential technological change.
* These Curves of exponential change especially describe information technology.
* These Curves of information technology increasingly enable new Curves of exponential change in other fields, especially genetics, robotics and nanotechnology.
* All of these Curves of exponential change have major social, cultural and value impacts.

Critical uncertainties:

* Are The Curves of exponential change smoothly accelerating, or will they display unexpected slowdowns, stops or reversals?
* Are these Curves of exponential change under the control of society's culture and values, or are they impervious to human intervention?

- Will software improve at a rate as great as hardware, or will human ingenuity be stymied by the sheer size, complexity and bugginess of the software required?

- If these Curves are predetermined, must they result in the creation of infinite change?

- If these Curves are predetermined, must they result in greater-than-human intelligence?

- If infinite change or greater-than-human intelligence is inevitable, will this happen soon—that is, before 2030?

Embedded assumption:

- The only event that can alter this path is a cataclysm that will ruin civilization, such as nuclear war.

CHAPTER FOUR

Heaven

*Nature, Mr. Allnut, is what we are
put in this world to rise above.*

—Katharine Hepburn, in John Huston's
The African Queen, 1951

We are as gods and might as well get good at it.

—Stewart Brand, in the statement of purpose of
The Whole Earth Catalog, 1968

WELLESLEY, MASSACHUSETTS, is a crucible of the past and future. This stately village is not far from the gritty cities of Rhode Island and Massachusetts where modern America began. In those places, starting in 1793, waterfalls powered the continent's first textile mills, launching the Industrial Revolution in the New World.

Wellesley today is just outside the fabled Route 128 corridor, where, in the 1950s and 1960s the Information Age arose. That circulator highway curves west around the great research centers of Boston and Cambridge—especially the Massachusetts Institute of Technology. In low-rise office campuses that MIT researchers built for their companies along Route 128, the world's high-tech revolution blossomed. In fact, at one point, if one of these fellows came to the First National Bank of Boston to ask for money and he had a government contract associated with a high-technology scheme, the loan officers had standing orders: No matter how crazy either this person or his idea seemed, neither could be turned down without authorization from a senior vice president.

Wellesley is also just south of Walden Pond, where, starting in 1845, Henry David Thoreau lived for two years and two months in a 10-by-15-foot hand-crafted cabin, considering the prospects for human transcendence. In his book *Walden, or Life in the Woods,* he came to the studied conclusion that "men have become the tools of their tools."

Now the Route 128 area is as much an incubator of biotech as of information technology, creating new opportunity and wealth. That's why the village of Wellesley today is the sort of idealized spot where the 18th century meets the 21st. Wellesley is a place where emporia don't so much sell sandwiches as offer "savories." Shop windows are full of marvelously slinky dresses, without price tags, aimed at those who timed their stock options right. There are an impossible number of art galleries and sophisticated craft shops with names like The Gifted Hand. Wellesley is the kind of place where the sidewalks are accented by brick and are cleaned more often than some kitchen floors. The tavern of the old Wellesley Inn on the Square is called Brix° 576. Brix° with that little degree symbol is a measure of the sugar content in the juice of wine grapes before it is fermented. In parts of Silicon Valley, too, such as Burlingame, you can see a lot of this crafted balance between old and new—between greensward and gigapixels. Wellesley has taken to heart the words of Alan Kay, the father of the mouse-driven, point-and-click computer desktop, when he said, "The best way to predict the future is to invent it yourself."

Not far from this village, predicting the future by inventing it himself is clearly what Ray Kurzweil is up to. A two-lane blacktop named Walnut Street, as it nears an on-ramp to Route 128, crosses the Charles River. In a dozen miles the broad Charles will define Boston and Cambridge. But here it is easily spanned by an ancient stone bridge as it burbles on its way to a waterfall that once powered one of those venerable textile mills. Between the bridge and the dam today is an isolated office building with marvelous river views from three sides. The office building demonstrates that the future still needs a little work. It offers a shockingly butt-ugly example of how badly precast concrete exteriors can age. This is the headquarters of Kurzweil Technologies.

In the lobby of the second floor, where Kurzweil has his office, is a museum-worthy Ediphone, created by Kurzweil's hero, Thomas Edison. Billed as a "Voice Writer," the Ediphone was produced by the Edison Company in the 1920s. It has a tube you speak into the size of a powder horn. It was the world's first mass-market business dictation device, transforming speech into electrons and then back into speech. As Edison famously said, "I find out what the world needs. Then I go ahead and try to invent it."

Not far away is an equally museum-worthy Kurzweil Reading Machine from the 1970s. It goes the Ediphone one better. It can read out loud any book on your shelf. It looks like a photocopier that can talk, which is basically what it is. You lay the open book down on the scanner. Presto—the mechanical voice reads it to you. It is the first consumer device to use artificial intelligence. To be able independently to read books not published in Braille was a huge breakthrough for the blind. Kurzweil sold the first one to Stevie Wonder.

A minion demonstrates the old machine to me, laying a printed copy of the Gettysburg Address on its platen. "Four score and seven years ago," the mechanical voice promptly starts. Then it gets to the word *liberty*. "Li-bah-dee," it says. Still a few bugs in the system—the machine has a Boston accent.

When Kurzweil was in high school in 1965, he appeared on the hit television show *I've Got a Secret,* in which celebrity panelists attempted to puzzle out a guest's mystery. Kurzweil sat down on an old upright piano and played a composition. Then he whispered to Steve Allen, the show's host, "I built my own computer." "Well, that's impressive," Allen replied, "but what does that have to do with the piece you just played?" Kurzweil then whispered the rest of his secret: "The computer composed the piece I just played."

This primitive machine, which analyzed the work of famous musicians and then created new melodies in the same style, was Kurzweil's first foray into that core part of human thought so difficult for computers to match: pattern recognition.

Soon after college, Kurzweil developed three technologies. He invented the first practical flatbed scanner, launching a multibillion-dollar industry. He invented the character recognition device that could read any typeface, not just those weird ones that were designed specifically for computers. And he invented the first full text-to-speech synthesizer. Together, these brought the Kurzweil Reading Machine to life. When he unveiled his device, revered news anchor Walter Cronkite, on the air, had the machine read his signature sign-off: "And that's the way it was, January 13, 1976."

Kurzweil has created nine technology companies that continue to be leaders in their fields. One was built around the Kurzweil 250. Invited by Stevie Wonder to deliver an encore to his reading machine, Kurzweil invented the first music synthesizer that could match the response of the grand piano and other orchestral instruments. It is so realistic that most musicians can't tell the difference. Because it can create the sound of a universe of instruments, the 250 allows a teenager in her bedroom to imagine music for a rock band or an entire philharmonic, hear what her composition sounds like, and instantly revise it if she doesn't like it. It is now marketed in more than 40 countries.

Another company is Kurzweil Applied Intelligence, which was the first to commercially introduce large-vocabulary speech-recognition systems. By the early 21st century, doctors in 10 percent of American emergency rooms were already talking into them and seeing their notes transformed into print. Kurzweil's latest venture is the whimsically named FAT KAT (Financial Accelerating Transactions from Kurzweil Adaptive Technologies). It applies pattern recognition to stock market decisions. Its goal is to create an artificially intelligent financial analyst—which certainly sounds like an improvement.

In the course of all this Kurzweil has received a raft of decorations. He has been granted the National Medal of Technology, the nation's highest such award. He's been inducted into the U.S. Patent Office's National Inventors Hall of Fame. He was named Inventor of the Year by MIT and the Boston Museum of Science. Three U.S. presidents have honored him. The music industry and the handicapped have showered him with prizes. He's received eleven honorary degrees.

But that's not what made Kurzweil notorious.

His books are what really curl people's toes. In *The Age of Intelligent Machines,* Kurzweil in 1989 predicted the World Wide Web, the taking of the world chess championship by a computer, and the dominance of intelligent weapons in warfare. All of this seemed pretty far-fetched at the time. Yet in 1999, in *The Age of Spiritual Machines:When Computers Exceed Human Intelligence,* which has been translated into nine languages, Kurzweil went much farther than that, predicting that in a very few decades, because of The Curve discussed in the previous chapter, many intelligences will roam the earth that are not traditional humans. His 2005 work, *The Singularity Is Near: When Humans Transcend Biology,* tops that, however. It lays out the case for the imminent and cataclysmic upheaval in human affairs associated with The Singularity.

But even that isn't the audacious part.

What makes Kurzweil an outrage to some and an inspiration to others is that he is relentlessly and fiercely optimistic about these futures. He uses charts and graphs to systematically portray a near future that to some seems indistinguishable from the Christian version of paradise. On top of everything else, he is convinced that medicine is moving sufficiently fast that any person who can stay healthy for the next 20 years may so benefit from the explosion in biological technology as to be immortal. He lays out an extensive scientific, nonreligious, non–New Age case for personally planning to live for a thousand years. When challenged, he doesn't retreat from his logic at all. Once, to rattle his cage, I paraphrased the author Arthur C. Clarke's renowned line, "Any sufficiently advanced technology is indistinguishable from magic." I asked him, So, Ray, what are you saying? That any sufficiently advanced technology is indistinguishable from angels? He did not back off an inch. "Depends on what you mean by angels," he replied.

I call his scenario Heaven.

When we met in his office one bright, brisk New England April day, I delicately inquired, So, Ray, Do you have enough cash to last the next thousand years?

"Probably not," he replied unflappably, and without a trace of irony. "I can support my family and some businesses. I mean, capital to an entrepreneur is kind of like clay to a sculptor. So I can experiment with inventions and ideas, which is a nice freedom to have. But I think I'll need a bit more for a thousand years, particularly with the way things are going to speed up."

Raymond C. ("for Clyde—please don't put that in") Kurzweil is a first-generation American, born in 1948. His parents were refined Viennese Jews. His mother's father was a leading physician, a friend of the Freuds. He got the family out in 1938, one step ahead of Hitler. Kurzweil's mother, Hannah, is a gifted artist. A remarkable painting of hers, of flowers, hangs on Kurzweil's office wall, right next to a remarkable painting created by a computer, also of flowers. His late father, Fredric, a brilliant musician, had been Hannah's music teacher in Vienna. Just before the Holocaust, a wealthy American society woman who was a patron of the arts brought him out of danger. He joined the U.S. Army and ended up as musical director at Fort Dix, where he and Hannah had a reunion, a quick courtship and a military wedding.

Until Kurzweil was six, his family lived the life of struggling artists in one of America's first "garden apartments," a walk-up in Jackson Heights, in western Queens, New York. Jackson Heights was then a lower-middle-class Irish and Jewish quarter, different in flavor from today's polyglot neighborhood of Indians, Bangladeshis and Pakistanis as well as Colombians and Ecuadorans. Those immigrants are overlaid on the community of elderly Jews still living there. All are being joined by the urban hip moving into those same old newly trendy garden apartments.

Fredric Kurzweil became conductor of the Bell Symphony, performing at Carnegie Hall, and was chairman of the music department at Queensborough College. (Kurzweil recalls that when his father wrote a symphony, he had to rent a hall, hire all the musicians and write out all the sheet music for each of them just to hear what he needed to change. The son wishes his father had had a Kurzweil 250.) The family eventually moved to a nice brick middle-class house. Ray Kurzweil attended Martin Van Buren High School, a vast public school with 2,000 students per grade and a science program that produced more Westinghouse award winners than any other high school in the country at the time—even surpassing the Bronx High School of Science, no small boast. Kurzweil didn't take school terribly seriously, but he fondly recalls several mentors who validated his notion that imagination was not incompatible with science.

Several forces shaped his attitudes toward human frailty. One was the tragedy born of human nature that was the Holocaust, uprooting his family as well as destroying much of his wife's. Another was his father's medical history. The elder Kurzweil had a major heart attack when he was 51—four years younger than was Kurzweil at the time of our conversation. The elder Kurzweil died at 58. Kurzweil himself was diagnosed with Type II diabetes when he was in his 30s. He brought it under control, he says, by a very aggressive program he researched and designed himself, involving lifestyle, nutrition and supplements. He now takes 250 pills a day. When Kurzweil was younger, he grew accustomed to being told he resembled the musician Paul Simon, of Simon and Garfunkel fame. He has seriously reduced his body fat since then, so at the time of the interview, he had the sort of deeply lined, sad and wizened visage that you might expect of a man in his mid-50s whose face used to be much fuller. He looked like a cross between Paul Simon and a basset hound.

What Kurzweil really worships is ideas. "Religion in my household was ideas and knowledge," he says. "I remember my grandfather telling me that he was allowed into a special part of a library and was actually able to handle some of da Vinci's original documents. He described it in reverential terms, like it was a religious experience, like he handled the Ten Commandments. And what is the significance of da Vinci's documents? They were ideas. Ideas that were different, that defied the current common wisdom. He just gave free rein to his imagination. Things like the airplane."

Kurzweil's formal religious upbringing was eclectic. He still belongs to a Reform temple. "I see Judaism as a cultural identification. It's one I'm proud of, because Jews have been pioneers, from Bob Dylan to Einstein to Freud. Jews are very often in the forefront of democratic movements and the rights of women, and things like that."

His parents, however, "wanted to avoid the provincialism of the narrow religions," as he says they saw it. So the family joined a Unitarian church. "In keeping with the power of ideas, my own religious education was in a Unitarian Sunday school. The theme of this educational program was 'many paths to the truth.' We would spend six months studying a particular religion—let's say Buddhism. We would go to those services, we'd study those books and bring Buddhist religious leaders into our discussion group. Some of the kids took it more seriously than others. But I actually was one of the more involved."

What he learned was that "seemingly different stories are really speaking the same truth. The theme was tolerance of different ways of looking at things, and that was also very influential. The real truth of the matter is something that can transcend an apparent contradiction. We see that in physics. Is light a particle or a wave? Both are true, even though they seem to contradict each other."

Today his is a "Buddhist's view of God—as the sort of life force, the force of creativity. As opposed to a specific cranky personality that makes deals with humanity and gets mad and exacts vengeance."

This worship of a life force fuels his optimism about the coming transcendence of human nature. "What we see in evolution is increasingly accelerating intelligence, beauty. We find evolving organisms, like humans, that are capable of higher emotions like love. I mean, if you go to the point where there were just reptiles, there was no love. They don't have much emotional intelligence. They don't have art, music. So part of the evolutionary process—and this has continued with our technological growth of human cultural and technological history—is an increase of those higher emotional, intelligent functions. We see exponentially greater love.

"Even 200 years ago, 98 or 99 percent of human beings lived lives of utter desperation. Extreme poverty. Extreme labor. Spending all their time to prepare the evening meal. Extremely disaster-prone. No social safety nets. Now at least an increasing portion of human civilization is free of that level of desperation. So our ability to appreciate arts and music and to have stable relationships is increasing. That was relatively difficult to do even 200 years ago, let alone thousands of years ago."

The core element of Kurzweil's thinking is that The Curve of exponentially increasing technology is rising smoothly, as if on rails. It is in command, in his view, and unstoppable. Everything flows from that. He sees The Curve as a force of nature. He sees it as an extension of evolution. He does not particularly see The Curve as something humans chose to create. Like evolution, it is simply a pattern of life to be recognized, the outcome of billions of small actions. He calls it "The Law of Accelerating Returns." In his view, nothing any one country or collection of countries can do will deflect it or even slow it down. Forget oil shocks or climate change. The only possible limit he sees is a complete and catastrophic collapse of civilization or the extinction of the human species, worldwide, and he only inserts that as something of a rhetorical footnote.

"Exponential progress, in recent times, has marched right through" disasters such as the Depression and World War II, he observes. "It really is an evolutionary process. Biological evolution is full of unpredictable events like visitors from outer space in the form of meteors and asteroids. But nonetheless, out of that chaos comes a very smooth curve. Now the progress is so rapid that The Curves are on a very fast track. But they still emerge from an evolutionary process that is full of disruption. I mean, a lot of people ask me, 'Well, now with 9/11 things must be different? Or with the high-tech recession and the meltdown of communications and Internet stocks, surely that has disrupted these curves?'"

Well, no they haven't, Kurzweil says.

"These meltdowns were only meltdowns in the capital markets—they don't affect the fundamental issues that are driving this progress. We get more powerful tools. We use the more powerful tools to create the next generation. That is built into our economic expectations. A company, even to survive or tread water, has to move ahead. The perception that Wall Street had, that telecommunications and the Internet were revolutionary, was correct. But that does not mean there is not a pace to these things. Investment got ahead of itself. They overinvested in one aspect of the technology, like fiber, and underestimated other aspects. So the capital market suffered. But the actual technologies have proceeded. You can see very smooth exponential growth with the boom or the bust, even though every point on those curves is a human drama of competition and marketing programs and bankruptcies and initial public offerings. So you'd think it would be very erratic. But it's not. It is very smooth. Now, those curves are moving very quickly, so the underlying chaos is happening very quickly."

I ask Kurzweil what he thinks of scenarios that are very different from his, such as the ones in the ensuing chapters of this book: Hell, in which the technology is used for extreme evil, threatening humanity with extinction, and Prevail, in which humans shape and adapt it in entirely new directions.

He can't see these scenarios. "I mean, you go to these academic conferences and they talk as if we could get some consensus and stop these processes. 'Let's keep those good technologies, but those dangerous ones, you know, like nanotechnology—let's just not do those.' How unrealistic that is. Nanotechnology is not just one thing or three things. It is really the end result of miniaturization, which is pervading all of technology. Most technology will be nanotechnology in the 2020s. You would have to relinquish all of technology."

Kurzweil shrugs at the controversy over the use of stem cells from human embryos, for example. "All the political energy that has gone into this issue—it is not even slowing down the most narrow approach." It is simply being pursued by others outside the United States in places such as China, Korea, Taiwan, Singapore, Scandinavia and Great Britain, where scientists will probably get there first, he notes. He compares political obstacles to boulders thrown into a stream—the water just flows around them, although he does acknowledge that such political boulders can divert the flow. In the case of stem cells, the ethical problems involved in destroying human embryos, he believes, may hasten research that will allow any cell—such as your own skin cell—to be redirected to form any other kind of cell, maybe a nerve cell. "I'd rather have my own DNA anyway," he says.

Kurzweil has heard all the arguments. What about the notion that people will simply recoil from a technoid future? He's impatient with that. "Your quintessential grandmother uses her cell phone and she calls up British Airways and makes reservations with a virtual reservation agent"—a computer that understands your speech and talks back intelligently. "We don't have to be experts."

Perhaps most important to his notion that The Curve is unstoppable is that he sees it as a voyage of tiny advances. "Each application is a relatively small, noncontroversial, benign step. If you have a better way of diagnosing something, or more effective treatment, it's not controversial. It is just readily accepted. We get from here to these more revolutionary scenarios through a hundred steps like that."

Kurzweil is incensed that more scientists and technologists are not taking seriously their moral responsibility to raise their hands if they think they are looking at developments that will add up to profound transformation.

"Scientists are absolutely trained not to think about this. It is just part of the scientific style, or ethic, to be very conservative about what you are trying to do. A typical scientific speech is, 'We know very little about XYZ, we have just begun to scratch the surface, our ignorance is far more than our knowledge, we don't know this and we don't know that, and maybe, with luck, we'll be able to shed some light on what I am now working on, and anything beyond that is sheer speculation.'

"Well, that was all very well and good when a generation of that kind of technology was 20 years. But now when a generation is more like 2 years, society does have an interest in knowing what is going to happen 3, 4, 5, even 10 generations away. Yet scientists typically are trained not to speculate beyond maybe one or two generations of technology and science. So, hence, they very often take a ridiculing attitude towards the future and bring up wrong predictions from the past—that were made without any methodology. They just dismiss the whole idea of looking ahead. But that is not of service to society. I mean, even though we can't be absolutely sure what is going to happen and how these scenarios will play out, these basic trends are so strong—we need to appreciate their existence. They really do reflect a future that is going to be very different. I have tried to bring some disciplined approach to studying these trends, and it is something where I have actually some track record."

Kurzweil, for example, points to his 1989 prediction in *The Age of Intelligent Machines* of the collapse of the Soviet Union. It may seem obvious in hindsight, but in 1988, the U.S. Central Intelligence Agency averred that that regime would be America's principal foe through the lifetimes of our children and our children's children. Kurzweil, by contrast, saw the Soviets' dilemma. "Either they would support these decentralized technologies, which would ultimately destroy how tight a control they would have, or they would avoid them—as they tried to relinquish, say, photocopiers—in which case they would destroy their economy and ultimately fail. I actually predicted that they would do a little bit of both and both would do them in, and I think that is exactly what happened."

Even the rapid, peaceful, and complete collapse of the Berlin Wall and the five-decade-long Cold War is less breathtaking than Kurzweil's predictions for the near future. Those are about technology entering our bodies and our brains, fundamentally transforming us.

In *The Age of Spiritual Machines,* written in 1998 and published in 1999, Kurzweil went way, way out on the proverbial limb. He did something popular Nostradamus-style "futurists" always avoid: He predicted not only what he thought would happen but also when. He did so unambiguously, with specific numbers, and in print. He made clear his methodology in ways others could replicate and critique—which they certainly did. In 2001 he opened an award-winning Web site, www. kurzweilai.net, complete with a daily e-mail newsletter on which is posted papers and news reports that address the possibilities that his predictions might be right or wrong. He even collaborated on a book of essays by some of his most scathing critics—*Are We Spiritual Machines? Ray Kurzweil vs. the Critics of Strong AI,* albeit one in which he gets to comment on their challenges. Kurzweil majored in creative writing as well as computer science at MIT. (By the end of his sophomore year, he'd taken all the computer science classes MIT had to offer. There weren't that many of them in 1967.) One suspects that the skills he picked up helped gain him a much wider audience than his peers who languish in the sometimes difficult-to-plow-through ghetto of science writing. Kurzweil became the John the Baptist of the technologists, offering broad and striking visions of the future. Even those who disagree with him are forced to acknowledge their debt. As you will see in the next chapter, he cleared the way for Bill Joy, the former Sun chief scientist, to get so agitated that Joy felt compelled to publish his Hell Scenario.

In his narrative describing the end of the first decade of the 21st century, Kurzweil portrays a world in which individuals primarily use computers that are portable, as many as a dozen of which are commonly worn as wristwatches, rings, earrings, eyeglasses and clothing. They provide services previously associated with cell phones, personal digital assistants and credit cards. But they also provide directions, play games and allow you into secure buildings. Keyboards still exist in Kurzweil's 2009, but the majority of text is created by devices that recognize speech. Many commercial transactions—such as purchasing gasoline at a 7-Eleven—are conducted with an animated personality. That is to say, the gas pump has a face and a voice. Supercomputers match the estimated hardware capabilities of the human brain—20 million billions of calculations per second. An ordinary $1,000 personal computer can perform about a trillion calculations per second. Children learn to read on their own, using their personal computers, before grade school. Synthetic voices sound fully human. Persons with disabilities—the deaf, the blind and the lame—are rapidly overcoming their handicaps through intelligent devices. Telephones that translate languages are common, though still buggy. Sexual experience at a distance—with other flesh-and-blood people as well as synthetic computer-generated virtual partners—is emerging but is not great. Intelligent assistants with animated personalities routinely assist with finding information, answering questions and conducting transactions. Both males and females prefer female personalities for their computer-based intelligent assistants. "The two most popular are Maggie, who claims to be a waitress in a Harvard Square café, and Michelle, a stripper from New Orleans. Personality designers are in demand, and the field constitutes a growth area in software development," Kurzweil writes.

Work groups are dispersed. People are successfully working together despite living in different places. The average household has more than a hundred computers—which is to say chips—most of which are embedded in appliances, entertainment and telephone systems, automobiles and even key chains. (Remember, Kurzweil is predicting this for 2009. Count the number of tiny colored lights in your household.) Household robots have emerged but are not yet fully accepted. Intelligent roads are in use, Kurzweil predicts, primarily for long-distance travel. They allow you to sit back and relax after your car's smart cruise control locks onto the control sensors. Local roads, though, are still predominantly conventional. Privacy has emerged as a major political issue. There is a growing anti-technology neo-Luddite movement. There is continuing concern with an underclass that has been left behind by a lack of skills, although the size of this underclass appears to be stable, since most people seem to be keeping up. Most visual art—especially performance art, photography and video—is a collaboration between humans and their software. Human musicians routinely jam with computer-generated musicians. Those who wish to create music no longer absolutely need fine finger coordination, the kind you see in traditional pianists and guitarists. There is a surge of interest in new "air" controllers, in which you create music by moving your hands, feet, mouth and other body parts. Virtual realities are convincing in terms of what you see and hear, although transmitting the sense of touch is still rocky. In fact, the creation of virtual reality environments is a form being addressed by artists. Wars are decided by those who have the best and most secure computers, Kurzweil predicts. Most conflicts are not between nations but between nations and terrorists. The greatest threat to national security comes from bioengineered weapons.

Cancer and heart disease mortality is being reduced by bioengineering. There is serious speculation as to whether computers will ever be conscious, like humans.

Phew.

Okay, then comes Kurzweil's narrative for the end of the second decade of the century—2019. In this prediction, based on The Curve, computers are largely invisible. They are embedded everywhere—in your walls, tables, desks, clothing and body. People generally communicate with them the way they might with a human assistant—through speech, gestures and facial expressions, which the computers recognize and respond to, as in the Steven Spielberg film *Minority Report*. Keyboards are rare, as are connecting cables. Phone calls routinely include high-resolution three-dimensional moving images piped directly into the eye. They can fool people into thinking a person is physically present. Automated language translation is high-quality and common. Some manufacturers use nanotechnology to assemble objects one atom at a time, like masons laying bricks, but that process is still not widespread. Paper books and documents are rare, since handheld displays are about the size, weight and resolution of a magazine. Adult human workers spend the majority of their time acquiring new skills and knowledge.

In Kurzweil's 2019, it is difficult to tell if a person might be blind, deaf or paraplegic because of the way computers have been so completely integrated into the body. As a result of sophisticated brain scanning, significant progress has been made in understanding how the human brain works, and that has been translated into how machines work. The price of a computer with the same raw power as a human brain is $4,000 in 1999 dollars. Of the total computing capacity of the human race—all human brains, plus all the technology the species has created—10 percent is nonhuman.

You can do virtually anything with anybody regardless of physical proximity. The technology to accomplish this is easy to use and ever-present, since it does not require any gear that is not normally worn or implanted. Household cleaning robots finally work. The vast majority of commercial transactions include a simulated person with high-quality animation and understanding of language. Often, no human is involved in this transaction at all; it's common for the human to have asked his automated personal assistant to, say, book the reservation. Transportation fatalities are rare because of the intelligence in both cars and highways. Violent crime is also down because of tiny cameras everywhere. Privacy is a difficult thing to come by. Kurzweil is reminded of Phil Zimmerman's line. Zimmerman created the leading encryption software of the early 21st century, Pretty Good Privacy: He once said, "In the future, we'll all have fifteen minutes of privacy." This turned on its head Andy Warhol's line from the 1970s that in the future, "every person will be world-famous for fifteen minutes."

The personality of their intelligent assistant is something people spend a lot of time crafting. You can model it on actual persons, including yourself, a rock star or the secretary of state. Or you can create a personality that, like characters in a work of fiction, combines traits. In fact, people are beginning to have relationships with their automated assistants, who serve as companions, teachers, caretakers and lovers. These assistants are not yet regarded as equal to humans in the subtlety of their personalities, although there is disagreement on this point. After all, automated personalities are superior to humans in some ways—they have very reliable memories, and their responses are not governed by wildly unpredictable hormonal fluctuations.

In 2019, "an undercurrent of concern is developing with regard to the influence of machine intelligence," Kurzweil writes. "There continue to be differences between human and machine intelligence, but the advantages of human intelligence are becoming more difficult to identify and articulate. Computer intelligence is thoroughly interwoven into the mechanisms of civilization and is designed to be outwardly subservient to apparent human control." Humans are required by law to take responsibility for the actions of their machine-based intelligences. On the other hand, few decisions are made without consulting them. The life processes encoded in the human genome are now understood, including those that underlie aging, cancer and heart disease. As a result, life expectancy, which has marched steadily upward since the 1800s, continues to rise. It is now over 100. Virtual artists in all of the arts are emerging and are taken seriously. These machine artists, musicians and authors usually collaborate with humans who have contributed to their knowledge base and techniques. However, interest in their output has gone beyond the mere novelty of machines being creative.

The primary threat to security comes from small groups combining human and machine intelligence using unbreakable encrypted communications. Their weapons of choice are software viruses and bioengineered disease agents. Most flying weapons are tiny—some the size of insects. Kurzweil imagines having a conversation about security with a 2019 person, Molly. Molly reports, "It's true that the century so far has seen much less bloodshed. But the other side of the coin is that the technologies are so much more powerful today. If something did go wrong, things could spiral out of control very quickly. With bioengineering, for example, it feels a little like all ten billion of us are standing in a room up to our knees in a flammable fluid, waiting for someone—anyone—to light a match."

Kurzweil replies, "But it sounds like a lot of fire extinguishers have been installed."

"Yeah, I just hope they work," Molly replies.

Now comes Kurzweil's 2029. This is about the time that those who follow The Singularity expect it to kick in, and Kurzweil comes close. But note how Kurzweil's predictions, for all their breathtaking change every ten years, are still essentially incremental. They all seem natural and obvious to those who are experiencing them.

In 2029, says Kurzweil, a $1,000 unit of computation—one hesitates to call it a computer, because how can you tell where one leaves off and another begins?—has the hardware capacity of 1,000 human brains. (That is 20,000,000,000,000,000,000 calculations per second—20,000 million billion.) Of the total computing power of the human race—all human brains plus all the technology that the species has created—more than 99 percent is nonhuman. Remember, this is a mathematical projection of The Curve into the near future. It is easily within the lifetime of most people reading this book.

The architecture of the human brain has been much more difficult to decode than had been expected at the turn of the century, says Kurzweil about this 2029. Nonetheless, many areas of the brain are now understood. This knowledge has been used to create the computers of the day. That's why networks modeled on the brain's neural activity are substantially faster, more powerful and more capable of remembering than is the human brain. People have implanted in their eyes—either permanently or in something like contact lenses—the ability to see three-dimensional computer-generated images in addition to whatever is in front of them (handy for conference calls). Ear implants commonly provide communications both ways to the Web, in addition to helping people hear. Mind implants allow you to trade thoughts with computers. You can buy all the long-term memory and reasoning you want. It is not yet possible to download knowledge directly, however. Learning still requires time-consuming human experience and study. This is how humans spend most of their day. Automated agents, meanwhile, are learning on their own without being spoon-fed by humans. Computers have read all available human and machine-generated literature. Significant new knowledge is created by machines, which, unlike humans, easily share knowledge with each other.

It is as hard to tell if a person is physically handicapped as it is to guess his original hair color. Everything from nerve stimulation to intelligent prosthetics makes blindness, deafness and paralysis as quaint as whooping cough. In fact, most people use devices originally aimed at the disabled. Neural implants allow family members to sit around the living room enjoying one another's company without being physically proximate. The human population has leveled off at 12 billion people. The economy continues to boom. Basic necessities of food, shelter and security are available for the vast majority of the human population. There is almost no human employment in manufacturing, agriculture and transportation. Human and nonhuman intelligences are focused on the creation of knowledge. The largest profession is education.

Aging has slowed dramatically. Most diseases can be prevented or reversed. Drugs are individually tailored to an individual's DNA, so there is nothing like the 100,000 annual deaths even from properly used prescription drugs that had been common in the United States. Robots the size of blood cells—nanobots, as they are called—are routinely injected by the millions into people's bloodstreams. They are used primarily as diagnostic scouts and patrols, so if anything goes wrong in a person's body, it can be caught extremely early. The nanobots also serve as an early warning system against bioengineered pathogen attacks. To a more limited extent, they are used as repair agents in the bloodstream and as building blocks for organs built from scratch. Life expectancy is now around 120 years. Fifty-year-olds look like 35-year-olds did in the year 2000. Significant attention is being paid to the psychological ramifications of a substantially increased human life span. Continuing extensions to the human life span will involve further use of bionic organs, including portions of the brain.

A sharp division no longer exists between the human world and the machine world. Machine intelligence is derived from the design of human intelligence, and human intelligence is enhanced by machine intelligence. Through neural implants of computer hardware into the brain, humans have greater understanding, memory and perception than ever. At the same time, many machines have personalities, skills and knowledge derived from the reverse engineering of human intelligence. In either case, the implants mean that human intelligence is popped directly into machines and vice versa.

Defining what constitutes a human being is a significant legal and political issue. It is difficult to cite human capabilities that machines can't match. For reasons of political sensitivity, machine intelligences generally do not press the point of their superiority. The rapidly growing capability of machines is controversial, but there is no effective resistance to it. Humans realize that disengaging the human-machine civilization from its dependence on machine intelligence is not possible. Since machine intelligence was initially designed to be subservient to human control, it has not presented a threatening face to the human population. Discussion of the legal rights of machines is growing. Those not embedded in a human brain are independent. Many of the leading artists, for example, are machines. These artists say they are conscious and have as wide an array of emotional and spiritual experiences as their human progenitors. These claims are largely accepted.

Molly is Kurzweil's foil. She is his alter ego—"the woman I would want to be," he says. Molly reports from the year 2029 that she is having an affair with her personal assistant. "George is a different person every day," she says admiringly. "He just grows and learns constantly. He downloads whatever knowledge he wants from the Web and it becomes part of him. He's so smart and intense, and very spiritual." It also turns out that George, because of mind-machine sensory inputs, can make Molly smile in every possible way.

"I'm awfully happy for you," the Kurzweil character says. Then he thinks of Molly's husband. "But how does Ben feel about you and George?" he asks.

"He wasn't too crazy about it, that's for sure," replies Molly.

"But you've worked it out?"

"We've worked it out, all right. We broke up three years ago."

Put Kurzweil's 2029 in context here. Suppose you decide to go out and buy a new home today. The year 2029 will come well before you pay off your 30-year mortgage. If you have a newborn child, this is the world Kurzweil says she will be entering before she is old enough to get her medical degree.

After Kurzweil finishes describing the world of 2029, he really torques out. He takes his inexorable Curve logic out to 2099. If nothing else, his predictions demonstrate what happens when you claim exponential acceleration will never level off. In this view, the change encountered over the course of the 21st century has been like ten centuries in one.

In this world, humans exist who are machine-based. They no longer have neurons, flesh or blood. They don't have "wetware." Instead, their brains are based on electronic and photonic equivalents. That's because the human brain has been reverse-engineered. It has been fully scanned, analyzed, understood and translated into machine analogues.

In Kurzweil's 2099, these "software-based humans" vastly exceed in number those still using native neuron-based brains for their thinking. Such a software-based human is not dependent on a physical body. Whenever and however they want, they can create a physical presence either through virtual reality or by creating the kind of reconfigurable nanobot swarms that Michael Crichton wrote about in his novel *Prey*, set in his usual day-after-tomorrow near-present. In this world, even those human intelligences still using carbon-based neurons plentifully augment how they think, feel, see and sense. "Humans who do not utilize such implants are unable to meaningfully participate in dialogues with those who do," Kurzweil writes. People in the early 21st century without e-mail got left out of a lot of conversations, too.

In his dialogue with the Molly of 2099, Molly acknowledges, "The spiral of accelerating returns lives on."

"I'm not surprised," responds the Kurzweil character. "Anyway, you do look amazing."

"You say that every time we meet."

"I mean you look twenty again, only more beautiful than at the start of the book."

"I knew that's how you'd want me."

"Great, now I'm going to be accused of preferring younger women."

"My body right now is just a little fog swarm projection. Neat, huh?"

"Not bad, not bad at all. You feel pretty good, too."

"I thought I'd give you a hug, I mean the book's almost over."

"This is quite a technology."

Molly has transcended. She has become essentially indistinguishable from the Christian portrayal of angels. She no longer has a permanent physical body. She can, however, project a material or virtual body at will, as she did to hug the Kurzweil character and dazzle him with her beauty. Bit by bit, over the years, everything that makes Molly herself was scanned into the larger knowledge base of human creations. There is no distinction any longer between humanity and its creations. As a result, Molly is immortal. She has, in Kurzweil's words, gone over to the other side. She has become pure knowledge.

Kurzweil introduces the idea of MOSHs—Mostly Original Substrate Humans. The drollery here is that *substrate,* for the last third of the 20th century, usually referred to silicon—the stuff that computer chips were built on. In Kurzweil's 2099, it means meat. A Mostly Original Substrate Human is one of those people who, for esthetic or ethical or moral reasons, insists on primarily using the flesh that humans were heir to for her living and thinking. MOSHs are honored and legally protected by "grandfather legislation," in Kurzweil's projection. They deserve it for being the progenitors of what humankind has become. But they cannot compete with those who have gone over to the other side. For those like Molly, it happened one incremental step at a time, as they chose to move on. Shedding the physical body was like shedding a large house that was a nuisance to maintain and no longer matched people's lives. It simplified things.

Work is still the point of human life in 2099. But those on the other side can do things of which MOSHs never dreamed. Molly is creating symphonies with frequencies, tempos and musical structures that MOSHs are not capable of understanding. All humans are now entrepreneurs, creating new knowledge, for which other people are willing to pay. A lot of time is filled by family, of which there are now many generations. To get to a point of peace and serenity, Molly meditates. She lives for moments of spiritual experience—"an artful expression, a moment of serenity, a sense of friendship."

Back in the early 21st century, in his office near Wellesley, I ask Kurzweil how he originally got into this prediction business. "Fairly conservatively," he replies. "I realized I needed to do some better modeling of technology trends to be a better inventor and entrepreneur. The world really was different by the time I finished a project. Early conceptions didn't necessarily pan out. Either they were too fast or too slow. I really needed to understand where technology would be—what the enabling forces would be. That's part of the craft of being an inventor. So that's how I got into this. Then it did take on a life of its own. This is what comes out of a scientific examination of technology trends. You come up with these perspectives and try to examine what that means for human civilization. It does create a philosophy. I try to actually overcome some of the older ways of thinking that are more death-oriented.

But I came to it through the scientific route. I wasn't trying to reverse-engineer a religious vision."

T HE ESSENCE OF The Heaven Scenario is stealing fire from the gods, breathing life into inert matter and gaining immortality. Our efforts to become something more than human have a long and distinguished genealogy. Tracing the history of those efforts illuminates human nature. In every civilization, in every era, we have given the gods no peace.

Efforts to transcend our origins begin in the most primitive of times. Sorcerers would create a likeness of a living thing and, with the rituals of magic, seek to animate it. In our earliest epic, the Sumerian tale of Gilgamesh, the climax is the king seeking, finding and losing the secret of immortality. Barely three pages into Genesis, the serpent is telling Eve that she doesn't have to worry about losing immortality by tasting of the fruit of knowledge. "No! You will not die! God knows in fact that on the day you eat it your eyes will be open and you will be like gods," he says. (Not coincidentally, one of the biggest attractions of Christianity, even as an upstart religion, was its promise of eternal life.) Ancient Greece was full of heroes harassing the deities. Prometheus not only created humans, teaching them many of their useful skills, but he filched fire for them. Daedalus confounded King Minos by crafting wings of wax and feathers to flee Crete. His son Icarus, of course, flew too close to the sun, giving us one of our earliest warnings against taking presumptuous pride in our technologies. But remember, Daedalus did succeed—his mythic wings worked, and ancient Greece gave us the tools of logic, skepticism and natural philosophy that became the underpinnings of science. The market for harvest and fertility goddesses has never been the same.

The cultural humanism of the Renaissance pushed ancient pieties aside. *Make something of yourself!* was its message to mankind. Human nature was not predetermined by anybody's secondhand image and likeness, in this view. We could shape ourselves to make the world better. Pico della Mirandola's 1486 *Oration on the Dignity of Man,* the manifesto of the Italian Renaissance, eloquently centers all attention on human capabilities. In it, says God to Adam: "We give you no fixed place to live, no form that is peculiar to you, nor any function that is yours alone. According to your desires and judgment, you will have and possess whatever place to live, whatever form, and whatever functions you yourself choose."

Indeed, in 1580, the kabbalistic Jews of Prague imagined creating a Golem—an artificial man made from clay—who would protect them from persecution. Galileo, as he laid the foundations of modern science, believed that for peering directly into the mind of God, there was nothing like a telescope. It was more profitable to study the deity's handiwork than it was to study scripture. "Philosophy," he wrote, "is written in this grand book, the universe, which stands continually open to our gaze."

The Age of Enlightenment elevated scientists over priests. The notion was that the logically deduced laws that governed human behavior were an even purer expression of God's law than that which could be gathered from scripture. Since God's law will always work to good ends, the same must be true of the natural laws governing our individual lives, this hypothesis concluded. Francis Bacon was among the first to see critical reasoning as a means of finding the destiny and nature of man. "The formation of ideas and axioms by true induction is no doubt the proper remedy to be applied for the keeping off and clearing away of idols," Bacon wrote in *Novum Organum* in 1620. He saw it as a new grounding for morality and, indeed, the perfection of society. The idea of a rationally discovered natural law to achieve a heaven on earth would fuel both the American and French revolutions. Is the universe like clockwork? Then we, too, could be universal clockmakers. One of the wonders of the Paris salons of the 1730s was a complex mechanical duck created by the ingenious engineer Jacques de Vaucanson. It was able convincingly to waddle, eat fish and poop. It was the incarnation of René Descartes' idea of nature as machine. It was all the rage.

"Unlimited progress" was the future seen by Marie-Jean-Antoine-Nicolas de Caritat, Marquis de Condorcet, in his *Sketch for a Historical Picture of the Progress of the Human Mind*. A world in which "death would result only from extraordinary accidents" was foreseen by Condorcet, who was a friend of Voltaire. "No doubt man will not become immortal, but cannot the span constantly increase between the moment he begins to live and the time when naturally, without illness or accident, he finds life a burden?" This was published in 1795.

In 1780, Benjamin Franklin wrote to the chemist, biologist and minister Joseph Priestley, "The rapid progress true science now makes, occasions my regretting sometimes that I was born too soon. It is impossible to imagine the height to which may be carried, in a thousand years, the power of man over matter. We may, perhaps, deprive large masses of their gravity, and give them absolute levity, for the sake of easy transport. Agriculture may diminish its labor and double its produce: all disease may by sure means be prevented or cured (not excepting even that of old age), and our lives lengthened at pleasure, even beyond the antediluvian standard. Oh that moral science were in as fair a way of improvement, that men would cease to be wolves to one another, and that human beings would at length learn what they now improperly call humanity."

If wresting power from the gods and seeking to transcend the human condition is fundamental to who we are, being terrified by the implications of our audacity is equally primal. Are we alarmed by the prospect of computers becoming human? Could that come from guilt? From fear? What happens if we create a monster? Will our silicon successes punish us? That is the horror and the dread. Call it "The Frankenstein Principle." The history of that idea begins in the summer of 1816, when Mary Wollstonecraft Shelley, while still a teenager, came to her spectacular leap of imagination. More about guilt, fear, horror and dread in the next chapter, on The Hell Scenario.

To continue tracing the history of stealing fire from the gods, however, in 1859, Charles Robert Darwin published *On the Origin of Species by Means of Natural Selection, or the Preservation of Favoured Races in the Struggle for Life*. It changed everything. By placing humans in a history of millions of years in which all living things are connected and nothing is constant, it transformed the way people thought about God and themselves. Human nature is not etched in stone. This insight became the dominant metaphor of our age. The still-controversial *Origin of Species*—described by the philosopher Daniel Dennett as the greatest idea ever to occur to a human mind—became the most lastingly influential book of the modern era, surpassing even Freud and Marx.

A golden age of fiction about the future followed, much of it starting out quite utopian. With his series about extraordinary voyages, the Frenchman Jules Verne caught the enterprising spirit of the age, along with its uncritical fascination with inventions and scientific progress. *A Journey to the Center of the Earth*, in 1864, and *From the Earth to the Moon*, in 1865, rode the sense of optimism and godlike vistas humankind saw opening up for itself. The universe was our dominion, the environment ours to be mastered, from the planet's core to its orbit. In *Twenty Thousand Leagues Under the Sea*, Verne introduced one of the world's first superheroes, Captain Nemo, with his baroque, extravagant submarine, the *Nautilus*. That story so captured the sense of adventure and limitless possibilities of the age that Disney made it into a movie starring Kirk Douglas in 1954. The same year the United States Navy christened the first nuclear submarine the *Nautilus*. In 1958 that vessel performed the mind-boggling, top-secret, hitherto considered impossible feat of sailing to the North Pole. (The Soviets were not amused.) Verne's *Around the World in Eighty Days* was made into a film in 1956 starring the aggressively international cast of David Niven, Cantinflas, Charles Boyer and Jose Greco.

A few decades after Verne, the Englishman Herbert George "H. G." Wells, in his more than 100 books, produced even more shocking, overwhelming divinations. His 1895 *The Time Machine* anticipated Einstein's theories in postulating time as the fourth dimension. Even eerier to a 21st-century reader is his 1896 *The Island of Dr. Moreau,* in which a mad scientist uses bioengineering to transform animals into human creatures. In movie versions, Dr. Moreau inspired such diverse actors as Charles Laughton, Burt Lancaster and Marlon Brando. Wells' *The Invisible Man* in 1897 was a Faustian story of a scientist who tampered with human nature in pursuit of transcendent, superhuman powers. *The War of the Worlds,* published in 1898, was so powerful a tale of Martians invading the earth that when Orson Welles' news-bulletin-style Mercury Theater radio dramatization was broadcast on October 30, 1938, widespread panic ensued.

H. G. Wells' prescience was extraordinary. In *The World Set Free* he wrote, "Nothing could have been more obvious to the people of the early twentieth century than the rapidity with which war was becoming impossible. And as certainly they did not see it. They did not see it until the atomic bombs burst in their fumbling hands." He wrote that in 1914. In 1920, he described human history as becoming "more and more a race between education and catastrophe."

By the middle of the 20th century, after the Industrial Revolution had delivered one miracle after another—electricity, the telephone, radio, the airplane—a remarkable change occurred in our culture and values. People no longer found startling the idea that man had acquired godlike powers to reshape himself and his environment. (The first Superman comic book appeared in 1938.) They did, however, begin to anticipate The Hell Scenario. They questioned whether technological change really equaled "progress."

Meanwhile, in the late 20th century, fiction continued to be the key forum pushing the boundaries of our imagination about what could happen to the future of human nature. Novels stretched our conceptions of human-created Heaven, kick-starting our thinking about what was possible, forcing us to change our perception of what was serious. Arthur C.

Clarke's *2001: A Space Odyssey*—vividly brought to the screen in 1968 by Stanley Kubrick—introduced us to HAL, the intelligent computer. The biochemist Isaac Asimov shaped an entire generation of roboticists with his stories, starting with *I, Robot.* What would machines that had ethics behave like? He imagined what he called the Three Laws of Robotics: "The First Law—A robot may not injure a human being or, through inaction, allow a human being to come to harm. The Second Law—A robot must obey orders given it by human beings except where such orders would conflict with the First Law. The Third Law—A robot must protect its own existence as long as such protection does not conflict with the First or Second Law." Later, in his Foundation series, Asimov explored the notion that the social sciences and history had fundamental laws as immutable and calculable as those of mathematics or physics and that these truths, properly understood, could help perfect society. These stories molded the scenario-planning industry.

Popular culture continued the quest to imagine transcendence. Michael Crichton's 1972 novel, *Terminal Man,* became the television series *The Bionic Man.* The hottest writer in Hollywood in the early 21st century, Philip K. Dick, had been dead for two decades. The scruffy, poverty-stricken, five-times-married, clinically paranoid Dick, a garage philosopher, autodidact and conjurer of false realities, never cared much about robots or space travel. He was, however, fascinated by hallucinatory worlds in which there is a blur between what is real and what is not. Needless to say, that has become a very hot subject as we try to process living in a time that includes telekinetic monkeys. Dick's work has become the stuff of big budgets, exalted directors and illustrious actors, such as Ridley Scott's *Blade Runner,* starring Harrison Ford; Paul Verhoeven's *Total Recall,* with Arnold Schwarzenegger; Steven Spielberg's *Minority Report,* starring Tom Cruise; and John Woo's *Paycheck,* with Ben Affleck and Uma Thurman. Dick's sensibility is all over *The Matrix.*

The impact of such storytelling on global culture should not be underestimated. In my travels I found it hard to find a cutting-edge researcher today who, when asked about the inspiration for his creations, did not reply by simply pointing to a shelf in a place of honor. There invariably sat expensively collected fables of the future that shaped his youth. For Kurzweil, it was the Tom Swift Jr. series.

"My personal philosophy," Kurzweil says, "is there is no problem or challenge that comes along that there isn't an idea to overcome that problem. I first encountered this idea in the Tom Swift series, which I have over there. Actually, those books on the left is the Tom Swift Sr. series. Those are original books, almost 100 years old." The ones on the right were the books from the Tom Swift Jr. series of 1954 to 1971 that Kurzweil devoured when young, such as the 1956 *Tom Swift in the Caves of Nuclear Fire.*

"I've been collecting them. I had the whole series, but I've been re-collecting them. If you remember how those books all progressed, Tom Swift Jr. would get into some jam, along with his friends and usually the whole human race. He would kind of go into his lair. That's what kept you reading—how is he going to get out of this jam? But it was always an idea, some clever insight. Really, the moral of each of these books was the power of the idea. There is no problem that you can get into where some well-leveraged idea wouldn't save the day."

Thence come our ideas of transcending origins and becoming something more than we are. Thence comes our demonstration that stealing fire from the gods is at the heart both of human nature and of The Heaven Scenario.

To expect the unexpected shows a thoroughly modern intellect.

—Oscar Wilde

"Why make people inquisitive, and then put some forbidden fruit where they can see it with a big neon finger flashing on and off saying 'THIS IS IT!'?," says the angel Aziraphale.

"I don't remember any neon," replies the demon Crowley.

"Metaphorically, I mean. I mean, why do that if you really don't want them to eat it, eh? I mean, maybe you just want to see how it all turns out. Maybe it's all part of a great big ineffable plan. All of it. You, me, him, everything. Some great big test to see if what you've built all works properly, eh? You start thinking: it can't be a great cosmic game of chess, it has to be just very complicated Solitaire."

—Neil Gaiman and Terry Pratchett, *Good Omens*

Scene: Alan Turing in the Bell Labs cafeteria, 1943

His high pitched voice already stood out above the general murmur of well-behaved junior executives grooming themselves for promotion within the Bell corporation. Then he was suddenly heard to say: "No,

I'm not interested in developing a powerful brain. All I'm after is just a mediocre brain, something like the President of the American Telephone and Telegraph Company."

—Andrew Hodges, *Alan Turing: The Enigma of Intelligence*

———w———

WASHINGTON IS A tribal town. One of its rituals involves code words. These evoke vast and nuanced concepts to those in the know, while sometimes appearing meaningless to outsiders. One of these words is *serious*, as in "He is a serious person," or not, or "That is a serious idea," or not. *Serious* does not necessarily have anything to do with whether the person or idea is correct, important or valuable, no matter what insiders would have you believe. *Serious* means that the notion or individual has been cleaned up, molded and adopted by sponsors already vetted as serious. It implies that the idea or person is deemed ready for admittance to the sacraments of authority—such as congressional hearings—which can lead to that Holy Grail, federal funding. It suggests a certain WASP respectability. It basically means housebroken.

That is why it is interesting to watch The Heaven Scenario begin to be taken seriously in parts of Washington. Two very serious organizations are the National Science Foundation and the United States Department of Commerce. They became regarded as serious in part because of their sober reputation for not taking injudicious chances. They also have wildly separate constituencies. Thus it is pretty remarkable to see them team up to produce a Washington policy document saying, "It is time to rekindle the spirit of the Renaissance" to achieve "a golden age that will be a turning point for human productivity and quality of life."

The 415-page, three-pound pronouncement entitled *Converging Technologies for Improving Human Performance* comes complete with Power-Point-ready bullet points laying out the radical evolution of humans for those with the attention span of a Cabinet secretary. Note how many exceed even some of Kurzweil's more startling predictions. Note also, because it is a Washington policy document, how quickly it cuts to the chase, aiming at constituencies in defense, in industry and in management of the economy, while genuflecting in the direction of many potentially important adversaries.

The bullet points in part say that in the next 10 to 20 years the advantages of radical evolution will include the following:

- Direct connections between the human brain and machines "will transform work in factories, control automobiles, ensure military superiority, and enable new sports, art forms and modes of interaction between people."

- Wearable sensors will "enhance every person's awareness of his or her health condition, environment, chemical pollutants, potential hazards, and information of interest about local businesses, natural resources, and the like."

- Teams will be able to "cooperate profitably across traditional barriers of culture, language, distance, and professional specialization."

- "The human body will be more durable, healthy, energetic, easier to repair, and resistant to many kinds of stress, biological threats, and aging processes."

- "Machines and structures of all kinds, from homes to aircraft, will be constructed of materials that have exactly the desired properties, including the ability to adapt to changing situations, high energy efficiency, and environmental friendliness."

- Technologies will "compensate for many physical and mental disabilities and will eradicate altogether some handicaps that have plagued the lives of millions of people."

- "National security will be greatly strengthened by lightweight, information-rich war fighting systems, capable uninhabited combat vehicles, adaptable smart materials, invulnerable data networks, superior intelligence-gathering systems, and effective measures against biological, chemical, radiological and nuclear attacks."

- "Anywhere in the world, an individual will have instantaneous access to needed information."

- "Engineers, artists, architects, and designers will experience tremendously expanded creative abilities," in part through "improved understanding of the wellsprings of human creativity."

- "The vast promise of outer space will finally be realized by means of efficient launch vehicles, robotic construction of extraterrestrial bases, and profitable exploitation of the resources of the Moon, Mars, or near-Earth asteroids."

- "Average persons, as well as policymakers, will have a vastly improved awareness of the cognitive, social, and biological forces operating their lives, enabling far better adjustment, creativity, and daily decision making."

* "Factories of tomorrow will be organized" around "increased human-machine capabilities."

* "Agriculture and the food industry will greatly increase yields and reduce spoilage through networks of cheap, smart sensors that constantly monitor the condition and needs of plants, animals, and farm products."

* "Transportation will be safe, cheap, and fast."

* "Formal education will be transformed" by an understanding of "the physical world from the nanoscale through the cosmic scale."

It concludes: "The twenty-first century could end in world peace, universal prosperity, and evolution to a higher level of compassion and accomplishment." It may be "that humanity would become like a single, distributed and interconnected 'brain.'"

Mihail C. Roco and William Sims Bainbridge of the National Science Foundation are the co-authors of this document. Roco is a compact man who speaks English with a thick accent that even some of his colleagues assume is Russian. This turns out to be a very big mistake. He was born in Bucharest, Romania, and is incensed and incredulous that anyone might be ignorant enough to make an error about this.

Roco is senior adviser for nanotechnology at the National Science Foundation, chair of the National Science, Engineering and Technology Council's subcommittee on nanoscale science, engineering and technology, chair of the President's National Science and Technology Council's Interagency Nanoscience, Engineering and Technology working group, and chair of the National Science and Technology Council's subcommittee on nanoscale science, engineering and technology. He is credited with 13 inventions.

Roco's version of The Heaven Scenario is driven by much more than The Curve. He sees transcendence resulting from no less than the convergence of all human wisdom. "The hallmark of the Renaissance was its holistic quality," he and Bainbridge write, "as all fields of art, engineering, science and culture shared the same exciting spirit and many of the same intellectual principles. A creative individual, schooled in multiple arts, might be a painter one day, an engineer the next, and a writer the day after that. However, as the centuries passed, the holism of the Renaissance gave way to specialization and intellectual fragmentation." Today, that's over, he says. "It's time to rekindle the spirit of the Renaissance."

The GRIN technologies—genetics, robotics, information technology and nanotechnology—drive a unified theory of everything, in this view. The information technology is the sire of them all. Here's how they intertwine.

One of the most challenging thinkers about the future of human genetics, the G in the GRIN technologies, is Gregory Stock. He is director of the Program on Medicine, Technology, and Society at the School of Medicine of the University of California at Los Angeles.

Stock's version of The Heaven Scenario departs from Kurzweil's. He doesn't think humans will transcend because of computers. He thinks humans will transcend because of genetic engineering. Such biological remodeling is "a plausible way for people to overcome their bodily frailties, but a larger game is afoot," he says. It is "biology's bid to keep pace with the rapid evolution of computer technology."

"No one really has the guts to say it, but if we could make better human beings by knowing how to add genes, why shouldn't we?" he approvingly quotes James Watson, co-winner of the Nobel prize for discovering the structure of DNA, as saying. The titles of several of Stock's books display his position. One is called *Redesigning Humans: Our Inevitable Genetic Future.* An earlier one is called *Metaman: The Merging of Humans and Machines into a Global Superorganism.*

"To not be human in the sense we use the term now" is the fate of our descendants, Stock says. We will soon see humans as physically and intellectually divergent as "poodles and Great Danes." But the passing of people like us is hardly a tragedy, he believes. "Unlike the saber-toothed tiger and other large mammals that left no descendants when our ancestors drove them to extinction, *Homo sapiens* would spawn its own successors by fast-forwarding its evolution."

That's not far off, says Stock. It's a whole lot closer than the "distant space travel we see in science fiction movies." He sees it as the inevitable outcome of the decoding of the human genome. "We have spent billions to unravel our biology, not out of idle curiosity, but in the hope of bettering our lives. We are not about to turn away from this." Genetic transcendence "does not hinge on some cadre of demonic researchers hidden away in a lab in Argentina trying to pick up where Hitler left off. The coming possibilities will be the inadvertent spin-off of mainstream research that virtually everyone supports. Infertility, for example, is a source of deep pain for millions of couples. Researchers and clinicians working on in vitro fertilization (IVF) don't think much about future human evolution but nonetheless are building a foundation of expertise" that will "one day be the basis for the manipulation of the human species," he says matter-of-factly.

There are two kinds of genetic engineering. The first is called somatic gene therapy. The phrase comes from the Greek word *soma,* meaning "body." It's intended to fix genes gone bad in one person. It treats wrenchingly awful diseases such as cystic fibrosis, immune deficiency disorders, sickle-cell anemia and hemophilia. Somatic gene therapy usually is not terribly controversial—except when it kills the person on which it is being tested, and especially when that person was not adequately informed of the risk, as happened in the 1999 death of Jesse Gelsinger. The reason somatic therapy is not hugely debated is that it only swaps out bad genes in specific areas—the lungs, the liver—of the person being treated. Changes are not passed on to succeeding generations. Thus it is usually viewed as not tremendously different from surgery. Even the Amish use it.

Where the balloon goes up is over germ-line interventions. *Germ* comes from the Latin word for "root." Germ-line engineering changes the genetic makeup of the embryo at the very start, altering the child's every cell. Thereafter, it changes every grandchild and great-grandchild propagated, forever. This is where the controversy over "designer babies" begins. It is the vision in which people start by selecting the color of their kid's eyes, and the next thing you know, you have "clusters of genetically enhanced superhumans who will dominate if not enslave us," as Stock puts it. Yet he dismisses such horror stories. "No one understood the powerful effects of the automobile or television at its inception," he notes. "Our blindness about the consequences of new reproductive technologies is nothing new, and we will not be able to erase the uncertainty by convening an august panel to think through the issues. No shortcut is possible. As always, we will have to earn our knowledge by using the technology and learning from the problems that arise."

If it's safe, it ain't sex. But risks in reproduction, we're wary about. That's why Stock is betting that we'll like artificial chromosomes. Right now, we have 23 chromosome pairs, with the chromosomes numbered 1 through 46. Messing with them is tricky—you never know when you're going to inadvertently step on unanticipated interactions. If, however, we add a new chromosome pair (numbers 47 and 48) to the embryo, the possibilities are endless. Think of it as scaffolding. It doesn't change anything itself. It just holds plug-in points where you can stick gene modules and their controls. "The auxiliary chromosome would be a universal delivery vehicle for gene modules fashioned by medical geneticists throughout the world," Stock says. He sees it as the safest way to substantially modify humans. It would minimize unintended consequences. On top of that, the insertion sites could have an off switch activated by an injection if we wanted to stop whatever we'd started. This would give future generations a chance to undo whatever we did, if they choose. It would also allow us to stick modifications into our kids that they could choose to turn on or off later in life, as they become adults. "When children who have received auxiliary chromosomes to improve some mental or physical characteristic grow up, they may want to give their own child the same advantage. They won't, however, want to pass on the outdated auxiliary chromosomes they received a generation earlier, any more than a middle-aged father today would try to give his Internet-savvy college-bound daughter that state-of-the-art typewriter he used for his term papers," Stock says. "Parents will want the most up-to-date genetic modifications available. Were these prospective parents' own modifications scattered through their chromosomes, cleaning them out and upgrading them would be tricky, but with changes confined to an auxiliary chromosome, a parent could simply discard the entire thing and give his or her child a newer version." He sees this as disarming the ethical argument, offered by the Council for Responsible Genetics, that germ-line engineering should be unconditionally banned because future generations screwed up by wrongful or unsuccessful germ-line modifications would have no control over the matter.

What would the impact of all this be? Genetic research is one of the ways the NSF report sees us quickly getting to "vastly improved awareness of the cognition, social, and biological forces" operating our lives. James Watson, the provocative co-discoverer of the structure of DNA, talks about molecular biology curing stupidity. "If you are really stupid, I would call that a disease," the Nobel laureate says. He also has few qualms about engineering beauty: "People say it would be terrible if we made all girls pretty. I think it would be great."

Stock—who with his deep-set, penetrating eyes, carefully barbered beard and youthfully bushy head of gunmetal-gray hair could moonlight as a local TV anchor—personally prefers the idea of becoming genetically altered flesh, rather than Kurzweil's kind of beyond-human machine. But he also sees his own Heaven Scenario as "tame" and near-term compared to Kurzweil's. Rudimentary artificial chromosomes already exist. Within a few years "we should have a fair idea of the size of the task facing future genetic engineers," he says. By then, "traditional reproduction may begin to seem antiquated, if not downright irresponsible." He sees his projections as not at all out of touch with reality, compared to Kurzweil's, which he characterizes as "far-fetched," "techno-exuberance" and a "huge leap of faith."

Perhaps that is another useful definition of human nature—the species that punctuates the most far-reaching discussions of its future with asides about how competitors stink.

Nanotechnology, the *N* in the GRIN technologies, means manipulating the unimaginably small. A nanometer is one billionth of a meter. It is the length of five carbon atoms in a row or the distance your fingernail grows in one second. If a nanometer were the size of your nose, a red blood cell would be the size of the Empire State Building, a human hair would be more than two miles thick, your finger would span the United States from the Atlantic to the Pacific, and a person would be taller than six planet Earths stacked one on top of the other. Just as there are two kinds of genetic engineering, there are two kinds of nanotechnology, one more far-reaching and controversial than the other. The first kind is the one already coming into existence. It reduces big things to sizes so astonishingly little that their behavior changes dramatically—transistors, for example, or the active ingredients in sunblock. By 2003, hundreds of tons of nanomaterials were being made in U.S. labs and factories. Such nanotechnology is expected to be a $1 trillion business by 2015—comparable to the gross national product of Canada.

The other kind of nanotechnology is the one its proponents say promises godlike powers, immortality and unimaginable wealth. Its detractors say it will doom us and every living thing on the planet. This form of nanotechnology is intended to work the other way around from the first. It involves taking individual atoms and stacking them into any large thing we want, from diamonds to spacecraft.

The person credited with inspiring nanotechnology is the late Richard Phillips Feynman. He became widely known as an American original for his wildly best-selling 1985 autobiography *Surely You're Joking, Mr. Feynman!* in which he discusses, in his relentless pursuit of knowledge, gambling with Nick the Greek, painting a naked female toreador, and accompanying a ballet on his bongos. Before that, Feynman was merely the renowned Caltech physicist who won the Nobel Prize for his work on quantum electrodynamics.

On December 29, 1959, at a meeting of the American Physical Society, Feynman gave an after-dinner talk entitled "There's Plenty of Room at the Bottom." At a time when the audience freshly remembered computers that used vacuum tubes the size of candy bars, he talked about storing "all the information that man has carefully accumulated in all the books in the world" in a cube "one two-hundredth of an inch wide—which is the barest piece of dust that can be made out by the human eye. So there is *plenty* of room at the bottom!"

In his lecture, Feynman described a world in which you "give the orders and the physicist synthesizes it. How? Put the atoms down where the chemist says, and so you make the substance." Make computers much smaller and therefore faster, he said. Make "mechanical surgeons"—nanobots—that could travel to trouble spots inside the body. To get things going he offered two prizes: $1,000 to the first person to make a working electric motor no bigger than 1/64 of an inch in any dimension, and another $1,000 to the first person to shrink text such that the entire *Encyclopaedia Britannica* would fit on the head of a pin. The prize for the motor was awarded almost immediately, in 1960. The text prize was awarded in 1985.

Much of nanotechnology at the beginning of the 21st century involved zapping relatively large amounts of things until they become small enough to acquire unusual powers. For example, the marvelously surnamed Richard E. Smalley of Rice University shared the 1996 Nobel prize for chemistry for hitting a batch of pure carbon with a special laser beam until the atoms rearranged themselves into a previously unknown molecule—a ball made of 60 atoms that looked like the kind of geodesic domes pioneered by Buckminster Fuller. The molecule is nicknamed the "buckyball" in Fuller's honor. Buckyballs and their cousins, the nanotube fibers, have many intriguing properties, including 60 times the strength of steel, the weight of plastic, the electrical conductivity of silicon, the heat conductivity of diamond and the size and perfection of DNA.

The U.S. government is throwing big bucks at jump-starting this sort of nanotechnology, because, among other things, it sees national security implications. Boeing is working on cutting the weight of rockets, aircraft and satellites. Asian and European governments are also frantically attempting to achieve a lead in what is hyped as overwhelmingly The Next Big Thing. General Electric, Motorola, DuPont, Lucent and Kodak are pursuing nanotech. The first nano products trickling to market were not terribly impressive—self-cleaning windows and non-staining trousers. It is the promise, however, that drove Congress wild. What might you do with nanotechnology? Defeat biological warfare. Defeat all disease, in fact. Nano may be the basis of half of all pharmaceuticals by 2010. Heal the environment—lock up carbon dioxide in the atmosphere, consume toxic waste. End our dependence on oil—produce solar panels embedded with nanocrystals so efficient and flexible that they can be used in everything from roof shingles to clothing. Conquer space. The U.S. National Aeronautics and Space Administration loves what you could do with enough carbon nanotubes—the strongest, lightest material ever made. You could stretch a cable 22,347 miles to a stable point in space. You could then run electromechanical vehicles along it full of people and payloads. What you'd have is a space elevator. Five-hour ride, one-way. Forget those preposterously expensive chemical rockets that haven't changed much since the 1960s. Forget those painfully fragile landing craft so susceptible to burning up on reentry. This could be the breakthrough to the stars, NASA says.

Mind you, that's the mild version of nanotechnology. That's not where you get the real excitement. The real excitement explodes from the second kind of nanotech. It starts with the work of K. (for Kim) Eric Drexler. In the 1970s, as an MIT undergraduate, Drexler noted that all life in every jot of its riotous variety is created by little biological machines like those in photosynthesis that manipulate basic elements of matter. Why not adapt those methods to build nonliving things? he wondered. Why not build computer-driven machines with parts the size of molecules to haul atoms to precisely the correct locations to build anything you pleased? *Anything you pleased.* In 1986, he single-handedly launched and named the nanotech industry with his book *Engines of Creation: The Coming Era of Nanotechnology,* which explained how his ideas could truly change the world. With nanotechnology, you could grow a house, or a car, or a completely real, molecularly accurate T-bone steak, or a new heart, from little more than software instructions and some handfuls of dust. It would enable digital control of the very structure of matter.

Even as he approaches 50, Drexler maintains an uncanny ability to look like a graduate student. He's as gaunt as a stray puppy. You want to take him home and feed him. Sitting in Silicon Valley's Original House of Pancakes in Los Altos, his breakfast order seems promising—a robust plateful of ham and eggs. Then he lets them get cold as he quietly but intensely explains the vast implications of the world he sees us entering.

If anything, Drexler is even more optimistic than Stock and Kurzweil. He wears a medallion around his neck that asks the finder, in case of Drexler's death, to "Call now for instructions/Push 50,000 U heparin by IV and do CPR while cooling with ice to 10C/Keep PH 7.5/No embalming/No autopsy." He and others believe that robots smaller than a red corpuscle will soon work like Pac-Man, gobbling up diseased cells and keeping the human body perpetually in tune. If he were to meet an untimely end before that day arrives, however, Drexler plans on coming back from the dead. That's why he's wearing the medallion. He wants to get frozen right next to Ted Williams so that when the appropriate technology arrives, he can be thawed, have a nanotech workover and a slap on the butt and get on with his efforts.

Does he think this will make him immortal?

"Depends on what you mean by immortal," he says, still ignoring his eggs. "There is such a thing as proton decay."

Pause.

He's talking about the eventual collapse of subatomic particles in eons vastly beyond the time it will take for the solar system to die.

Okay, what about merely geological time? Hundreds of thousands of years?

"Oh yeah." He smiles. "That. For sure."

One problem with a genie like strong nanotech is what it might demand in return for granting your every wish. Drexler now calls his original vision of nanotech "molecular manufacturing" to distinguish it from what he views as the promiscuous debasement of the word *nanotech* to include anything small, including ordinary chemistry. The key idea about his vision of nanotech is the creation of "assemblers." Assemblers would be the molecular manufacturing machines that would seek out the appropriate atoms and put them in the right places. Of course, since it would take billions of assemblers to make a product of any significant size, the first thing an assembler would have to do is make a second assembler. Then each would make another one, and so on, up the exponential curve, doubling and redoubling their numbers. There are several pause-giving aspects to this. One is that nobody knows how to make an assembler—they are hypothetical. (In fact, those seeking venture capital for their near-term visions of nanotechnology dismiss Drexler as at best a daydreamer and at worst a "scarer of children.") A second is that were the "assembler breakthrough" ever to start, it's hard to know how to steer. The world might be transformed so utterly and abruptly by such a breakthrough as to resemble The Singularity. Another problem is that even if you got them started and steered, you'd better be awfully sure you know how to make them stop. If they can't be made to stop, in principle they could blithely reproduce themselves until they've consumed every bit of energy on the planet, wiping out all life. This is the "gray goo" plot that is a critical element of The Hell Scenario.

Nonetheless, the promise is staggering. Nanotech is where the NSF gets the idea that the human body will soon start conquering age, the energy crisis will be solved, the environment will be cleaned, aircraft will be more adaptable than birds, and the American armed forces will be invulnerable.

As in genetic engineering and nanotechnology, computer intelligence, the *I* in GRIN, has two meanings. There is strong artificial intelligence and weak machine intelligence. *Weak* essentially means that it does a pretty good job of imitating humans. *Strong* means that it *is* intelligent in important ways. It *is* conscious. It *is* a challenge to the identity of humans and their human nature. Weak machine intelligence is the United Airlines reservation computer that hears what you're saying on the phone when you ask whether your plane will be on time, and responds to you usefully in a mechanical voice. Strong artificial intelligence, if it ever comes to be, is you not being able to tell whether or not you're talking with another flesh-and-blood person. Weak machine intelligence can and does fly a jumbo jet 10,000 miles across the Pacific, setting it down in Hong Kong with such soft precision—and without the pilot touching the controls—that the passengers applaud the landing. Attempts at strong artificial intelligence cannot yet achieve the common sense we expect of a three-year-old when she dresses herself.

Marvin Minsky is the grand old man of the notion that a successful computer model of the brain will be able to explain how it thinks. Young researchers now regard as quaint some of his views that consciousness can be described as machinelike. Nonetheless, Minsky is revered. In 1959 he co-founded what would become the MIT Artificial Intelligence Lab. (In fact, he was present when the term *artificial intelligence* was invented in 1960.) He is one of the last living links to the early pioneers of the Information Age, under some of whom he studied. His work on artificial intelligence is still regarded as the pushing-off point for discussions of machine intelligence. His students fill the field. He remains as inquisitive as a squirrel with a new bird feeder.

It's a treat to see Minsky perform. This winner of the Japan Prize—that nation's highest honor in science and technology—thinks nothing of showing up at august academic forums in a palm-tree-decorated Hawaiian shirt. His very bald head is now marked by light liver spots and a wild fringe of long baby-fine white hair near his neck. His smile is as wide as a frog's. He is a great waver of his arms over his head as he makes pronouncements like "There is nothing certain but taxes"—conspicuously excluding death. He sports a credit-card-sized digital camera on a silver cord around his neck, black leather sneakers with Velcro fasteners on his feet, and a telescope in his pocket with which he recognizes people blocks away. He wears a belt woven of 8,000-pound-test Kevlar filament that can be quickly unraveled into a single cord with which you can pull your car out of a ravine. On this belt hangs a pouch full of tools, including lasers and cutting implements. Seems reasonable to him. "Who would fix the plane if it got into trouble?" he asks.

Minsky believes it is important that we "understand how our minds are built, and how they support the modes of thought that we like to call emotions. Then we'll be better able to decide what we like about them, and what we don't—and bit by bit we'll rebuild ourselves. He views this project of understanding intelligence as comparable to "only two earlier inventions: those of language and of writing." He sees it as a matter of urgency. "The more we can learn about how human minds work, the better we will be able to guide the development of our genetic successors or of beings whose making we'll supervise," he says.

"Why should we alter ourselves instead of forever remaining the same? Because we have no alternative. If we stay unchanged in our present state, we are unlikely to last very long—on either cosmic or human scales of time. In the next hundred or thousand years, we are liable to destroy ourselves, yet we alone are responsible not only for our species' survival but for the continuation of intelligence on this planet, and quite possibly in this universe." He takes The Hell Scenario seriously. He recites a long list of ways that we can become extinct, from global warming to rogue asteroids. His solution? "We've seen many suggestions about dealing with these, but none of them yet seem practical. A more practical course might simply be," he says, to focus "instead on finding ways to make ourselves more intelligent."

As with strong nanotechnology, strong artificial intelligence has not yet been achieved. Nonetheless, advances in machine intelligence are what the NSF has in mind when it discusses direct connections between you and your car, factory work, environment and weapons, as well as the creation of new arts, sports and ways to transcend cultural and language barriers between people.

Robots, meanwhile, have come a long way from Karel Čapek's 1921 play *R.U.R.*, for "Rossum's Universal Robots," whence comes the word *robot*. The Japanese—especially Sony and Honda—love to create two-legged humanoid mechanical creatures. In the time-honored tradition of nearly life-sized Bunraku stage puppets, Honda has fielded a squad of Asimo robots, the company's walking, talking, child-sized pseudohumans. The company says its long-term aim is to create "a partner for people." Sony's "corporate ambassador," meanwhile, is a two-foot-tall humanoid called Qrio. Qrio is produced by Sony's Entertainment Robot Company, which already sells Sony's robot dog, Aibo. The company's president, Satoshi Amagai, says that Japanese Aibo owners have a more emotional relationship with their mechanical pets than do Americans or Europeans. The AIBO ERS-7, at $1,599, promises to have six emotions—happiness, anger, fear, sadness, surprise and discontent. Pat one on the head and it appears to become happy enough to do tricks. Whack it on the nose, and it not only appears hurt, it learns not to repeat certain behaviors while prompting that profound philosophical question: Is it wrong to kick a robotic dog?

In much of the rest of the world, meanwhile, robots—defined as digitally driven creatures that can sense and move—have taken on new forms of life. "We are trying to build robots that have properties of living systems that robots haven't had before," says Rodney Allen Brooks, chairman and chief technical officer of iRobot, the company that brought to market the Roomba—America's first cheap, practical, sweeping and vacuuming robot. It looks like a portable CD player the size of a toilet seat. By all accounts, it really does clean the floors. Even under the bed. More than half the owners of these robots give them names. Finally in Brooks we have a roboticist whose achievements we can take seriously.

Oh yes, Brooks has a few other credentials as well. He is director of the MIT Artificial Intelligence Laboratory, Fujitsu Professor of Computer Science at MIT, and author of *Flesh and Machines: How Robots Will Change Us*. (Did you say the name of his company, iRobot, reminded you of the title of the Isaac Asimov book mentioned above that influenced a whole generation of roboticists? Told you.)

Actually, the person who led the robot charge toward The Heaven Scenario was Hans Moravec of Carnegie Mellon University in Pittsburgh. In 1988 this pioneer published *Mind Children: The Future of Robot and Human Intelligence*. In its first few sentences, it painted a picture in

which humans were swept away. We "have produced a weapon so powerful it will vanquish the losers and winners alike," Moravec said. But even
though the Cold War was still on, he didn't mean nuclear weapons. He
meant the intelligent children of our minds, his beloved robots. By 2008,
he thought, they would be able to perform like humans on the gear con-
tained in a $10,000 personal computer. By 1998, in *Robot: Mere Machine to
Transcendent Mind,* Moravec acknowledged that his time frame was a trifle
ambitious. (At the same time, his computer-price projections turned out
to be laughably conservative.) Nonetheless, he predicted the near arrival
of general-purpose robots that could figure out where they were in a
room with the precision of a lizard, followed by robots with the adaptability of a mouse, followed by robots with the imagination of a monkey,
followed by those with human reasoning. His predictions carefully copied
the biological evolutionary ladder up the scale as computers became
more and more powerful. No more will that proud race called robot be
sadly stuck welding Corolla fenders day after day. Moravec's vision is one
in which robots can be autonomous, responding innovatively to complex
and unanticipated situations. Real robots roam.

Indeed, by the 100th anniversary of the Wright brothers' 1903 flight,
at least 32 countries were developing more than 250 models of flying robots, with 41 countries already operating them. (In the last week of the
first Gulf War, five Iraqi soldiers waved white flags at a U.S. Pioneer unmanned air vehicle—the first time in history somebody tried to surrender to a robot.) Already in the air were reconnaissance robots the size of
sparrows, such as the Black Widow. The next generation would be the
size of insects. Such robots will find markets among firefighters, environmentalists, border patrols, film producers and farmers. The U.S. Department of Defense is expecting pilotless aircraft the size of F-16 jet fighters
to suppress enemy air defenses and sensors by 2012. Cargo-lifting pilotless
craft originally designed to supply soldiers were expected one day to replace FedEx vans. The story in aviation circles was widespread that in the
future, passenger plane cockpits would only have a pilot and a dog. The
pilot's job would be to watch the robot's computer screens. The dog
would be there to bite the pilot if he tried to touch anything.

But Brooks goes much farther than that. "We're trying to build robots
that can repair themselves, that can reproduce (although we're a long way
from self-reproduction), that have metabolism, and that have to go out
and seek energy to maintain themselves." Forget the image of a man in a
can. "Our theme phrase is that we're going to build a robot out of Jello.

We don't really mean we're actually going to use Jello, but that's the image we have in our mind. We are trying to figure out how we could build a robot out of 'mushy' stuff"—like flesh—"and still have it be a robot that interacts in the world."

Is Brooks worried about this all going too far? Well, he admits that Bill Joy—who in the next chapter details his fears about technology overrunning mankind—has stimulated him. "I have several interesting robotics projects underway," he says. "One of the robots I must say was inspired by Bill Joy, probably to his dismay. We have a robot now that wanders around the corridors, finds electrical outlets and plugs itself in. The next step is to make it hide during the day and come out at night and plug itself in. I'd like to build a robot vermin," Brooks says. He loves the idea of building a robot that could be killed only by locking it in a room with no power outlets or by going after it with a hammer. "I'm trying to build some robots like that as thought-provoking pieces—and just because Bill Joy was afraid of them," he says. Just what the world needs—a roboticist with a sense of humor.

Brooks is also trying to build robots made out of Legos that can make copies of themselves out of more Legos. There is a deep mathematical question of whether this is possible, but basically he says that's what machine self-reproduction will look like. After all, he points out, biological molecules can do that. Meanwhile, his iRobot company is big into PackBots. These are emergency and military robots weighing some 50 pounds that can be carried by one person. They can be dropped from a reasonably low-hovering helicopter and do search and rescue, disarm bombs, or perform reconnaissance in caves into which humans don't want to stick their heads. (DARPA loves the experimental versions being pioneered at the University of California at Berkeley that want to climb straight up walls and hang from ceilings, like those gecko lizards with suction pads on their feet after which they are modeled. DARPA also gets dizzy at the thought of air-dropping entire networks of tiny "throwbots" that can work with each other to saturate a battlefield.) Making these bots impervious to the abuse and neglect of a 19-year-old American is a very big challenge. But the troops so far deem them worth humping—no small compliment.

Brooks is both enthusiastic and sanguine about The Heaven Scenario. "Fairly soon, we may have to start banning kids with neural Internet connection implants from having them switched on while taking the SATs," he says. Shortly after that, he thinks, such implants will be mandatory after the SATs are revised to accommodate such people. As a result, "we will become a merger between flesh and machines. We will have the best that machineness has to offer, but we will also have our bioheritage to augment whatever level of machine technology we have so far developed. So we (the robot-people) will be a step ahead of them (the pure robots). We won't have to worry about them taking over."

This merger of flesh and machine is what the NSF is thinking of when it talks about machines operating "on principles compatible with human goals, awareness, and personality." That's also what it's talking about when it contemplates "construction of extraterrestrial bases, and profitable exploitation of the resources of the Moon, Mars, or near-Earth asteroids."

Who knows if all these GRIN technologies collectively will produce "world peace, universal prosperity, and evolution to a higher level of compassion and accomplishment," as the National Science Foundation would have us believe. If a fraction of these technologies has non-negligible possibilities, however, even that will have an overwhelming impact on humanity.

───※───

H OW WOULD WE KNOW if we were achieving The Heaven Scenario? I ask Kurzweil. How would we know if we were transcending human nature? What would the milestones be?

"Well, it depends on how you define human nature," he replies.

Kurzweil defines humans as "that species that seeks to extend its own horizons—that represents the cutting edge of evolution." It embodies "evolution towards ever greater knowledge and intelligence and creativity." If you buy that definition, he holds, "then there is no transcending it. It is, ultimately, the cutting edge of transcendence."

If, however, you "define it in more limited ways, such as human beings are a certain biological species that has brains organized in this certain biochemical way, then we will transcend that. Since the technology is going to grow exponentially, ultimately the non-biological portion will dominate."

One nice thing about our becoming less biological, he says, is that we'll be easier to upgrade. The way we do things now is a nuisance. For example, for most of human and prehuman history, every last calorie was precious—you never knew when famine would hit, so it was better to hold on to all those calories while you could, as the extra fat might be crucial to your survival.

"But now we live in a completely different period—at least I do—in a period of abundance," Kurzweil says. "We don't need all those calories. I have a low fat percentage, but I have to do that by controlling my eating.

Ultimately we'll separate that. We've actually identified the gene that does that."

Is that the marker of transcendence? When we can eat all we want and never be fat? "That's going to be one step, among many," Kurzweil replies. "When we do that, it will be considered a conservative, sensible thing to do. After you have done hundreds and thousands of those changes, we will be very different than we are today."

At a conference at Boston University, I once was told that we'll know we have transcended when The Enhanced start having difficulty telling the difference between Naturals and dogs. We will have evolved so far that all those carbon-based life forms will tend to blur. Is that what Kurzweil is talking about?

"Okay, the MOSHs," Kurzweil replies. "The Mostly Original Substrate Humans. Well, I think enhanced humans will become greatly enhanced compared to MOSHs. But I think there will be a continued respect for human life. I think that is built into our species. That is not necessarily an ultimately moral position. It's a species–centric position. That one is deeply rooted in our nature."

Really? I ask. What about our predecessor species? I don't see any Cro-Magnons or Neanderthals left—with the possible exception of a couple of editors.

"My view is that there will be respect for our species and where we have come from. It will take an absolutely trivial portion of the intellectual output of enhanced humanity to meet all of the material and other needs of the rest of humanity. So from the perspective of unenhanced humans, these new nonbiological entities will appear to be their transcended servants. But from the perspective of nonbiological intelligence, they are devoting a very trivial portion of their intellectual output to providing that service."

Do you believe in a God you plan on meeting when you die? I ask.

"I'm not planning to die," Kurzweil responds. "I expect to use the power of ideas. I am a survivor as an entrepreneur and a human being. It's my plan to be involved in this next phase of humanity where we get past some of the frailties of these Version 1.0 bodies we have. The way to 'meet our maker,' so to speak, is, in fact, by staying alive. We will be part of this very rapid explosion of intelligence, and beauty, and a very rapid acceleration of this evolutionary process. And that, to me, is what God is. Evolution, I think, is a spiritual process because it moves closer to what we have considered God. It moves closer to infinity."

I am nonplussed by this answer. He is not talking about us someday meeting God.

He is talking about us *becoming* God.

Aware of how this sounds, he rephrases a little. "I don't think we actually ever become God. But we do become more God-like," he clarifies.

He then barrels right back to a grand view of The Heaven Scenario. "I see it, ultimately, as an awakening of the whole universe. I think the whole universe right now is basically made up of dumb matter and energy and I think it will wake up. But if it becomes transformed into this sublimely intelligent matter and energy, I hope to be a part of that."

He is talking about participating in the creation of Heaven.

The Heaven Scenario

Ray Kurzweil and others in his camp do not consider their description of the future a scenario. They consider it a prediction. That is, they do not see the future they describe as one logical possibility among several for which it would be prudent to prepare, as is the case in scenario planning. They see the logic of their position as patently inescapable and their version of the future as inevitable. They frequently argue against the possibility that the other two scenarios described in this volume might ever occur. Nonetheless, since others disagree, and those with different versions of the future frequently end up debating each other, the vision of the future that in this volume is called Heaven, in which The Curve is inevitable and the outcome good, is treated here as a scenario.

Predetermined elements:

- There are Curves of exponential change governing technology.

Critical uncertainties:

- Are The Curves of exponential change accelerating smoothly? If so, how fast? Or are they displaying unexpected slowdowns, stops or reversals?

- Are The Curves of change leading to progress, disaster or both?

Embedded assumption:

- Technology drives history.

Early warning signs

- Almost unimaginable are entering The Heaven Scenario:
  conquering of disease things are happening, including the
  wisdom, love, truth and ty, but also an increase in beauty,
- Predictions that recently se
  exceeded.
- Even in the face of such wonder since fiction are routinely
  spectators. Technology seems in co
- The phrase "The Singularity" enters c ore or less
  phrase "global warming" at the turn of
  did the

Early warnings that we are not entering The Heaven Scenario:

- The growth of complexity slows or starts proving erratic.

- Almost unimaginably bad things start happening. (This would be an early warning that we are entering The Hell Scenario.)

- Culture and values gain control of technology such that events that once seemed inevitable are now consciously being avoided over significant periods of time, and by everyone on the planet. (This would be an early warning that we are entering The Prevail Scenario.)

Figure 3.1

Early warnings that we are not entering The Heaven Scenario:

The growth of complexity slows: stars proving events

Almost unimaginable bad things start happening. (This would be an early warning that we are entering The Hell Scenario.)

Culture and value gain control of technology: such that events that once seemed inevitable are now consciously being avoided over significant periods of time, and by everyone on the planet. (This would be an early warning that we are... entering The Prevail Scenario.)

CHAPTER FIVE

Hell

Technological progress is like an axe
in the hands of a pathological criminal.

—Albert Einstein, letter to a friend, 1917

Human *nature will be the last part of Nature to surrender to*
Man. The battle will then be won. We shall . . . be henceforth free
to make our species whatever we wish it to be. The battle will
indeed be won. But who, precisely, will have won it? For the power
of Man to make himself what he pleases means, as we have seen,
the power of some men to make other men what they please.

—C. S. Lewis, *The Abolition of Man*, 1944

THE PILOTS WHO ENJOY flying the United run into Aspen, Colorado, tend to be former Navy aircraft-carrier fighter jocks. They greatly admire bucking the downdrafts swirling around the young, incisorlike peaks very deep in the Rocky Mountains. On approach, they get to evade the fang of Triangle Mountain as it fills up the windows just off their left wing. Then, when the tiny island of flatness that is the airport finally appears at the bottom of the very narrow Roaring Fork Valley, they drop on it like a raptor. To snag the end of the runway, they've got to lose 4,400 feet in three minutes. That plunge never fails to thrill the passengers, especially when it is followed by reverse thrust that shudders the plane to a stop just before the asphalt ends. In a snowstorm. In May. The pilots love this—reminds them of their youth.

Aspen is by no means the most isolated settlement in the Intermountain West. But no matter which parlous path you use to make your way to this town, on the ground or in the air, there is no way to avoid feeling vulnerable to the towering forces that surround you. The mountains stretch out endlessly, dusted with snow like powdered sugar on a chocolate cake. Coming into these mountains—on which trees as tall as 10-story buildings seem like no more than beard stubble—it's hard to escape thinking about how fragile the human situation is, how we cheat death every day.

Compared to other outposts in the formidably empty land between Denver and Salt Lake, Aspen is urbane. From frontier days, it always served the needs of wealthy entrepreneurs. From 1877 to 1893, it was a silver boomtown. In fact, locals insist, you can still get from the mountains on one side of town to those on the other side through the mine shafts deep beneath your feet. Now Aspen's main offering is white gold—boundless snow and 300 days a year of sunshine. Its antique Western brick and sandstone buildings with their pressed tin ceilings are full of $100 wines sold in shops covered with fairy lights. Women with long curves of blond hair, wearing snug white spangly jumpsuits and bejeweled chokers, sweep into memorable restaurants that feature old-timey heavy green drapes held back by gold swags. Chaps with British accents serve smoothies costing more than five bucks at a coffee joint called Zélé, across from the Louis Vuitton shop, which is next to Christian Dior, which is next to Gucci. The mountains are so close—they so loom over the rooftops, blocking the sky—that you have to crane your neck to examine the clouds being trapped between their peaks. They are so close that offices near the center of the business area are only a five-block stroll from the ski lifts that take off dramatically from the edge of town.

It is in this rarefied place that you'll find Bill Joy, "the Edison of the Internet," when he's willing to be found. His office has a six-inch rule: Six inches of powder and the office closes. Skis and ski boots hang by the door. The skiing is not really why he's here, though, he says. And it certainly isn't Aspen's glitz—if anything, Joy finds that an annoyance. Joy is in Aspen because in 1989 he decided to get away from everything he hated about Silicon Valley, about cities, about California—the maniacal ambition, the traffic jams, the earthquakes. He could afford to search for serenity just about anywhere. Sun Microsystems, of which he was co-founder, could whistle up for him a Citation X whenever he needed one. At the time it was the world's fastest private jet, capable of flying from Silicon Valley to Washington at Mach .92 in under four hours. He settled on Aspen, population 5,914, he says, because it was at the end of the road. That was important to him. There's nobody just passing through. It is far removed from the madding crowd. But most critically, he says, this was the most remote town in the Intermountain West with a decent bookstore.

Joy is not really part of the Aspen community, he admits. He was the commencement speaker at the local high school once. But mention his name to the locals, and generally you draw a blank. His house is nothing particularly special in Aspen terms. It is only in the neighborhood of 5,000 square feet—comparable to America's first supermarkets. In Aspen, he wryly explains, a house is not considered significant unless it has a 2,000-square-foot master suite. Some houses run to 40,000 square feet. His is nothing like that. In fact, what he likes best about it is the crabapple tree in his little backyard, under which, in the fall, a bear sleeps.

But Aspen's not secluded enough for him. What genuinely lights him up is his real retreat, called Three Meadows. It's a ranch sufficiently farther west into the Colorado wilds that it looks like Wyoming—mostly sagebrush, a few aspen trees, elk, deer, coyote and more bear. His cabin there is tiny. His bedroom is barely bigger than the bed into which he tucks his gaunt 6-foot-3-inch frame. But it surveys enough of a spread that he doesn't talk about how many acres it adds up to. There, Joy is unlikely ever to see a human being who is not a guest. It is not even on civilization's utility grid. Joy must employ a fair amount of his technological ingenuity just to make it habitable. His phone there is the size of a brick. The receiver has to beam directly to orbiting satellites, for there are no phone lines anywhere nearby. No power lines go to his cabin. He generates his own electricity with a wind turbine and solar panels connected to batteries almost the size of file cabinets; in Japan, such batteries power passenger trains. To get to the Internet, he fires up a satellite dish. Don't tell Bill Joy no man is an island. He will never send to know for whom the bell tolls. He will never hear any steeple bell.

The Edison of the Internet. Joy has gotten inured to being called this, but such adulation still makes him cringe. "We're both from Michigan," he says of the comparison to the father of the electrical age, the Wizard of Menlo Park. "We like to stay up late at night, and we've done some things that have succeeded and some that have been spectacular failures. But I'm not yet deaf and I don't work as hard as he did."

Yet the comparison is not all wrong. Joy's had a hand in some of the most important aspects of the Net. In 1978, while still a grad student, Joy became the principal programmer for a Defense Advanced Research Projects Agency project inventing the Berkeley Systems Distribution (BSD), the first operating system linking computers over this newfangled thing that would come to be called the Internet. ("It was fun," he says, predictably enough.) In the early 21st century, the BSD architecture was still the main rival to Microsoft's server system, being the basis of Apple's OS X operating system and Sun's machines, and an underpinning of Linux. In 1982, Joy married that system to a cheap but powerful computer called the S.U.N. workstation, after the Stanford University Network. This is how he wound up as chief scientist of Sun Microsystems, until he resigned the post in 2003. In 1984, he and a colleague developed a revolutionary new reduced-instruction-set computing (RISC) processor with extraordinary performance.

When others were just beginning to think that stand-alone desktop personal computers might someday be a big deal, Joy imagined an entirely different world. In it, intelligence would be embedded in everything from telephones to shoes to doors to eyeglasses, all talking and thinking with each other in a fertile jungle of information, often without the intervention of humans. A camera, for example, might talk to a printer without having to go through a PC. Everything was simple, like using a lamp or a telephone. The complexity—as with the electricity or telephone systems—was managed by the network Sun served. In 1988 this crystallized in Sun's enduring slogan, "The network is the computer."

Joy has developed themes that have shaped his life. His watchword is "simple"—avoid the puffed-up and esoteric. Another important concept for him has been openness. He has found great power in letting the world freely understand the underpinnings of his work in order to communally and cooperatively tinker with and improve on it. This open-source philosophy contrasts markedly with the proprietary and jealously guarded approach of outfits such as Microsoft. Finally, there is the importance of networks. Nourish the network. Trust the network. The network collectively can do things beyond the power or even imagination of any individual.

Joy enjoys a reputation in Silicon Valley as thoughtful and level-headed. "Nobody is more phlegmatic than Bill," says Stewart Brand, the Internet pioneer. "He is the adult in the room."

That's why it came as such a shock in March 2000 when this godfather of the Information Age predicted "something like extinction" of the human race within the next generation. Most extraordinarily, he blamed it on the accelerating pace of technological change he had helped create. He intended his warning to be reminiscent of Albert Einstein's famous 1939 letter to President Franklin Delano Roosevelt alerting him to the possibility of an atomic bomb.

In a vast, 24-page spread in *Wired* magazine entitled "Why the Future Doesn't Need Us," Joy announced that, to his horror, he found advancing technology poses a threat to the human species. "I have always believed that making software more reliable, given its many uses, will make the world a safer and better place," he wrote in the article, on which he worked for six months. "If I were to come to believe the opposite, then I would be morally obligated to stop this work. I can now imagine that such a day may come."

At the peak of the Internet boom, when technologists of Joy's stature were beating movie stars to the covers of magazines, his conversion went off like a thunderclap on a sunny day.

"I think it is no exaggeration to say we are on the cusp of the further perfection of extreme evil," he wrote. He was horrified by "a surprising and terrible empowerment of extreme individuals." Joy reviewed the prospects for the genetic, robotic, information and nano technologies and was aghast. Genetic technology, he noted, makes possible the creation of "white plagues" designed to kill widely but selectively—attacking only people of a targeted race, for example. It also holds out the possibility of engineering our evolution into "several separate and unequal species . . . that would threaten the notion of equality that is the very cornerstone of our democracy." Robots more intelligent than humans could reduce the lives of their creators to that of pathetic zombies. "A robotic existence would not be like a human one in any sense that we understand," Joy wrote. "The robots would in no sense be our children . . . on this path our humanity may well be lost." Nanotechnology holds out the possibility of the "gray goo" end-of-the-world scenario, in which devices too tough, too small, and too rapidly spreading to stop, suck everything vital out of all living things, reducing their husks to ashy mud in a matter of days. "Gray goo would surely be a depressing ending to our human adventure on Earth," he writes, "far worse than mere fire or ice, and one that could stem from a simple laboratory accident. Oops."

"Most dangerously, for the first time, these accidents and abuses are widely within the reach of individuals or small groups," he wrote. "Knowledge alone will enable the use of them." Nuclear weapons require heavy industrial processes and rare minerals. They take the resources of, at the very least, a rogue state. With a variety of GRIN technologies, by contrast, one bright but embittered loner or one dissident grad student intent on martyrdom could—in a decent biological lab, for example—unleash more death than ever dreamed of in nuclear scenarios. It could even be done by accident. Joy called these "weapons of knowledge-enabled mass destruction." What really alarmed him about these GRIN weapons was their "power of self-replication." Unlike nuclear weapons, these horrors could make more and more of themselves. Let loose on the planet, the genetically engineered pathogens, the superintelligent robots, the tiny nanotech assemblers and of course the computer viruses could create trillions more of themselves, vastly more unstoppable than mosquitoes bearing the worst plagues.

This belief—that one strike and you're out—is, in short, what I have come to call The Hell Scenario. "The only realistic alternative I see is relinquishment," he concluded, "to limit development of the technologies that are too dangerous, by limiting our pursuit of certain kinds of knowledge."

"There's nothing optional about this, you see," Joy says, staring stoically across his long conference table in an unadorned room in the outback of the American West. "It's like death and taxes. We have this problem. It will get us." Joy is The Hell Scenario's most celebrated apostle.

William Nelson Joy was born on November 8, 1954, in what then were the ragged fringes of metropolitan Detroit. The area around Fourteen Mile and Orchard Lake Roads in which he grew up was then called Farmington. Now more tony, the area is pleased to be called Farmington Hills. But at the time, you could get a four-bedroom two-story Colonial with a basement and a humidifier on a quarter-acre lot for $29,000, which is what his family did. His dad, William, was a counselor and vice principal in the Detroit schools who taught business courses such as accounting and typing, as did his mother, Ruth. By the time he was three, Bill was already reading. His dad took him to the elementary school, where he sat on the principal's lap and read him a story. As a result, Joy started school early. Then he skipped a grade. He drove adults nuts, asking lots of questions. He escaped into books. He wanted to have a telescope to look at the stars. Since there was no money for him even to make one, he checked out library books on telescope making and read them instead.

His mother was of Swedish extraction, so he was raised Lutheran, but to no lasting effect. "The scars are invisible," he jokes. As a teenager, he wished to be a ham radio operator. It was a kind of predecessor to the Internet, he remembers—very addictive, and quite solitary. His family didn't have the money to buy him the equipment, but cost aside, his mother was appalled. Joy—who was not exactly living up to his name even then—was not going to spend his life in the basement with headphones on. "I was antisocial enough already," he recalls her feeling. "Oh yes, I looked like Clark Kent with my hair greased back and the glasses. But I didn't have the alter ego." He did not go to his high school prom.

Ruth Joy died when he was 18. Her death was particularly difficult on his younger brother and sister. A few days after his mother died, his father said something that stuck in his mind: "Sorry you kids have to go through this." Joy still thinks about that, because unlike his father's generation—which had grown up before antibiotics and which had been sent off to war—he is impressed by how many young people today have no firsthand experience with death or disease. He believes they are unprepared for the evils that he fears will come.

Growing up, Joy had few close friends, but he did discover the great writers who speculated about the future. He especially remembers Robert A. Heinlein's *Have Spacesuit—Will Travel,* and, like so many others of his temperament, Isaac Asimov's *I, Robot.* Thursday nights his parents went bowling and the kids stayed home alone. That was the night of Gene Roddenberry's original *Star Trek.* It made a big impression on Joy. He seized on the notion that humans had a future in the cosmos, and it was like a Western, with big heroes and big adventures. He devoured Roddenberry's vision of the centuries to come as one of strong moral values, embodied in codes such as the Prime Directive:

> As the right of each sentient species to live in accordance with
> its normal cultural evolution is considered sacred, no Star Fleet
> personnel may interfere with the healthy development of alien life
> and culture. Such interference includes the introduction of superior
> knowledge, strength, or technology to a world whose society is
> incapable of handling such advantages wisely. Star Fleet personnel
> may not violate this Prime Directive, even to save their lives and/or
> their ship unless they are acting to right an earlier violation or
> an accidental contamination of said culture. This directive takes
> precedence over any and all other considerations, and carries with
> it the highest moral obligation.

"This had an incredible appeal to me," he recalls. "Ethical humans, not robots, dominated this future, and I took Roddenberry's dream as part of my own." Now, in The Hell Scenario, the possibility looms that humans already may be violating the Prime Directive—on their home planet.

At the age of 16, in 1971, Joy embraced the advanced courses of math majors at the University of Michigan. When he discovered computers, he never turned back. The computer had a clear notion of correct and incorrect, true and false, and Joy found this "very seductive." He liked clear answers about what was right and wrong. He took few liberal arts courses.

Joy got a job programming early supercomputers and discovered their amazing power to simulate reality. In grad school at Berkeley in the mid-1970s, he stayed up late, inventing new worlds. He recalls "writing the code that argued so strongly to be written." In his novel *The Agony and the*

Ecstasy, Irving Stone described Michelangelo as feeling he was only releasing the statues from the stone, "breaking the marble spell." Joy vividly remembers exactly this kind of ecstatic moment. The software emerged the same way. It was as if the truth, the beauty, was already there in the machine, waiting to be freed. Staying up all night seemed a small price to pay. He recalls it as the "rapture of discovery."

Joy tells these stories to make clear he is not a Luddite. He insists he has always had a strong belief in the value of the scientific search for truth. The Industrial Revolution, in his view, immeasurably improved everyone's life; he always saw himself as part of this inevitable march of truth and progress. Sun hardware powers a significant portion of the Net to this day. Joy's RISC processors are still at the heart of the company's $10 billion server business.

The turning point for Joy came on September 17, 1998, at the Resort at Squaw Creek, near North Lake Tahoe, an exquisite mountain locale in northern California's Sierras half a day from Silicon Valley. There he ran into Ray Kurzweil, whom up until then Joy had known only as a famous inventor. He and Kurzweil were speakers at different sessions of something called the Telecosm Conference, which at the time was one of the hottest tickets in geekdom. Financiers with what seemed like unlimited funds flocked to this gathering to worshipfully gather crumbs of wisdom from digital visionaries.

At this conference Joy first heard Kurzweil's prediction that computers will supersede humans in the next step of evolution. Joy was in the bar of the hotel with John Searle, a professor of philosophy of mind at the University of California at Berkeley, who scoffs at Kurzweil's idea that machines can become conscious. Kurzweil wandered in, and he and Searle picked up their wrangle. As Joy listened to them debate whether The Heaven Scenario was feasible, a possibility occurred to Joy that was entirely different from the one those two were discussing.

Joy buys the first half of The Heaven Scenario. He views it as obvious that the fundamental reality shaping our future is The Curve. That Indian summer day, however, Joy remembers thinking—oh my God, this could go just the opposite way. The outcome could be terrible. Instead of The Curve going straight up to Heaven, it could be the mirror image and go straight to Hell. This was the moment of conception for what ultimately would be Joy's "Why the Future Doesn't Need Us."

Later, reading page proofs of Kurzweil's *Spiritual Machines,* he came to the passage in which Kurzweil, discussing the new Luddite challenge, quotes another author's scenarios:

First let us postulate that the computer scientists succeed in developing intelligent machines that can do all things better than human beings can do them. In that case presumably all work will be done by vast, highly organized systems of machines and no human effort will be necessary. Either of two cases might occur. The machines might be permitted to make all of their own decisions without human oversight, or else human control over the machines might be retained.

If the machines are permitted to make all their own decisions, we can't make any conjectures as to the results, because it is impossible to guess how such machines might behave. We only point out that the fate of the human race would be at the mercy of the machines. It might be argued that the human race would never be foolish enough to hand over all the power to the machines. But we are suggesting neither that the human race would voluntarily turn power over to the machines nor that the machines would willfully seize power. What we do suggest is that the human race might easily permit itself to drift into a position of such dependence on the machines that it would have no practical choice but to accept all of the machines' decisions. As society and the problems that face it become more and more complex and machines become more and more intelligent, people will let machines make more of their decisions for them, simply because machine-made decisions will bring better results than man-made ones. Eventually a stage may be reached at which the decisions necessary to keep the system running will be so complex that human beings will be incapable of making them intelligently. At that stage the machines will be in effective control. People won't be able to just turn the machines off, because they will be so dependent on them that turning them off would amount to suicide.

On the other hand it is possible that human control over the machines may be retained. In that case the average man may have control over certain private machines of his own, such as his car or his personal computer, but control over large systems of machines will be in the hands of a tiny elite—just as it is today, but with two differences. Due to improved techniques the elite will have greater control over the masses; and because human work will no longer be necessary the masses will be superfluous, a useless burden on the system. If the elite is ruthless they may simply decide to exterminate the mass of humanity. If they are humane they may use propaganda or other psychological or biological techniques to reduce the birth rate until the mass of humanity becomes extinct, leaving the world to the elite. Or, if the elite consists of soft-hearted liberals, they may decide to play the role of good shepherds to the rest of the human race. They will see to it that everyone's physical needs are satisfied, that all children are raised under psychologically hygienic conditions, that everyone has a wholesome hobby to keep him busy, and that anyone who may become dissatisfied undergoes "treatment" to cure his "problem." Of course, life will be so purposeless that people will have to be biologically or psychologically engineered either to remove their need for the power process or make them "sublimate" their drive for power into some harmless hobby. These engineered human beings may be happy in such a society, but they will most certainly not be free. They will have been reduced to the status of domestic animals.

Joy remembers nodding in unease as he read this. But then he flipped the page. His anxiety changed to horror. He discovered he was reading the words of Theodore Kaczynski in *The Unabomber Manifesto*. Kaczynski was everything Joy found despicable. "I am no apologist for Kaczynski," Joy writes. "His bombs killed three people during a 17-year terror campaign and wounded many others. One of his bombs gravely injured my friend David Gelernter, one of the most brilliant and visionary computer scientists of our time. Like many of my colleagues, I felt that I could easily have been the Unabomber's next target. Kaczynski's actions were murderous and, in my view, criminally insane. He is clearly a Luddite, but simply saying this does not dismiss his argument; as difficult as it is for me to acknowledge, I saw some merit in the reasoning in this single passage. I felt compelled to confront it."

It's eerie the extent to which Joy's Hell Scenario mirrors Kurzweil's Heaven Scenario. Both are driven by The Curve. Joy readily anticipates astonishing increases in the information drivers of the GRIN technologies. "A calculation that, on a normal computer, would take the age of the universe, on a quantum computer it would take, like, a second. It would be exponentially better," he says. Both see the increase in information power clearly driving vast change in our ability to become masters of genetics, robotics and nanotechnology. In fact, when it first came out, Joy devoured Eric Drexler's book announcing the possibility that nanotechnology could be made to work. He was a speaker at one of the first gatherings of Drexler's organization, the Foresight Institute, in 1989.

Where Kurzweil and Joy totally and diametrically diverge is what happens to humanity as a result. Like Kurzweil, Joy can't see any possibility for the future other than the scenario he lays out. "If we create widely dispersed technology for this kind of mass destruction, we would have grave consequences. It's almost a tautology. How could that be wrong? If you gave a million people their own personal atomic bombs, would some of them go off?" he asks.

If anything, Joy's gloom has deepened in the years since "Why the Future Doesn't Need Us" first appeared. When the *Wired* piece detonated, New York publishers threw money at him to turn it into a book. Some would-be authors might be thrilled to put a gloss on what they had already written, spending a few months padding and thickening it sufficiently to justify a $25 cover price. Not Bill Joy. When he gets down to business, he gets down to business. So here we are years later and still no book, though not for lack of effort on his part. He has entire rooms full of research material. "You know, there is this old quote that says, 'Extraordinary claims require extraordinary proof,' " he says.

He's written three drafts, but events keep overtaking him. At first, he says, his book was aimed at overcoming people's denial. So for over a year his research assistant rounded up all the reasons to think there were really bad people in the world with really lethal ideas. On September 10, 2001, Joy was unpacking the resultant library of extraordinary proof in the place he had rented in Manhattan only blocks from the World Trade Center. "I was alphabetizing all my books on the plague and books on loose nukes and chem and bio and anthrax. I didn't have any of the anarchist cookbooks. But I think if, on the morning of the 12th, someone had come to my hotel room, I'd have probably been in jail."

Bam—as quickly as you can fly airplanes into towers, the need for the warnings of that book disintegrated. Within hours, it seemed to him, "basically a thousand people—that seems like a large number, but I bet it would be provable"—started reporting on his subject matter, blasting it out from every medium.

So Joy abandoned that plan for the book and started reorganizing it around the next important thing to which he needed to alert people: the mortal threat from the development of self-replicating man-made anything. Things that can make more of themselves might not be eradicable. In the original *Wired* piece, Joy had called attention to the specter of self-replicating robots and nanotechnology. His research, however, persuaded him that biotech was the much more imminent threat. Indeed, during this period, one of the reports that most agitated intelligence agencies in Washington was the hint that Saddam Hussein's Iraq was working on a designer pathogen of genetically combined pox virus and snake venom. A 1986 study in the *Journal of Microbiology* had reported that fowlpox spread faster and killed more chickens in the presence of venom extract. Investigators heard that Iraq sought to splice together pox and cobra venom. Such an artificial life form—created by inserting genetic sequences from one organism into another—is called a chimera, after the fire-breathing monster of Greek mythology commingling lion, serpent and goat. The investigators pursuing this weapon after the second Gulf War discovered that Iraq's technical capabilities to produce any such thing had been crippled by bombings, embargoes and UN inspections. But Joy's alarm is underlined by the fact that such a weapon in the hands of someone like Saddam Hussein was a serious worry.

Joy looks at the way in which jet travel allows a newly discovered virus such as avian flu—which can kill humans—to move from Southeast Asia to New York in hours. He looks at the way the industrialization of our food supply can distribute mad cow disease or resistance to antibiotics. He looks at how development pressure on wild areas seems to be connected to the spread of West Nile virus, Lyme disease and AIDS. He sees these suddenly emergent diseases as repeated wake-up calls from Mother Nature to human nature. Then he thinks about what an evil person could create intentionally. Imagine somebody opening a jar of chimera on Capitol Hill. Suppose it had the quick contagion and awful bloody flesh-melting aspects of Ebola, plus the long-lasting debilitating horrors of AIDS. Imagine someone unleashing flesh-eating self-replicating nanobots. After all, he says, it's astonishing what genetic engineers are doing by accident. Particularly terrifying is the Australian mouse pox incident.

Australia is one of those unusual isolated ecosystems where, when new species are introduced, they can run amok because they have no natural enemies. Mice are one of those species. Population explosions of mice—mice everywhere—just overrun Australia from time to time. Interest in figuring out how to control the buggers is keen. Toward this end, two Canberra researchers, Ron Jackson and Ian Ramshaw, in late 2000 were trying to create a new mouse contraceptive. Instead, they created a monster. They added one gene to a mousepox virus, and this new virus turned out to be 100 percent fatal. It killed every last mouse in the experiment in nine days. No survivors. This came as a complete surprise. Until then, researchers believed that changes in the genetic makeup of viruses invariably made them less lethal, not more. But even when this engineered virus was tested on mice that supposedly had been immunized, it killed half of them. "It's surprising how very, very bad the virus is," Ann Hill, a vaccine researcher from Oregon Health Sciences University in Portland, told *New Scientist*. Later work in the United States created a version that wipes out 100 percent of the mice even if they have been given antiviral drugs as well as the vaccine that would normally protect them. Cowpox virus also has been genetically altered in a similar way.

Mousepox does not affect humans. It is, however, a close relative of smallpox, which lethally does. Such a manipulation in the genetic makeup of smallpox, therefore, were it ever to be released into the population, might stand a decent chance of wiping out the human race. Nonetheless, after consulting with the Australian Department of Defence, the Australian researchers submitted their work for publication in the *Journal of Virology*. "We wanted to warn the general population that this potentially dangerous technology is available," said Jackson. "We wanted to make it clear to the scientific community that they should be careful, that it is not too difficult to create severe organisms."

All this—especially publishing the details of how to duplicate this evil—drives Bill Joy wild. He believes we simply have no idea what a modern plague, specifically one engineered to evade defenses, might be like. Have we learned nothing from the outbreak of SARS, which kills 10 or 20 percent of all infected humans and can be spread by sneezing? He has taken to quoting to anyone who will listen the first-century B.C. Roman poet and Epicurean philosopher Lucretius. This is his description of the plague of Athens:

At first, they'd bear about
A skull on fire with heat, and eyeballs twain
Red with suffusion of blank glare. Their throats,
Black on the inside, sweated oozy blood;
And the walled pathway of the voice of man
Was clogged with ulcers; and the very tongue,
The mind's interpreter, would trickle gore,
Weakened by torments, tardy, rough to touch . . .
From the mouth the breath
Would roll a noisome stink, as stink to heaven

Rotting cadavers flung unburied out . . .
Night and day,
Recurrent spasms of vomiting would rack
Alway their thews and members, breaking down
With sheer exhaustion men already spent . . .
The inward parts of men,
In truth, would blaze unto the very bones;
A flame, like flame in furnaces, would blaze
Within the stomach.

This is what Joy says he is worried about. This is The Hell Scenario.

──────※──────

THERE ARE REASONS for both hope and desperation regarding the unprecedented terrors of The Hell Scenario. The good news is that end-of-the-world predictions have been around for a very long time, and none of them has yet borne fruit. The gloom concerning The Hell Scenario is that it is distinct. It is based on the unprecedented geometric increase of The Curve. Another cause for melancholy is that many suggested ways of avoiding it are unconvincing.

The seeds of Bill Joy's alarm land on the most thoroughly fertilized ground of our imagination. Of all the scenarios we can conjure, oddly enough the ones we find easiest to embrace are the visions of disaster. No matter how much better off we may be than our ancestors, the easiest sell in the world is doom. Perhaps it is because we feel latent guilt about our achievements. Perhaps it just makes for a more exciting story. (Alternatively, if you find yourself bogging down in despair in this chapter, skip ahead to the next one, "Prevail." I promise you there is an alternative future.)

Whatever the case, there is no dearth of examples matching that of Paul R. Ehrlich. In 1968, he published *The Population Bomb*. He predicted "certain" world mass starvation by 1975 unless the world's population growth was halted. "The battle to feed all of humanity is over," Ehrlich's first sentence read. "In the 1970s and 1980s hundreds of millions of people will starve to death in spite of any crash programs."

At best, he predicted, America and Europe would have to undergo "mild" food rationing within the decade, even as starvation and riots swept across Asia, Latin America, Africa and the Arab countries. At worst, the turmoil in a foodless Third World could set off a series of international crises, leading to thermonuclear war. His misjudgment sold over 3 million copies.

Armageddon has a long and distinguished history. Theories of progress are mirrored by theories of collapse, Arthur Herman notes in *The Idea of Decline in Western History*. If success scenarios are usually tied to a wonder at the way humanity is stealing fire from the gods, calamity scenarios are tied to a conviction that we will pay mortally for our sin.

The Bible gets at it quickly. Eat of the tree of knowledge, and here's the price:

> Accursed be the soil because of you. With suffering shall you get
> your food from it every day of your life. It shall yield you brambles
> and thistles, and you shall eat wild plants. With sweat on your brow
> shall you eat your bread, until you return to the soil, as you were
> taken from it. For dust you are and to dust you shall return.

In Greek myth, Pandora, the first woman, displays curiosity and opens a box sealed by Zeus. She accidentally releases every kind of evil, including sickness. Only hope remains inside. The Hindus, too, have a saga where a golden age is replaced by misery, as do the Confucians, the Zoroastrians, the Aztecs, the Laplanders, numerous North American tribes and the earliest peoples of Iceland and Ireland.

> To whom can I speak today?
> The iniquity that strikes the land
> Has no end.
> To whom can I speak today?
> There are no righteous men,
> The earth is surrendered to criminals.

That was written circa 2000 B.C., in Egypt's Middle Kingdom.

In Greek myth, Cassandra, the daughter of Priam and Hecuba, was given the gift of prophecy by Apollo, who lusted for her. When she spurned his advances, Apollo retaliated by decreeing that no one would believe her forebodings, no matter how accurate. Today, the word *Cassandra* conjures up a nag with nothing to offer but premonitions of ruin. That's because divining how we might *avoid* doom is the far more formidable challenge.

What's interesting is how quickly and thoroughly our visions of doom became intertwined with our uneasiness about technology. To "make a name for ourselves," as it is recounted in Genesis, we decided to build a tower to heaven. Yahweh took one look at it and surmised, accurately,

"This is but the start of their undertakings! There will be nothing too hard for them to do." The tower ended up being called "Babel, therefore, because there Yahweh confused the language of the whole earth. It was from there that Yahweh scattered them over the whole face of the earth." Despite such dramatic intervention, Babel may have slowed us down, but obviously it has not stopped us yet.

Through the Middle Ages, barbarism was so close a companion that anything smacking of civilization seemed like a good idea. When the Enlightenment held out the prospect of self-government or liberty as preordained by God's natural law, it seemed superb. It was hard to argue against rationality, the arts, science, literature and poetry, compared to fighting to the death over who gets to gnaw which bone. The idea of progress was inseparable from the idea of civilization. "Dependency, especially on political and religious authority, is the distinguishing mark of a barbarous and primitive society, while autonomy—liberty—is the mark of a modern and civilized one," Herman writes. The perfection of human reason seemed to equal transcendence.

In 1627, Francis Bacon, one of the founders of modern science, published his fable *The New Atlantis*. In it, he anticipated aspects of The Hell Scenario. He describes a society much like ours, with the benefits and challenges of advanced science and technology, living with its burdens and its blessings. In it, the scientists meet to decide which of their inventions to make public and which to keep secret. This is to protect humankind from the dark side of their discoveries. But also they are trying to protect themselves from the backlash of society. They have no illusions that their beneficiaries would always be grateful.

In 1788, as the American republic arose, Edward Gibbon, in the most famous work of Enlightenment history, *The History of the Decline and Fall of the Roman Empire,* laid out a theory of societal self-destruction. All great empires reach a point of no return, he argued. "Virtue and truth produced strength, strength dominion, dominion riches, riches luxury, and luxury weakness and collapse—fatal sequence repeated so often," as the historian John Anthony Froude phrased it.

The Industrial Age created the ancestors of today's Hell Scenario. In 1794, the Rev. Robert Malthus made a curved projection from the rise of affluence and population that he observed, and concluded that there was no way civilization could continue to feed itself. The result would be starvation, destitution and ruin. He was spectacularly wrong, of course. His error, however, offers a lesson that echoes to this day: Just because the problems are increasing doesn't mean solutions might not also be increasing to match them. (Malthus, by the way, may have been the first to note that change was occurring at "accelerated velocity" and that this was exhilarating and disorienting to those caught up in it.) In 1808, Johann Wolfgang von Goethe published the first part of his monumental work immortalizing one of the most durable legends of Western folklore. It is the story of Dr. Faustus, the German wizard and astrologer, who sold his soul to the Devil in exchange for knowledge and power.

The most breathtaking Romantic depiction of technology's impact on human nature, however, came from Mary Shelley.

In 1816, Mary Wollstonecraft Shelley, while still a teenager, came to a sensational leap of imagination. She wrote a novel in which her protagonist famously recalls his precise moment of damnation: "It was on a dreary night of November that I behold the accomplishment of my toils. It was already one in the morning, the rain pattered dismally against the panes, and my candle was nearly burnt out, when, by the glimmer of the half-extinguished light, I saw the dull yellow eye of the creature open; it breathed hard, and a convulsive motion agitated its limbs. . . . His yellow skin scarcely covered the work of muscles and arteries beneath; his hair was of a lustrous black, and flowing; his teeth of a pearly whiteness; but these luxuriances only formed a more horrid contrast with his watery eyes, that seemed almost of the same colour as the dun-white sockets in which they were set."

Her novel, the first and still most celebrated work of fiction examining the impact of Industrial Age technology on the human mind, probed the psyche of a scientist obsessed with creating life who then rejected his creation, even though it could think and feel. She called her story *Frankenstein: The Modern Prometheus* and later described the arrival of her vision as almost miraculous. Speaking of her own dreams, she would later write:

When I placed my head on my pillow, I did not sleep, nor could I be said to think. . . . I saw—with shut eyes, but acute mental vision,—I saw the pale student of unhallowed arts kneeling beside the thing he had put together. I saw the hideous phantasm of a man stretched out, and then, on the working of some powerful engine, show signs of life, and stir with an uneasy, half vital motion. Frightful must it be; for supremely frightful would be the effect of any human endeavor to mock the stupendous mechanism of the Creator of the world. His success would terrify the artist; he would rush away from his odious handywork, horror-stricken. . . . He sleeps; but he is awakened; he opens his eyes; behold the horrid thing stands at his bedside, opening his curtains, and looking on him with yellow, watery, but speculative eyes. I opened my eyes in terror.

Shelley said that in *Frankenstein* she "endeavoured to preserve the truth of the elementary principles of human nature." To this day, Shelley's title is used as an incantation and an amulet. Whenever any powerful new technology disturbs our ideas of what it means to be human—interspecies organ transplants, genetic engineering, cloning—such research is rocked by that one-word hiss: *Frankenstein*.

A century later, Shelley's fears had suffused society. By the first third of the 20th century, it was commonplace to question whether technological change really equaled progress. Writers warned us against any technological utopias promised by supposedly far-seeing elites. Indeed, they saw these as hell on earth. Utopias became dystopias.

In 1932, Aldous Huxley addressed our fears of a world driven by mass production and mindless pleasure in *Brave New World*. The date is 632 A.F. (After Ford). Ford is explicitly portrayed as a god. Where once there was Christ and his Cross, in the Brave New World there is Ford and his Flivver. Psychological conditioning, genetic engineering and a perfect pleasure drug called soma are the cornerstones of the new society. Reproduction has been removed from the womb and placed on the assembly line. Workers tinker with the embryos to produce grades of human beings like so many Lincolns and Mercurys. They range from the superintelligent Alpha Pluses down to the dwarfed semi-moronic Epsilons. Each class is conditioned to love its type of work and its place in society—Epsilons are supremely happy running elevators. Outside of their work, people spend their lives in constant pleasure—buying new things whether they need them or not, participating in elaborate sports and universal uninhibited sex. Love, marriage and parenthood are viewed as obscene. The World Controller explains that contentment is more important than freedom or truth.

George Orwell's *1984*—written in 1948—emerged from one of mankind's most perilous decades. The promises of any coming utopia, such as those pledged by Hitler and Stalin, were smashed by a cataclysm of warfare and slaughter. Orwell vividly explained who these monstrous dictators were, and the nature of their hold on others. The book still resonates in our nightmares and language. It reflects the dawning of the Information Age understanding that those who controlled knowledge could be as gods. In *1984*, Big Brother can create a world in which everything means the opposite of what it is called. The Ministry of Love delivers torture, the Ministry of Truth delivers lies, the Ministry of Peace engages in war, and the Ministry of Plenty creates starvation. It's all about information. "To know and not to know, to be conscious of complete truthfulness while telling carefully constructed lies, to hold simultaneously two opinions which cancelled out, knowing them to be contradictory and believing in both of them," is at the core of totalitarianism, Orwell writes in *1984*. Big Brother, the Thought Police and the Memory Hole remain part of our vocabulary. *Orwellian* has become the gold standard adjective to apply when measuring the gulf between political language and moral reality, as Glenn Frankel noted in *The Washington Post* on the occasion of Orwell's 100th birthday.

Greatly fueling the idea that human nature could be reduced to machinelike soullessness was the work of Burrhus Frederic "B. F." Skinner, who by the middle of the 20th century had become the most famous psychologist since Sigmund Freud. His work on operant conditioning demonstrated that rewarding any desirable act could readily shape the behavior of pigeons and rats. Extrapolating from these to humans, Skinner saw no value in understanding the human psyche. He believed there was no such thing as the human mind. Human behavior was just the sum total of the effect of external forces. Shape the reward system and you've shaped human nature. In 1971 he wrote *Beyond Freedom and Dignity*, which advocated abandoning individual freedoms to further the goals of an ideal society. By then his ideas had so evoked *Brave New World* and *1984* as to lose traction. The regimes of Mao and Pol Pot cast doubt on the fruitfulness of this approach. So did the inclination of humans generally to behave in ornery and unpredictable ways. Even the lab animals had a nasty habit of not getting with the program. Skinner's students once formulated the Harvard Law of Animal Behavior: "Under controlled experimental conditions of temperature, time, lighting, feeding, and training, the organism will behave as it damned well pleases."

One of the more hopeful aspects of The Hell Scenario is that it might be so persuasive, and so terrifying, as to become an inoculant. We still so fear the 20th century's fictional dystopias that we invoke *Brave New World* and *1984* when we see developments we abhor. They vaccinated us against the futures they described. The atomic nightmare movies that populated the Cold War offered similar medicine. Take *On the Beach*, the

Nevil Shute novel made into the 1959 film starring Gregory Peck, Ava Gardner and, oddly enough, Fred Astaire. In it, the Northern Hemisphere has been wiped out in a nuclear exchange, and the cast awaits their extinction in Australia as the fallout creeps slowly, inexorably south. Then there is Stanley Kubrick's 1964 *Dr. Strangelove, or How I Learned to Stop Worrying and Love the Bomb,* the greatest black comedy ever made. Slim Pickens, as bomber pilot Captain "King" Kong, riding the bucking bomb to oblivion, is a pop culture icon to this day.

These and hundreds of similar works effectively established such a horror and taboo against the use of nuclear weapons that we live in one of the least likely futures anyone at midcentury could imagine: Humans have not used a nuclear weapon in anger for well over half a century. This, of course, could all change tomorrow. But as of this writing, the truly unimaginable has happened: Nuclear war—by at least nine orders of magnitude the most elaborately planned apocalypse in the history of man—has not yet come to pass. We may dream that The Hell Scenario provides as great a service to humankind.

Rachel Carson's *Silent Spring*—which convincingly envisioned a vernal equinox so poisoned by pesticides that it would not even be heralded by the song of birds—is another example of an alarm that resulted in great social good. Informing all of Carson's work was the idea that although human beings are part of nature, we are distinguished by our power to alter it irreversibly. "The 'control of nature,'" she wrote, "is a phrase conceived in arrogance, born of the Neanderthal age of biology and philosophy, when it was supposed that nature exists for the convenience of man." Her 1962 work led to the rise of the environmental movement, which squarely challenges three centuries of thought about the idea of "progress." Its bedrock observation is that scientific change is specifically *not* progress if it destroys the nature of which humans are a part. In the 1970s, the historic preservation movement built on this, with its core belief that change is specifically *not* progress if it destroys the community of which humans are a part. The measure of these movements is that two generations ago, not one person in a hundred had ever heard the word *ecology.* The public had never contemplated a photo from space of the whole Earth, swirling blue and green and white like a beacon of life in the utterly black void. It would have been fantasy to think that the nation's rivers—then open sewers—would ever be so clean as to spark a real-estate boom along their banks.

This is small comfort to those who tie global environmental degradation to technological "progress." As if the Three Mile Island and Chernobyl nuclear disasters, ozone loss and acid rain were not enough, now looms the specter of abrupt climate change. It could cause the extinction of a quarter of the planet's species, a report in the journal *Nature* suggests, not to mention mass migrations of humans as Germany, Britain, Scandinavia and Russia are returned to the Ice Age. Such scenarios are frequently linked to atmospheric change caused by burning carbon-based fuels. Hence the environmental movement's profound distrust of technology and the corporations that produce it.

For many, this specter is just a foretaste of The Hell Scenario.

<hr>

IN DANTE'S *Inferno*, the ninth and final circle of hell—the frozen lake Cocytus—is reserved for those who betray love and trust. Those who writhe in icy agony farthest from the warmth of redemption are those who have sold out the ones to whom they should ever be faithful: family, country, God.

For Francis Fukuyama, one of America's most thoughtful and challenging public intellectuals, the lowest ring of hell will be reserved for those who, through biotechnology, dream of leaving behind human nature. Unlike Bill Joy, he worries less about physical extinction. Fukuyama systematically attacks the ethics, morals and economics of those who would transform the human species. "Human nature exists, is a meaningful concept, and has provided a stable continuity to our experience as a species," he writes. It is, "with religion, what defines our most basic values." Messing with our minds, memories, psyches and souls, he fears, risks "leading to a brave new world." In Huxley's hell, he notes, we "no longer struggle, aspire, love, feel pain, make difficult moral choices, have families, or do any of the things that we traditionally associate with being human." As a result, we "no longer have the characteristics that give us human dignity." He quotes the bioethicist Leon Kass: "Unlike the man reduced by disease or slavery, the people dehumanized à la *Brave New World* are not miserable, don't know that they are dehumanized, and, what is worse, would not care if they knew. They are, indeed, happy slaves with a slavish happiness." Fukuyama thinks that this could easily be our future and he wants to stop it now.

This is no small contention. Fukuyama first achieved renown for authoring *The End of History and the Last Man*. In that 1989 work he argued that the end of the Cold War and the collapse of communism signaled the emerging triumph of democracy and capitalism around the globe. For this the Reaganauts lionized him. Now, however, he is a darling of the European left, especially those opposed to genetically modified crops, much less genetically modified critters, not to even countenance genetically modified human character. That's because a decade after *The End of History* Fukuyama realized that of all the myriad critiques of his work, the one he couldn't refute was the argument that there could be no end to history—no final triumph of democracy and capitalism as we know it—if there were no end to science. In fact, since human nature shapes the kinds of political organizations that are possible, technology could destroy democracy. Kiss "all men are created equal" goodbye, for example. That idea drives Fukuyama wild. "What will happen to political rights once we are able to, in effect, breed some people with saddles on their backs, and others with boots and spurs?" he asks. The result was his 2002 book, *Our Posthuman Future: Consequences of the Biotechnology Revolution.* With it, Fukuyama became the first fully credentialed "serious" person to introduce the word *posthuman* into the debate in Washington, D.C. He announced to the pinstripes that this was important. We are headed toward a transformation beyond the human species in our lifetimes.

When Fukuyama and I get together one mild and sunny November afternoon, I am reminded of the Washington of pulp fiction. Have you ever noticed that in every espionage thriller, whenever a clandestine meeting is required in the nation's capital, there always seems to be a tiny, out-of-the-way, European-style hostelry into which the protagonists can conveniently dive? That's because every espionage thriller writer in Washington seems to have spent time at the Tabard Inn on N Street. Nice place for lunch if you like overpriced organic grub and dark, mysterious, seedy ambiance. Not that Fukuyama and I have anything to be stealthy about. We were just looking for a place handy to Think Tank Row on Massachusetts Avenue, where can be found the formidable SAIS—the Paul H. Nitze School of Advanced International Studies of Johns Hopkins University—at which Fukuyama now occupies an endowed chair.

Picking at our crab cakes, we talk about human nature, what it would take to transcend it and why that would necessarily be such a terrible thing. By now my reporting has taken me to the studied conclusion that we are heading into a world in which we well might see three distinctly different kinds of humans, which I have come to call The Enhanced, The Naturals, and The Rest.

- The Enhanced are people who embrace the opportunities of the GRIN technologies. They love the idea of thinking faster, living longer, remembering everything, connecting to anything, being muscular, staying muscular, never worrying about fat, conquering disease, being sexy forever. They will pay almost any price for that kind of transformation. The Enhanced are defined as those who, through modifications to their minds, memories, metabolisms and personalities, can perform feats so unattainable by original-equipment human beings as to draw attention to themselves.

- The Naturals are those who have access to those opportunities but pass them by like fundamentalists eschewing modern pleasures or like vegetarians shunning meat. For esthetic, moral or political reasons, they recoil in horror from the consequences, especially the unintended ones. Naturals are original-equipment humans who have the opportunity to become Enhanced but have chosen to turn against it.

- The Rest are those who, for economic or geographic reasons, do not have access to these technologies. They envy and despise those who do. The Rest are original-equipment humans with no choice to become Enhanced.

In the early 21st century, there were no bright lines separating these three and it was an open question whether or how such stark separations might occur. The gaps, should they arise, are more a values proposition than a technological issue. If a person has a test-tube baby, is she an Enhanced? In the developed world, we don't recognize her as such. She is perhaps just a 50-something attending elementary school meetings. Her blessings are remarkable, but hardly seen as threatening. If her family travels to Africa, however, the locals may properly regard her with wonder, if they even believe her claim not to be the child's grandmother. But again, there is no bright line, either of technology or of acceptance.

If a person has a Viagra prescription, is he an Enhanced? That's more problematic. Viagra is a metabolic improvement that possibly allows a wealthy 50-year-old to compete for a young lovely against a 30-something or even a financially struggling 20-something. Does resentment ensue?

What about Barry Bonds? He has been implicated in the steroid trade. If he is an Enhanced, should he go around with an asterisk on his forehead for the rest of his life marking him as a kind of human different from his godfather, Willie Mays, whose home-run feats he surpassed?

These questions reflect primitive 20th-century technologies. What happens when The Curve allows a vast collection of formidable internal upgrades for those who can afford them? How will you someday tell, looking at an Enhanced, if she genuinely represents a transformation of the species—comparable to the difference between Neanderthals and today's humans?

I propose The Shakespeare Test.

(This deliberately avoids the swamp concerning how you would establish if a machine intelligence is truly conscious. That debate leads to thousands of excruciating pages of monographs. Besides, as Vernor Vinge notes, "self-awareness may be overrated.")

In The Shakespeare Test, as I imagine it, you take the being who has been so seriously modified that you are concerned. Forget the old bases for discrimination of the past. She is way beyond simple differences of class or race. She has a significantly transformed mind, memory, metabolism and personality. You're curious whether this has changed her immortal soul. So you stick her into your thought experiment's hypothetical time machine and dial this creature back to 1603 or thereabouts. You then present her to Mr. Shakespeare, who knew something about human nature and humans' reactions to outsiders, as he created both Othello and Caliban. You ask Mr. Shakespeare a simple question: "Do you recognize this creature as one of yours? Is she human?" If yes, then she passes the human nature test. If not, then no.

The Shakespeare Test is based on the observation that historically, human nature changes much more slowly than do our circumstances. In the 21st century we perform the ancient Greek plays all the time and recognize the characters as being like us. This is not to say that human nature is utterly unchanged. Some habits of millennia ago, such as eye gouging, are now regarded as pathological and monstrous. Similarly, although *Romeo and Juliet* still resonates, the violent aspects of Verona would not play today if set in a wealthy zip code such as Beverly Hills. Rich people today rarely slaughter each other like that, and when they do, it becomes a national morality play, as in the O. J. Simpson trial. To make sense, *Romeo and Juliet* now must be set in a neighborhood marginal to the human experience in the developed world, such as the South Central L.A.-like gangsta hood of the 1996 Leonardo DiCaprio film, or the Spanish Harlem of *West Side Story*.

Therefore, if you were to apply The Shakespeare Test to the cast of *Apollo 13*, they would pass effortlessly. Once Mr. Shakespeare got past the fact that the film involves a craft headed toward the moon—and it's nonfiction—he would have no trouble recognizing the astronauts as adventur-ers trying to make it home, exactly like those of *The Odyssey*. Similarly, most of the characters on the bridge of the original Star Ship *Enterprise*—Bones, Sulu, Uhura, Captain Kirk—would easily pass The Shakespeare Test. The ones Mr. Shakespeare might have trouble with are Spock and, in the later series, Data, and those dudes with the horseshoe crabs on their foreheads, like Worf. In addition to their appearance, their emotional re-sponses, or lack thereof, would definitely be suspect. It's easy to imagine these characters sparking a lively conversation with Mr. Shakespeare as to who should count as a human and why. I would especially like to know what Mr. Shakespeare would make of Lt. Commander Data.

Back at the Tabard Inn, Fukuyama wonders how Gestapo officers would fare when presented to Mr. Shakespeare—or sociopaths who lack the capacity for embarrassment. "Something that's very typically human is the ability to perceive how other people perceive you," he notes. "Autistic children do not have this. A source of their problem is that they shout out loud in class and they don't understand that people don't like them."

What Fukuyama's really concerned about, though, are not people like this—well out on the fringes of the bell curve. "The thing I'm worried the most about is the attempt to modify on a large scale some basic char-acteristics of human behavior in ways that will make us scarcely recogniz-able," he says. He loves the work of the anthropologist Donald E. Brown, author of *Human Universals*. In fieldwork with a hunting-gathering soci-ety, Brown couldn't help but observe the huge differences between them and Westerners. Yet, he notes, both societies have in common emotional responses that tie them to all human beings through tens of thousands of years. "It's instantly recognizable that this is another human being, despite that huge difference," says Fukuyama.

If there is prompt recognition even when a hunter-gatherer and an In-ternet maven meet, imagine a different, posthuman future, he asks. Sup-pose our grandchildren have been genetically manipulated to "improve" the species, and pharmaceuticals have taken all the rough edges off their personalities. "Is it possible to imagine that somehow we would be miss-ing some of those typical human characteristics, good or bad?" That's his big concern. He defines human nature as "the sum of the behavior and characteristics that are typical of the human species, arising from genetic rather than environmental factors." He goes to great lengths to establish that he is by no means suggesting that gays—or, for that matter, dwarfs or any other minority—are not human, even though they are not the statistical norm. Fukuyama says that when you strip away all of a person's accidents of birth—skin color, looks, social class, gender, culture and even talents—there is still some essential human quality underneath that is worthy of respect. That is the source of human dignity, he argues. That essence, whatever it is, he calls "Factor X." He wrestles with what that might mean. He does not, for example, insist that "Factor X" means a soul. He knows there are a lot of people who believe that matter and energy are all that is real. Yet he points out that there is some essential quality there that we obviously recognize, because "the idea that one could exclude any group of people on the basis of race, gender, disability or virtually any other characteristic from the charmed circle of those deserving recognition for human dignity is the one thing that will bring total obloquy on the head of any politician who proposes it." The resulting idea of a universal equality of human dignity "is held as a matter of religious dogma by the most materialist of natural scientists."

When pressed, Fukuyama acknowledges that human nature has not been static over all these millennia. It has evolved. Christianity, for example, was an effort at transcendence that made a difference. "Look at the difference between the Christian world and the Roman world in terms of things like sympathy, compassion, the prevalence of cruelty," he says. "I mean, the Romans were unbelievably cruel. Basically, if I'm powerful I'm going to get all the women I want and all the money I want and the hell with you."

He recalls a passage in *Democracy in America* not two centuries ago in which Alexis de Tocqueville quotes a letter from a French noble lady, Madame de Sévigné, dated October 30, 1675. The lower classes in Brittany have just staged a revolt against the imposition of a new tax, and they are being put down. Madame de Sévigné is very cultured; she sponsors a salon. She's writing to her daughter and says well isn't it marvelous that you seem to have kissed every lad in Provence, and then she immediately goes on to give the news from Rennes. It seems old men, women about to give birth and children are "wandering around and crying on their departure from this city, without knowing where to go, and without food or a place to lie in. Day before yesterday, a fiddler was broken on the wheel for getting up a dance and stealing some stamped paper. He was quartered after death," she reports cheerfully, "and his limbs exposed at the four corners of the city." Then she immediately goes on to say what a wonderful day she's had with Madame de Tarente. Soon after, she writes, "You talk very pleasantly about our miseries, but we are no longer so jaded with capital punishments; only one a week now, just to keep up appearances. It is true that hanging now seems to me quite a cooling entertainment."

Tocqueville says, "It would be a mistake to suppose Madame de Sévigné, who wrote these lines, was a selfish or cruel person; she was passionately attached to her children and very ready to sympathize in the sorrows of her friends; nay, her letters show that she treated her vassals and servants with kindness and indulgence. But Madame de Sévigné has no clear notion of suffering in anyone who was not a person of quality. In our time"—a century and a half later; he wrote this in the 1830s—"the harshest man, writing to the most insensitive person of his acquaintance, would not venture to indulge in the cruel jocularity that I have quoted; and even if his own manners allowed him to do so, the manners of society at large would forbid it. Whence does this arise?" Tocqueville attributes it to the transformative notion that "all men are created equal," especially as the idea evolved in American democracy. "In democratic ages men rarely sacrifice themselves for one another, but they display general compassion for the members of the human race," he writes. "At the time of their highest culture the Romans slaughtered the generals of their enemies, after having dragged them in triumph behind a car; and they flung their prisoners to the beasts of the Circus for the amusement of the people." Even Cicero does not see barbarians as belonging "to the same human race as a Roman. On the contrary, in proportion as nations become more like each other, they become reciprocally more compassionate, and the law of nations is mitigated."

This is one of Fukuyama's big arguments against changing human nature. It threatens the optimal "end of history" of modern capitalistic democracy that he firmly believes suits humans better than any other existing system.

Okay, so cultural evolution works, I tell him. We can learn from experience and pass that learning on to our descendants. Fukuyama acknowledges that. He even points out that humankind's constant effort to fix its shortcomings is what drives human history. So suppose we modify ourselves through technological evolution. The problem with that would be what?

It's wrong, he replies, to assume that technology always produces positive social outcomes. The cotton gin was bad, for example. In the late 1700s, slavery was becoming unprofitable in America. It might soon have waned. Eli Whitney's clever invention, however, made lucrative the use of slaves to harvest cotton. So bondage expanded. The ultimate result was the bloodiest conflict in American history, the Civil War. Today, "you can't have modern democracy," says Fukuyama, "unless you have this basic belief in equality, which means that you should empathize with suffering and feelings of other people and recognize their rights as equal to your own." His concern is that the divisions between The Enhanced, The Naturals and The Rest may be so profound as to make past ruptures over race and religion seem quaint and paltry. If wealthy parents figure out a way to increase the intelligence of all of their descendants, "we have the makings not just of a moral dilemma but of a full-scale class war," he writes. What's more, "human nature is what gives us a moral sense, provides us with the social skills to live in society, and serves as a ground for more sophisticated philosophical discussions of rights, justice, and morality." He's terrified that The Enhanced, in time, "will look, think, act, and perhaps even feel differently from those who were not similarly chosen, and may come in time to think of themselves as different kinds of creatures." He sees that separation as easily leading to people getting "off their couches and into the streets . . . actually picking up guns and bombs and using them on other people."

But suppose we end up using technology to expand our circle of empathy by increasing contact with people all over the world? I ask Fukuyama. Why are you so sure we can't improve on human nature?

"I think the answer is no, we're not sure that we can't. But I would say there's a high probability of our screwing this thing up," he replies. "Certainly no one would say that we want more hatred. But if you think about things like anger and the kind of violence and pride and the responses that lie behind a lot of acts of violence, it actually is all in the service of defending norms of communities. So the question is whether you can actually intervene to dampen that emotional response in ways that won't undercut your ability to actually defend your community. If you could get rid of just random, pointless violence but not directed, necessary violence—if you think that you're good enough to figure out what the sources of one are, and not the other—then be my guest. I just think it's so likely that we're going to screw this thing up. We just don't understand how complex these interdependent feelings are.

"Even something like the elimination of pain and suffering, you know. This is the argument that's the most difficult to make. But I think it's ultimately the most critical one. There's something about the experience of pain and longing and anxiety and all of these things that our therapeutic society is trying to get rid of. It is somehow necessary to our self-understanding of what we are as human beings. I mean, you can't have courage without risk. You can't have real compassion or sympathy without the personal experience of pain.

"I'm not sure that we would be better off as gods. For example, I presume that one of the attributes of being a god is not ever having to worry about your own mortality. That seems to me a perfectly good case of something that every individual would wish for but which is going to be disastrous for society as a whole." He denies the possibility that we might gain new wisdom with age. To the contrary, he doesn't think immortals will ever have a new idea again. He believes the only way new ideas get accepted is, "literally, people dying off." This is an intriguing hypothesis. If immortals are not capable of innovation, that at least does solve the problem of The Curve's ever-exponentially-increasing technological change—unless robots learn to dream.

Fukuyama fears a world in which "you've got two-thirds to three-quarters of your population beyond the age of sexual interest, very rigid views, kind of fundamentally unable to adapt in certain basic ways to changes in what's going on in the world." Fukuyama, who was born in 1952, can't see the possibility that people might enjoy having a series of new lives. He doesn't like the fact that with the current extension of age, a smaller and smaller portion of our lives is concerned with the raising and socialization of children. He thinks the more time we spend raising children, the better humans we become: "It plays a role in the socialization of the parents." He also doesn't like the idea that "sex becomes a fairly minor part of life. There's all this political correctness about there's no reason why people in their 70s can't be sexy. Total bullshit, you know. I would prefer to live in a more natural kind of society. I think it's nice to look at young, sexy women."

Apart from the utilitarian aspects of this outcome, Fukuyama sees a moral issue in immortality. "The deeper issue is, can people conceive of dying for a cause higher than themselves and their own fucking little petty lives? I mean, can they think of dying for God or for their country or for their community or anything beyond themselves? I think that the very aspiration is wrong—this aspiration to want to live forever—because your own petty life trumps all other values, and I think that any traditional notion of transcendence began with the notion that the continuation of your personal human life is not the highest of all goods. No animal is capable of formulating an abstract cause to die for."

You're not going to like the next 100 years of your life, I tease him.

"Well, no male in my family has ever lived past about the age of 75, and I'm not expecting to, either."

What if you get double-crossed by advancing technology and last for a very long time?

"I'm not sure that I'd be happy about that. My mother had a stroke about six or seven years before she died, and she was really never the same person after that. When my father died, he was in a hot bath in Japan and he was 73. Just had a heart attack. I would much rather go the way my father did than the way my mother did. I wouldn't want the extra 10 years."

Of course, the real problem Fukuyama has with The Hell Scenario is not whether it is persuasive and realistic and horrifying. That's not a hard case to make. But what do you do to prevent it?

"It has been a long time since anyone has proposed that what the world needs is more regulation," he writes. That is exactly what he proposes, however, going so far as to insist on a regulatory regimen covering the entire globe. He thinks little of scientific self-regulation, arguing that there are too many greedy people chasing too much money to leave it up to scientists to regulate themselves. "Science cannot by itself establish the ends to which it is put. . . . Only 'theology, philosophy or politics' can do that."

He dismisses the notion that the GRIN technologies are beyond control. To do it, though, he thinks we will need new institutions, reaching into the internal workings of China, India, Japan, Korea and Europe, as well as the United States, to bring the power of public opinion to bear on technologies he views as offering an unprecedented threat. That might happen, he says, only after enormous public outcry caused by, for example, hideously deformed babies—products of experiments gone horribly wrong. We might need to suffer as much revulsion as we did after Hiroshima, he believes.

———

BILL JOY AGREES with Fukuyama that a tragedy may be needed to bring action. His version of optimism is to hope that the wake-up call will be only a medium-sized catastrophe, killing millions. As opposed to some full-blown version of The Hell Scenario, such as a genetically altered virus eating the flesh of the entire species, or legions of tiny robots sucking the nutrients out of the entire biosphere, or there being worldwide class warfare between divergent kinds of humans, or super-intelligent machines arising that perceive humans either as pests to be destroyed or, perhaps worse, as pets.

But what exactly do you do in response to the agonizing end of the human race so vividly projected in The Hell Scenario? If you find this scenario utterly persuasive, what do you do to defeat it? That remains the key issue.

Joy has several valuable ideas. He believes scientists can and should regulate themselves, being deathly cautious about creating anything that can uncontrollably replicate itself. "Scientists do not believe they can do their work if they have to consider consequences," he says. "But such free passes are no longer sensible in the age of self-replication. Scientists and technologists must take clear responsibility for the consequences of their discoveries. That this will slow the pace of discovery is unfortunate. But there is no alternative. It is not sufficient to have great science, to have great and repeatable laboratory results. What we need is great results in the world." Joy believes that "market forces should replace regulation. Companies that wish to make personalized drugs can take legal and financial responsibility for the outcomes." He wants liability law to put a cost on catastrophic risks. "We don't need the costs to be perfect. If they are at least roughly proportionate to the magnitude of the true risk, some very dangerous things, to our great relief, will become uneconomic. There will still be rogues, but this is a game of risk reduction, not risk elimination, and the markets can provide us a great leg up." He also believes we need to recognize that information is now the same thing as a physical object. If you view an organism as so dangerous as to require P4 containment—the highest level, complete with airlocks, moon suits, double-door autoclaves and liquid waste sterilizers—then keep information about that organism under the same kind of wraps. "There is no reason to publish the plague genome to everybody. You can publish it to people who need to know it. You wouldn't sell them the material. If the material is the same as the information, then why give them the information? That's common sense. It's just in two different forms. The idea that everything ought to be available to everybody is foolishness. The reason is we're scientists and we've always published and if we don't publish, blah, blah, blah, blah. That's bullshit! A realistic view of the world would say that people are evil and people make mistakes and we ought to use common sense. These technologies were created by large groups. An individual couldn't have created bioengineering, and so the group of people who created it has a responsibility for the way in which it's used. A directed attack against a particular important node in the network can cause incredible havoc. Slow the thing down to give yourself some time. There are some messes we can't clean up."

Nonetheless, when pressed, no matter how convincingly Joy portrays his Hell Scenario, he finally dribbles off when it comes to conquering it. "We should relinquish stuff that we decide is too dangerous for us—

which is like a tautology. That's a definition of sanity—that you don't do what is fatal to you," Joy says.

He's right on several counts, one being that it is a tautology, a needless repetition of an idea in different words. How do you decide what's fatal? Who decides what's fatal? What happens if there is a disagreement? Especially if, say, Asians, Americans and Europeans have widely diverging views? Nuclear weapons are easy to control compared to the GRIN technologies. Few people want nuclear war. Many people, however, want healthy, smart, athletic, beautiful, long-lived children.

Even as distinguished a describer of The Hell Scenario as Martin Rees runs into problems figuring out what to do about it. Rees is the "astronomer royal" of the United Kingdom—sort of a poet laureate to the stars. From this lofty vantage point, Sir Martin has produced the most far-reaching version of The Hell Scenario imaginable. He can even see that "experiments that crash atoms together with immense force could start a chain reaction that erodes everything on Earth; the experiments could even tear the fabric of space itself, an ultimate 'Doomsday' catastrophe whose fall-out spreads at the speed of light to engulf the entire universe." In his book *Our Final Hour; A Scientist's Warning: How Terror, Error and Environmental Disaster Threaten Humankind's Future in This Century—On Earth and Beyond*, Rees vociferously makes the case that "technical advances will in themselves render society more vulnerable to disruption." He quotes the odds of our species surviving to the end of the 21st century as no better than 50-50.

Chapter by chapter, Rees catalogs everything that could go colossally wrong in this century. It's hard to believe he's missed anything. Climate change, asteroid collision, flesh-eating viruses assembled and released by madmen, genetics run amok, nanobots run amok, machine intelligences run amok—it's all in there. So are quite a few exotic catastrophes that could theoretically result from experiments in Rees' specialty, physics—such as the universe tearing mentioned above.

As expert and devastating as his recitation is, Rees has difficulty with the issue of what to do about these doomsdays. He does not like risk taking at all. Take our world today. If it were up to him, we wouldn't be living in it. Given the risks of a nuclear exchange during the Cold War, Rees flat out says he would "rather be red than dead," in the words of the old slogan. "I personally would not have chosen to risk a one in six chance of a disaster that would have killed hundreds of millions and shattered the physical fabric of all of our cities, even if the alternative was a certainty of a Soviet takeover of Western Europe," he says.

When it comes to remedies, Rees preaches the gospel of the "precautionary principle"—humans should not create something new unless they are reasonably certain something awful will not result. This is hard to argue with, in the abstract. The world might indeed be a better place if Queen Isabella had asked Christopher Columbus to file an environmental impact statement before setting sail. Columbus' heirs would probably be writing it still. But it seems to me that proponents of this principle bear a certain responsibility. Caution overcoming curiosity is not a conspicuous aspect of human history. As the renowned British satirist Terry Pratchett famously notes, "Some humans would do anything to see if it was possible to do it. If you put a large switch in some cave somewhere, and a sign on it saying 'End-of-the-World Switch. PLEASE DO NOT TOUCH,' the paint wouldn't even have time to dry." Nor does Pratchett attribute this simply to perversity. It may be hardwired. Of our species, he writes, "Perhaps it was boredom, not intelligence, that had propelled them up the evolutionary ladder." Humans seem driven by "that strange ability to look at the universe and think 'Oh, the same as yesterday, how dull. I wonder what happens if I bang this rock on that head?' "

Where you want Rees to explain how he proposes to constrain this behavior, however, he is disappointing. For example, he notes, "The surest safeguard against a new danger would be to deny the world the basic science that underpins it. . . . Should support be withdrawn from a line of 'pure' research, even if it is undeniably interesting, if there is reason to expect that the outcome will be misused? I think it should."

Okay, but you're going to do this how? On the very next page, he is intellectually honest enough to note that U.S. restrictions on stem cell research are resulting in a "brain gain" in places with more permissive guidelines, such as the United Kingdom and Denmark. "By offering a still more enticing regime to researchers and to their fledgling biotech industry, Singapore and China aim to leapfrog the competition," he adds.

Rees shoots a hole even in Fukuyama's courageous idea for a vast global regulatory scheme. "The difficulty with a dirigiste [government-controlled] policy in science is that the epochal advances are unpredictable. . . . X-rays were an accidental discovery by a physicist, not the outcome of a crash medical program to see through flesh. . . . In more recent times, the pioneers of lasers had little concept of how their invention would be applied (and certainly did not expect that one of the first uses would be for operations to repair detached retinas)."

All in all, *Our Final Hour* is a sober, sane, serious catalog of hideous threats, accessibly and compactly written, with useful and nuanced insights and even moments of unexpected humor about, say—anthrax. Rees writes: "To say that a few grams of an agent could in principle kill millions may be true, but it may also be misleading (just as it would be misleading to say that one man could father a hundred million children; spermatozoa are plentiful enough, but dispersal and delivery would be a real challenge)."

Nonetheless, given how broad his inventory of Medusa's snakes are, you find yourself asking Rees, "Why you telling me this stuff?" If we are shocked and convinced by his Hell Scenario, what should we do?

At the end of *Our Final Hour* he muses about one reasonably surefire way to preserve humanity: disperse it into space. At which point this advocate of the precautionary principle grows quite excited about a novel way to colonize Mars. He rhapsodizes about taking this virgin planet—on which no one can say for certain life does not exist—and suggests we precede its human habitation by landing on it a nuclear reactor bolted to a chemical processing factory.

What a human is Sir Martin.

One thing Rees' work does point up is that the road to hell is paved with a rich variety of scenarios. Hollywood has offered a nice array. In *Blade Runner* and *The Terminator,* the androids and cyborgs, respectively, are out to get us. In *The Matrix,* the machine intelligence is out to get us. And in *X-Men,* the genetically engineered are out to get us. (Actually, in the American Film Institute ranking of film's 50 greatest heroes and villains, *The Terminator* shows up on both lists.)

Baronness Susan Greenfield, a neuroscientist and the author of *Tomorrow's People,* provides another Hell Scenario. Greenfield is one of the exceedingly few women so far prominent in this debate. (There is also a conspicuous absence of blacks or Hispanics.) She speculates in her chapter "Human Nature: How Robust Will It Be?" that ultimately we will end up losing our individuality and identity in "a more atavistic state of consciousness where we permanently 'blow' our minds." She acknowledges that she started out trying to write her Hell Scenario as a science fiction novel featuring a ravishingly beautiful and brilliant female neuroscientist but couldn't handle the problems of plot, character and dialogue. So she tried to make her book nonfiction. Be that as it may, she consoles herself that her Hell Scenario may be grimly adaptive. Since "large-scale death and suffering if not global nuclear war" seem imminent to her, "then the materially comfortable, anodyne existence in which we lose the essence of our humanity, our human nature, might, amazingly enough, seem more desirable," she writes. She swears she is not a techno-phobe, yet in the end she quotes the philosopher and social critic Bertrand Russell: "Science has not given men more self-control, more kindliness, or more power of discounting their passions. . . . Men's collective passions are mainly for evil; far the strongest of them are hatred and rivalry directed toward other groups." In the end she argues that the ulti-mate priority should be "not just the preservation, but also the *celebration* of individuality." She also quotes the physicist and social activist Freeman J. Dyson: "The central problem of an intelligent species is the problem of sanity."

A more immediate version of a Hell Scenario comes from those who focus on the haves and have-nots. This is the case for the renowned digi-tal divide. It envisions a desolate future in which The Rest never have ac-cess to The Curve of exponentially increasing technology. They are left behind in irreparable misery, on a planet that is hotter, more crowded, and utterly degraded. "On a visit to North-East Asia, I saw this future," says Bob Carr, the premier of New South Wales in Australia, seeing in China a replay of the film *Blade Runner*. "The landscape was simple," he says. "There were clusters of shoebox-style tower blocks. They were linked by clogged expressways in a flattened, cleared landscape. It was so bleak, so denatured, that it could have been a place rebuilt after a nuclear blast. The air was heavy with smog. Acid rain fell. . . . This will be how more people will live in 100 years."

The critical uncertainty in the potential divide between The Enhanced and The Rest is access to the GRIN technologies—access to the ability to evolve. If developing peoples can progress—economically, socially and technologically—then this Hell Scenario may be less imminent. The good news is that such a future is not unimaginable. A have/have-later world would be considerably less scary than a have/have-not world.

In have/have-later, it goes without saying that the haves will get first crack at whatever is new. Rich people tend to get first pick in every soci-ety throughout all of history. As recently as 2002, the makers of films such as *The Bourne Identity* signaled that you were supposed to understand their characters as rich, powerful, worldly and sophisticated because they were peering into their tiny cell phones.

As a matter of fact, however, in the decade leading up to 2002, cell phones—the real personal computers of the early 21st century—increased phenomenally among people who are not rich, powerful, worldly or sophisticated. In developing countries the proportion of people with access to a phone grew an astonishing 25 percent in the 1990s, according to the Worldwatch Institute, an organization devoted to "an environmentally sustainable and socially just society." One in five of the world's population had used a mobile phone by 2002—up from 1 in 237 in 1992. Because of The Curve, the price of mobiles had dropped precipitously—below the cost of landlines. In 2002, for the first time, the number of global mobile phone subscribers (1.15 billion) was greater than the number of fixed-line connections. In 1999, Uganda became the first African nation to have more mobiles than traditional phones. Thirty other African nations followed by 2002. Judging from the billboards in the megacity of Lagos, Nigeria, cell phones were one of the three largest industries there, neck and neck with religion and nutritional supplements. This remarkable pattern fueled connections to the Net. In 1992, just 1 in 778 of the world's population had used the Internet. By 2002, 1 in 10 had. This has a tangible benefit for people's lives, Worldwatch reports. "By linking rural farmers to market information, craft workers to customers, patients to doctors, and students to teachers, the Internet can aid economic development," it says.

To be sure, in places capable of great technological sophistication, such as China and Russia, governments who fear their own dissidents—and thus try to control information—have attempted to intentionally slow the revolution, the RAND Corporation noted. Some Middle Eastern societies recoil at dissemination of Western ideas in general, and pornography in particular. Latin America is hampered by low literacy rates. There are some failed places on earth marked by such outrageous politics, pathetic infrastructure, abysmal annual incomes and few cities that it's hard to imagine how they will achieve any significant development. Singapore researchers examining Internet uptake in Asia pointed to a familiar list of failed suspects—Bangladesh, Cambodia, Kazakhstan, Laos and Myanmar, for example.

Nonetheless, the gap between the haves and have-nots has hardly proven to be hopelessly rigid, as the migration of software-writing jobs to India has demonstrated. The International Telecommunication Union, tallying broad measures of connectedness worldwide, including affordability, found Slovenia tied with France. Korea, Hong Kong and Taiwan were ahead of the United States. In the Caribbean basin, access for the Bahamas, St. Kitts and Nevis, Antigua and Barbuda, Barbados, Dominica, Trinidad and Tobago, Jamaica, Costa Rica, St. Lucia and Grenada was ahead of Russia. The Eastern European nations of Estonia, the Czech Republic, Hungary, Poland, the Slovak Republic, Croatia, Lithuania, Latvia, Bulgaria, Belarus and Romania were ahead of China. The Singapore researchers found that a lack of English-speakers did not necessarily correlate with poor technology pickup. In a postliterate world—in which the Internet increasingly becomes something you watch and listen to, rather than read—low literacy rates were less a barrier than one might expect, at least in Asia. The digital divide seems to be narrowing, a University of Toronto study says. The demographic lag between those who use the Internet in developing countries and those who use it in the United States was about five years, the Canadian researchers reported. This technology is getting to the masses a lot faster than did electricity, radio, washing machines, refrigerators, television, air conditioners and automobiles.

The big difference between the GRIN technologies and others separating the haves from the have-nots is price. Because The Curve rules, costs of the GRIN technologies drop dramatically. The transformative stuff quickly becomes affordable and ubiquitous, even in developing countries. This was not the case with older technologies. Heart transplants, for example, never became a bargain because transplantable hearts never became abundant. This may change. Growing replacement organs on demand is one of the promises of the GRIN technologies. It will be interesting to see what happens when genetic engineering becomes sufficiently sophisticated, cheap and unremarkable that it becomes common in a place such as India. The reincarnation implications alone are staggering.

Cost is hardly the only barrier to the global diffusion of the GRIN technologies. Probably more important is the "yuck factor"—the visceral rejection of technologies that are seen as anti-human. Headlines about human cloning produced one of the more vivid Hell Scenarios. That prospect caused Bill McKibben to wonder if there was any way to preserve human nature as it exists.

McKibben is most renowned for his environmental writing, especially his 1989 best seller, *The End of Nature,* the most influential early warning of the impact of global warming. It has been translated into 20 languages. In his 2003 book—greatly inspired by the Bill Joy alert that he calls "one of the great Paul Revere moments of our time"—he argues in favor of humans embracing their limits. He considered calling that volume *The End of Human Nature,* but discarded that title as too self-referential. Instead, he called it *Enough: Staying Human in an Engineered Age.* As a result, those people I call The Naturals—who have access to human enhancement but pass it up for moral, esthetic or political reasons, may very well end up calling themselves "the Enoughs." McKibben has written their manifesto.

McKibben begins with a tale of running the Boston Marathon at the age of 41. He finishes 324th, an hour and a quarter off the world record. Yet his saga is one of achieving his personal best. "When it was done, I had a clearer sense of myself, of my power and my frailty. For a period of hours, and especially those last gritty miles, I had been absolutely, utterly present, the moments desperately, magnificently clarified. As meaningless as it was to the world, that's how meaning*ful* it was to me. I met parts of myself I'd never been introduced to before, glimpsed more clearly strengths and flaws I'd half suspected. A marathon peels you down toward your core for a little while, gets past the defenses we erect even against ourselves. That's the high that draws you back for the next race, a centering elation shared by people who finished an hour ahead and two hours behind me." He feels connected to early humans, for few experiences are a more primal part of our nature than long-distance running. And yet, he fears, he will be among the last humans to be touched by this epiphany. He worries that enhancements will make such efforts meaningless. In his view, soon you will not know if anything you ever achieve will be because of your striving or because of your technology. "It's not the personal *challenge* that will disappear. It's the *personal*," he writes.

Yet the problem is not the plausibility and persuasiveness of his Hell Scenario. He hardly lacks eloquence in conjuring that fright. Again, the problem is what to do about it. McKibben's proposed solution is to renounce technologies that separate our essence, our beings, from our past.

He devotes a chapter—"Is Enough Possible?"—to how exactly you would do that. He offers three existence proofs that technological refusal is possible: the modern Amish, and the medieval Chinese and Japanese. Only people making the case exactly the opposite of McKibben's usually cite the last two. Yes, it is possible to renounce technology. The Chinese turned against global maritime exploration in the 1400s, and the Japanese gave up guns in the 1600s. But as a result, even he acknowledges, "the West, not the East, was to dominate world trade and come first to commercial and industrial might." He doesn't mention it, but Islam also ignored the global threats created by the scientific and industrial revolutions of the once-undeveloped Europeans. The Middle East, too, is still shaped by that decision. All of these are examples of once-great civilizations brought to their knees. They were forced to endure humiliation and exploitation by supposedly backward people who weren't overly scrupulous about the technologies they embraced. These are not confidence-inspiring illustrations of how to avoid The Hell Scenario.

In fact, if you want to pace the floor at night, playing with your own personal Hell Scenario, ask yourself a question. The first licensed gene therapy came out of China. The first cloned human embryos came out of South Korea. Suppose a machine suddenly achieved intelligence greater than that of humans. Suppose it continued to improve so rapidly as to evolve way past human capabilities, triggering The Singularity, as Vinge predicts. Would it matter if people from China originally shaped that intelligence? Would the super-intelligence share their cultural attitudes? Would it be different if the super-intelligence first arose in Europe? Or the United States? Or North Korea?

Suppose that super-intelligence first came out of the military, where the point is to win conflicts. Now suppose instead that the super-intelligence first comes out of some corporate back-end shop such as customer relations, where the point is to be very, very nice to us. Would that matter?

Such thought experiments show why the much more interesting case is that of the Amish, who do not reject all technology. In 1991, for example, a team of Amish carpenters gathered to build a clinic for the somatic gene therapy of children with an inherited illness prevalant among Amish and Mennonite children. The Amish use cars—if somebody else is driving them. They use phones—if they are not in the house. Their test is whether or not a technology serves the building of their community. That is a profoundly valuable analysis.

"Can we, even if we want to, actually rein in these technologies?" McKibben asks. "Can the opposition to them ever be more than academic?" Is it possible that "the visceral recoil from the loss of meaning that I've been describing will translate into effective political resistance"? Good questions.

There is great merit in McKibben's argument that we should pick and choose which technologies are right for us. For example, in the 1960s Americans rejected the technology then available to build supersonic airliners, on the grounds that it would damage the upper atmosphere. Their decision was vindicated in the early 21st century when the Europeans, who went ahead with the Concorde, allowed it to go out of service without a replacement because few could afford the tickets. We've also taken a pass on DDT, massive new hydroelectric dams and additional nuclear reactors.

It's not clear that such picking and choosing—"fine-grain relinquishment," it is called—would seriously deflect The Curve, however. After all, if people in one country such as the United States can't even agree on abortion and stem cells, it will be a formidable task to get a global consensus on the entire array of GRIN technologies coming at us ever faster. Nonetheless, when trying to figure out what to do about his Hell Scenario, McKibben admirably throws in his lot with democracy. He does not propose to hand the problem over to a carefully appointed commission of wise men. He's still willing to slog it out in the marketplace of ideas with that seething mass of the great unwashed, the voters.

He's certainly right that such a project may already be leading to a major reordering of our politics, scrambling any Industrial Age thinking about what constitutes left and right—who's got and who wants. The strange-bedfellows list when it comes to engineered evolution is intriguing. *Liberal* and *conservative* are awkward labels in this realm. Among those deeply skeptical of human enhancement, one will find prominent "bio-conservatives" such as Leon Kass, chair of the U.S. President's Council on Bioethics. But those finding some things about which to agree with him include Jeremy Rifkin's Foundation for Economic Trends, the Green Party, many who loathe globalism (especially the unregulated marketplace, the International Monetary Fund and the World Bank), people who don't believe we are descended from monkeys, off-shoots of the Christian anti-abortion movement, feminists from the Boston Women's Health Book Collective (including the editor of *Our Bodies, Ourselves,* whom McKibben describes as more pro-choice than the pope is Catholic), and Prince Charles. McKibben points out that environmentalists usually see organized science as their ally in the fight against climate change or for habitat conservation. Yet Friends of the Earth and Greenpeace USA have lined up against human cloning. William Kristol, editor of the conservative *Weekly Standard,* signed the same petition in this regard as did Tom Hayden and Todd Gitlin, stalwarts of the sixties left.

Gung-ho adherents of The Heaven Scenario offer a similarly fascinating spread. They include the staunchly libertarian individualists and market-driven entrepreneurs one finds so thick on the ground in Silicon Valley. But among these, you also universally get endorsement of open societies such as the Western social democracies and rejection of authoritarian or totalitarian systems, whether they be of the left or the right. When pressed, they of course will not abandon their position that government rarely does anything right. Yet they are grudgingly forced to acknowledge that the United States government created the Internet, and great gobs of technology funding is federal. In this political space, loathing of convicted monopolists such as Microsoft is common. In the Heaven enthusiasts' Web sites, it's easy to find denunciations of "racism, sexism, speciesism, belligerent nationalism and religious intolerance." Greens are imaginatively represented—especially in the Viridian movement—as looking forward to ecological fixes through technology. Disabled people, who are among the most technology-dependent humans on earth, had as their poster boy Christopher Reeve, the former *Superman* actor, who was determined to use the GRIN technologies to walk again. You can find feminists who welcome a vision of a posthuman future that does not require men for anything, including procreation. You will even see arguments among Heaven scenarists for a universal guaranteed human income—after all, it's the least all those robots who will produce our cornucopia can do for us, since we made them possible.

That's why the Heaven and Hell Scenarios, and those attracted to them, may be the new political divide, the defining political conflict of the 21st century. Optimists and pessimists cut across all lines. You can find them among cultural liberals and conservatives, economic liberals and conservatives, bioethical liberals and conservatives, and advocates of government big and small. Yes, among advocates of The Hell Scenario you will find social conservatives who have found great support in right-wing governments. But oddly, they are in the position of advocating massive government regulation to thwart individual choice, which was not a hallmark of conservative thought in the late 20th century. It may be more useful to refer to them less as conservatives and more as devotees of The Hell Scenario. By contrast, Nick Bostrom, co-founder of the World Transhumanist Association—a group that advocates The Heaven Scenario—can readily be described as a liberal. Not only is that his political history, but he embraces radical social change. Yet he is practically libertarian in his agenda, in that he proposes letting each individual decide for herself and her family which technologies they should embrace, how and when. So perhaps a more constructive label for him would be "Heaven Scenario devotee."

In imagining responses to The Hell Scenario, McKibben has two things he would like to see happen. The first is that we build up a network of taboos—as the Amish have done—that "keeps us more or less human." In this he is preaching the gospel of "voluntary simplicity," which is as sound an idea today as it was when advocated by Henry David Thoreau in *Walden*. As his friend Emerson noted, Thoreau "made himself rich by making his wants few."

Historically, however, voluntary simplicity has had as its most fervent adherents those who already have plenty. It would not be much of a surprise if many Naturals came from the ranks of the well-off and educated, similar to McKibben, who already makes his home close to nature in the mountains above Lake Champlain.

McKibben's second hope is more formidable. He would like it if "the rush of technological innovation that's marked the last five hundred years can finally slow." The practicalities of this, of course, have yet to be demonstrated. Rejecting the endless cycle of cheaper, faster, better, is probably an easier sell to McKibben's fellow faculty members at Vermont's Middlebury College than it is to those like my neighbor, whose maxim is "If Wal-Mart don't got it, you don't need it." The question is whether those who don't know where the next rent check is coming from—or worse yet, the billions who don't know where their children's next meal is coming from—will ever agree with McKibben that "the world has enough wealth and enough technological capability, and should not pursue more." Under what circumstances will those with little not seek more? When will they cry, "Enough"?

It will be particularly difficult to arrive at a consensus about what constitutes enough when we're discussing experimental therapies with parents who have a kid who is sick or perhaps dying. For them, there is probably no such thing as enough.

You might say that's a part of human nature.

A SMALL INDUSTRY has risen challenging aspects of both The Heaven and Hell Scenarios.

The controversy over The Heaven Scenario includes those who are so horrified by the very idea of transcending human nature that they see it as a Hell Scenario. But it also encompasses those who argue about whether The Heaven Scenario is possible.

Of those who argue that it can't happen—the Heaven Can Wait contingent, we may call them—the most thoughtful arguments frequently amount to modern versions of what used to be known as "vitalism." *Vital* is an interesting word. It comes from the Latin word meaning "to live," yet in English it means "essential." These critics profoundly question whether living things really are like machines. They suggest life has a vital spark that will be impossible for science to match.

This has not stopped J. Craig Venter from trying. Venter is the maverick genetic researcher who created Celera Genomics to beat government researchers to the goal of mapping the entire human genome. Now, funded by the U.S. Department of Energy, he is creating new microscopic life forms with special abilities, such as consuming toxic waste. His larger goal is to create from bits and pieces of DNA a single-celled organism with the minimum number of genes to sustain life.

Today's vitalists, however, are saying that playing God in biology will not be as technically easy as Victor Frankenstein or Craig Venter would have us believe. They think we will discover, one failure at a time, that there is something ineffable about life, some necessary, vital ingredient—perhaps consciousness created by the effects of quantum physics—that humans will have the Devil's own time matching.

John Searle is such a critic. Searle is a professor of philosophy of mind at the University of California at Berkeley. He points out that Kurzweil's Heaven Scenario assumes that when computers can perform sufficiently more operations per second than can our brains, they will inevitably seem conscious. That is, they will demonstrate true intelligence. Indeed, they will be "spiritual." That's what happened with the evolution of our brains, according to Kurzweil. Why shouldn't it happen with the evolution of computers?

Searle argues that we have no idea what consciousness is, and never will, because the very idea is utterly subjective. Therefore, who is to say that computers will ever achieve it? Kurzweil agrees that consciousness is utterly subjective. The only way you can guess whether another human being is intelligent is by the way she behaves. But, he says, since behaving intelligently is a good enough way to cause people to think you are conscious, so too will it be a good enough way to cause people to think that machines are intelligent. If it quacks like a duck convincingly enough, sooner or later someone will wonder how it might look in orange sauce.

Searle asks how anyone can be so sure consciousness can be reduced to zeroes and ones, that there is not some deep mystery behind it. Adds the scenario planner Kees van der Heijden, "They have to do a lot more than reverse-engineer the brain. They have to reverse-engineer the whole human body, with its complex sensing system, motor system and chemistry (emotions) in addition to its brain. All that is just as essential to human intelligence."

Kurzweil believes that "he can somehow do what philosophers have pondered for centuries: separate the body from the mind," writes Ellen Ullman, author of *Close to the Machine: Technophilia and Its Discontents,* and *The Bug.* He is "mistaking the machine and its principle of operations for the realities of humanity he intends to automate." Ullman is that unlikely combination, a software industry pioneer with the eye and voice of a poet. "Ray Kurzweil is a computer scientist. He has taken his knowledge . . . and superimposed it over everything he seeks to understand. . . . He has redefined the problem as something he knew how to engineer." He "would analyze the scan to derive the 'patterns' that constitute consciousness, what he calls the architecture and 'implicit algorithms' of the interneuronal connections. This analysis would involve differentiating between brain activities seen as essential parts of the mind, and those that 'do not directly contribute to its handling of information.'"

Ullman is applying to the essence of human nature the argument Robert Frost made when he said, "Poetry is what gets lost in translation." Ullman believes "we're beings who are suffused with error, dripping with imperfection, drenched in inefficiency . . . Ray Kurzweil would improve us. I don't know about you, but it always makes me nervous when someone wants to improve the human race."

Kurzweil's basic reply is that pattern recognition is his expertise, that this is the direction toward which he can clearly see that we're headed. Human intelligence and machine intelligence are becoming increasingly indistinguishable.

This is an argument that can be resolved only with time—although not very much time, because of the march of The Curve. Events will make it obvious to us right quick whether Kurzweil is a seer or a fool. It will all depend on how we see our lives merge with our machines.

Then there is Steven Pinker of Harvard. Pinker is the undisputed rock star of the human nature biz. Alarmingly, model-quality slim, with an immense mane of curly if graying hair from which a boyish lock falls over his forehead, he has blue eyes, high cheekbones, cheeks as dimpled as a fresh Titleist, a square jaw, androgynously full lips and a lead-singer voice. He has been known to show up at academic performances in a double-breasted black jacket, long collared blue-purple shirt, cranberry silk tie and tall, soft leather boots.

Pinker has massively annoyed the left with his best-seller *The Blank Slate: The Modern Denial of Human Nature.* In it, he argues that human nature is so complex and so based on genes (activated by upbringing) that it resists the most well-intentioned efforts to be molded by state-sponsored nurture, such as welfare. Nonetheless, a year after publication, Pinker was causing gas in the techno-libertarian right at what was billed as the first gathering of its kind: "The Future of Human Nature: A Symposium on the Promises and Challenges of the Revolutions in Genomics and Computer Science" at the Frederick S. Pardee Center for the Study of the Longer-Range Future of Boston University. Pinker claimed at the symposium that "not only is genetic enhancement not inevitable, it is not particularly likely in our lifetimes." His argument was that the genetic basis of human nature is not only so complex as to defy the best efforts of Mao or Pol Pot to shape it, but also very well might be so complex as to defy the best efforts of geneticists. Individual genes rarely produce beneficial effects, he says. There are no single genes for mathematical giftedness or athletic prowess. We shouldn't hold our breath for the musical talent gene—or for cures for schizophrenia, antisocial personality, or bipolar illness. It's more complicated that that, he says, with many genes interacting in feedback loops.

Pinker also points out that specific predictions about technologies as complicated as genetic enhancement are unreliable because so many things have to go right—technologically, psychologically, sociologically. How many people guessed correctly that the videophones of the 1960s would sink like a stone while text messaging from mobile phones would become a teenage craze? Real people weigh both the benefits and the costs of any innovation, he points out. If we're made nervous by genetically engineered soybeans, why do we think we will do it to our kids?

Other people don't think technology will evolve as quickly as Kurzweil does. There is a strikingly large number of bright people with impressive credentials who think that since it took us 200 years to get from the Industrial Age to the Information Age, it will take another 200 years or so to get to the Next Big Thing of human transcendence. Most of them aren't arriving at that conclusion in a convincing way, unfortunately, as I discovered in my travels. They simply fail to factor in the accelerating returns embodied in The Curve.

At that august conference at Boston University, where big names made deeply serious presentations in an impressive room decorated with classical columns and pediments, palm trees, and crystal and gold chandeliers, the organizers asked questions like "In 2200 will we recognize humans?" With a date that far out, the question became academic in the classic meaning of that phrase—of theoretical interest at best. Working the room, I asked why the participants thought we had that kind of time. This seemed a crucial question to me. Overwhelming change in 20 years yields a fundamentally different world than overwhelming change in 200 years. The more time culture has to adapt to and shape technology, the more likely it is that the transition will be softened. No one disagreed. But no matter how eagerly I sought out a thoughtful debunking of The Curve, that's not what I got. None of the organizers gave me a straight answer as to why he or she thought transcendence might be very far in the misty future. That is not to say they are necessarily wrong. But I left with the impression that they hadn't really grappled with the question.

A more sophisticated analysis may be the one popular in tech circles. It holds that predictions of immortality always occur among men in their 50s. They always say it is 20 years out, just when they're going to need it.

Bill Joy, meanwhile, derides The Heaven Scenario as selfish. "A traditional utopia is a good society and a good life. A good life involved other people. This techno utopia is all about 'I don't get diseases; I don't die; I get to have better eyesight and be smarter' and all this. If you described this to Socrates or Plato they would laugh at you."

In turn, Joy's peers acknowledge the seriousness of his Hell Scenario. What they take him to task about is his relentless pessimism regarding our ability to deflect it. More than 350 years ago, the poet John Milton wrote a pamphlet addressed to the Parliament of England with the title *Areopagitica*. He was arguing for the liberty of unlicensed printing, notes Freeman Dyson, the physicist and a professor at the Institute for Advanced Study at Princeton. Dyson has written many popular works on science and society, including *Infinite in All Directions*. He replied to Joy's provocative article in an essay titled "The Future Needs Us."

There is a connection between "the 17th-century fear of moral contagion by soul-corrupting books and the 21st-century fear of physical contagion by pathogenic microbes," Dyson says. In John Milton's 1644, England had just emerged from a long and bloody religious civil war, and Germany's was still in progress. In that century, "books not only corrupted souls but also mangled bodies," Dyson notes.

Milton talks about the difficulty of regulating "things, uncertainly and yet equally working to good and evil": "Suppose we could expel sin by this means; look how much we thus expel of sin, so much we expel of virtue: for the matter of them both is the same; remove that, and ye remove them both alike."

In arguing to free that earlier technology, capable of great good but also great evil, Milton expressed majestic confidence in our ability to prevail: "Lords and Commons of England, consider what Nation it is whereof we are, and whereof ye are the governors: a Nation not slow and dull, but of a quick, ingenious, and piercing spirit, acute to invent, subtle and sinewy to discourse, not beneath the reach of any point the highest that human capacity can soar to."

In such fashion did Milton seek to reconcile the enduring problems of individual freedom and public safety that we grapple with today. Can the most totalitarian regimes we can imagine today or tomorrow totally shut down hackers of genetic, robotic, information and nano engineering? If not, might the societies that thrive be the ones that widely disperse knowledge? Rather than closeting knowledge in a bureaucratic intelligence agency, might a citizen militia of smart people, armed with enough knowledge, be able to respond within minutes with brave solutions to hitherto unimaginable threats? Evidence exists that this works. On 9/11, the fourth airplane never made it to its target, and it's not the top-down military or the White House who can take credit for that. Empowered by their mobile phones, and without waiting for orders from any national leadership, the passengers aboard that aircraft within a matter of minutes diagnosed their society's problem and, with determination and at a great price, rolled.

An even more hopeful argument against the inevitability of The Hell Scenario, however, is "A Response to Bill Joy and the Doom and Gloom Futurists" by John Seely Brown and Paul Duguid. Brown was director of the Xerox Palo Alto Research Center back in the day when Xerox PARC was legendary in Silicon Valley. It was the birthplace of many of the things that allow a personal computer to feel friendly to ordinary human beings—from the computer mouse shaped to your hand to the icons you point at and click on your desktop screen. These breakthroughs eventually became the Apple Macintosh and then Microsoft Windows. Duiguid, Brown's regular co-author, is from the University of California at Berkeley. They wrote a very wise book on the intersection of our broadest culture and values, and how it influences our technology, reciprocally, called *The Social Life of Information*.

In their response, which appeared within weeks of Joy's article, Brown and Duiguid argue that Bill Joy's problem is that he is so tightly focused on technology, he can't see any other forces at work. He has tunnel vision, they argue. He can't see any controls because he can't see the broad workings of society and social systems. "Technological and social systems shape each other," they say. "Technologies—such as gunpowder, the printing press, the railroad, the telegraph and the Internet—can shape society in profound ways. But on the other hand, social systems—in the form of governments, the courts, formal and informal organizations, social movements, professional networks, local communities, market institutions and so forth—shape, moderate and redirect the raw power of technologies."

Genetic engineering of food presents the clearest example for them. It once seemed to be an unstoppable force. Now it is an object of consumer boycotts. When Joy wrote, both the promise and the threat of nanotechnology seemed immeasurable, for good reason: at the time the technology was almost wholly on the drawing board.

This is a good point. If someone in the 1970s had correctly forecast the ubiquitous presence in the 21st century of computer viruses viciously and constantly attacking the brains of our most sensitive systems, any sensible person would have concluded we were doomed. It might have seemed laughably Pollyannaish to believe that an immune system could co-evolve to match the problems.

Meanwhile, say Brown and Duiguid, "the thing that handicaps robots most is their lack of a social existence. . . . All forms of artificial life (whether bugs or bots) will remain primarily a metaphor for—rather than a threat to—society at least until they manage to enter a debate, sing in a choir, take a class, survive a committee meeting, join a union, pass a law, engineer a cartel or summon a constitutional convention. Those critical social mechanisms allow society to shape its future. It is through planned, collective action that society forestalls expected consequences (such as Y2K) and responds to unexpected events (such as epidemics)."

This is not to say that we will necessarily succeed at avoiding Joy's Hell Scenario. It takes work for defense and offense to co-evolve. Culture and values can be glacial. "We must shore up the foundations of civilization well in advance, much as the medieval cathedrals had flying buttresses retrofitted," says William H. Calvin, the University of Washington theoretical neurophysiologist and author of *A Brain for All Seasons: Human Evolution and Abrupt Climate Change.*

"Getting started now is important because political consensus takes so long to achieve. The development of atomic bombs required only a few years, yet Europe's impressive achievement of a common currency took fifty years. Action and effective societal reaction have very different time scales. While scientists can provide better headlights to spot the turnings and the washouts in advance, speeding up consensus-building requires a different set of skills. Only an effective combination of foresight and leadership stands a chance of building these flying buttresses that are needed to protect the cathedral of civilization from abrupt shocks. . . .

"The physician who waits until dead certain of a diagnosis before acting is likely to wind up with a dead patient. Sometimes things develop so rapidly that only early action—back when you're still somewhat uncertain—stands a chance of being effective, as in catching cancer before it metastasizes."

———

A FTER RETURNING FROM ASPEN, I reread Brown and Duguid's piece. I thought of Bill Joy on his mountaintop, as alone as he could get. In the profiles I'd read of him from as long ago as the 1980s, there had been many references to his "characteristic boyish grin." Granted, the topics we discussed had been grim. But over the course of a couple of days I would have thought I might have seen one or two characteristic boyish grins. I did not. Overall, his affect was markedly flat. Later that year, he got divorced, quit Sun, and put the book on hold.

Whenever Joy and I discussed the bullets humanity had dodged over the years, such as nuclear annihilation, he ascribed it to "luck." Computer viruses haven't destroyed the Information Age? Given Microsoft's "bug-ridden" software permeating the globe, "that's just serendipity." SARS? "We were just lucky." Whenever a social system works, for Joy that is just "luck." "The common defense is not in anybody's individual self-interest," he claims.

He may be right. The Hell Scenario may be inevitable. But the great irony is that Joy—the pioneer of networks, the pioneer of open source—believes we are doomed precisely because he no longer gives much credence to the power of either.

Brown would hardly argue with Joy's contention that "life involves risk. If it didn't involve risk, it would be inanimate. What you can do is responsibly reduce the risk." Nor would he argue with the proposition that Joy has done the world a great service with his warnings.

Brown's gentle and almost avuncular reminder to Joy, however, is that for hundreds of thousands of years, we humans have succeeded in making our own luck.

And no man is an island.

The Hell Scenario

Bill Joy and other advocates of The Hell Scenario see a future in which the advance of The Curve is unstoppable, and unspeakable evil ensues. In their conviction that their prediction is inevitable, they are the mirror image of those who promulgate The Heaven Scenario. Nonetheless, since others do not agree with this inevitability, here it is being treated as a scenario.

Predetermined elements:

- There are Curves of exponential technological change.

Critical uncertainties:

- Are The Curves of exponential change smoothly accelerating? If so, how fast? Or are they displaying unexpected slowdowns, reversals or stops?
- Are Curves of change leading to progress, disaster or both?

Early warning signs that we are entering The Hell Scenario:

- Almost unimaginably bad things are happening, destroying large chunks of the human race or the biosphere, at an accelerating pace.
- Horrors that recently seemed like science fiction are routinely exceeded.
- Even in the face of such disasters, no agreement is being reached to slow down or stop the spread of these technologies. Humans seem helpless before the onslaught.
- Technologies continue to accelerate as individual nations, continents, tribes or movements jockey for position in a hostile world.

Embedded assumption:

- Technology drives history.

Early warnings that we might not be entering The Hell Scenario:

- The pace of growth of complexity slows or starts proving erratic.

- Almost unimaginably good things start happening. This would be an early warning that we are entering The Heaven Scenario.

- Culture and values gain control of technology such that events that once seemed inevitable are now consciously being avoided over significant periods of time, and by everyone on the planet. This would be an early warning that we are entering The Prevail Scenario.

CHAPTER SIX

Prevail

I decline to accept the end of man. It is easy enough to say that man is immortal simply because he will endure: that when the last ding-dong of doom has clanged and faded from the last worthless rock hanging tideless in the last red and dying evening, that even then there will still be one more sound: that of his puny inexhaustible voice, still talking. I refuse to accept this. I believe that man will not merely endure: he will prevail. He is immortal, not because he alone among creatures has an inexhaustible voice, but because he has a soul, a spirit capable of compassion and sacrifice and endurance. The poet's, the writer's, duty is to write about these things. It is his privilege to help man endure by lifting his heart, by reminding him of the courage and honor and hope and pride and compassion and pity and sacrifice which have been the glory of his past. The poet's voice need not merely be the record of man, it can be one of the props, the pillars to help him endure and prevail.

—William Faulkner, Nobel Prize
acceptance speech, December 10, 1950

I N NEW MEXICO, El Camino Real de Tierra Adentro—the fabled Royal Road from Mexico City to Santa Fe—traces the course of the Rio Grande. Through the clear, dry air from El Paso del Norte toward the Sangre de Cristo—beyond the cotton, chilis and pecans of its fertile plain and below its gliding sandhill cranes—poke abrupt, strange mountain shapes. Spooky, spiky, funky and wild, they are hard not to imagine as Paleozoic reptiles awaiting the incantation that will bring them back to life. In this very ancient place, along this path of dreams, lies a small town called La Mesilla.

Today La Mesilla (the name means "the little tableland") is fewer than 30 miles north of the border between Mexico and the United States. But the boundaries move a lot in these parts. In 1598—9 years before the Cavaliers landed in Virginia and 22 years before the Pilgrims arrived at Plymouth Rock—the conquistador Don Juan de Oñate here led some of the first permanent European settlers into what would become the United States. For more than two centuries, Spain brought to this region the cross and the sword. The Pueblo Indians responded with mixed emotions. The Apaches and later the Comanches responded with great violence. In 1821, when it achieved independence, this empire became Mexico. Two and a half decades later, Americans, pursuing their Manifest Destiny, declared war and in 1848 began the process of turning the northern half of Mexico into all or some of California, Nevada, Arizona, New

Mexico, Colorado and Utah. Because of a surveyor's mistake, however, a band of territory including La Mesilla wound up still in dispute, claimed by both Old Mexico and the United States. The Yankees didn't resolve that border issue in their favor until just before the Civil War. Then came the Rebels, who grabbed La Mesilla in 1861 and claimed a vast dominion as part of the Confederacy. It took a year for the Union to reclaim it. (New Mexico became the 47th state only in 1912.)

After the borders got straightened out, you'd think things would have settled down in La Mesilla—or just Mesilla, as it is more familiarly known. But no. The area west of the Pecos was now called the New Mexico Territory, and it was as wild and lawless as ever the West was. People from as far as Chihuahua and Tucson came to Mesilla to attend bullfights, cockfights and dances called *bailes,* and visit the town's bars, pool halls and even a bowling alley. It was not uncommon for differences to be settled in the streets with gunfights. The children of Mesilla practiced the quick draw with their little wooden pistols, mimicking their teenage hero, William Henry McCarty, alias Kid Antrim, alias William H. Bonney, alias Billy the Kid, who was said to have killed more men than he had years of age. At the jail and courthouse on the southeast corner of the plaza, The Kid was tried and sentenced to hang in 1881 for the killing of Lincoln County Sheriff William Brady. When The Kid escaped after being transported to the northern part of the Territory to hang, Sheriff Pat Garrett tracked him down and shot him dead at the age of 19, 20 or 21—estimates vary, as his birth date is a surmise.

In the ensuing century, since the railroad and the Interstate passed it by, Mesilla has seen little change, some chroniclers claim. But that's not right. Granted, a lot of the residents are direct descendants of the rugged original settlers—Indian, Spanish, Mexican and Anglo. (Mesilla's still the kind of place where people make distinctions between being "Spanish" and being "Mexican." It's an ancient issue of Hispanic authenticity that hinges on how many centuries ago your ancestors got here, as well as who is browner than whom.)

One of the foremost connoisseurs today of whether Mesilla is closer to the ancient world or the universe of curving change is Jaron Lanier. Lanier, who was born in 1960, has vivid memories of how much Mesilla resembled a frontier when he was growing up there.

Jaron Zepel Lanier today is one of the world's more startling combinations of philosopher, creative artist and computer scientist. Learned journals have published his articles on the philosophy of consciousness and information—like how you might tell whether an intelligent entity is a zombie or whether there is actually somebody home inside. He says his book, *Technology and the Future of the Human Soul,* will be finished some-day, despite epic procrastination. He is a professional "new classical" recording artist who writes chamber and orchestral compositions, includ-ing a triple concerto commissioned by the National Endowment for the Arts and the American Composers Forum. In his Berkeley, California, home he maintains some 1,300 musical instruments, many of them ex-otic, all of them playable and all of which *he* can play. They include a glass harmonica invented by Benjamin Franklin, consisting of a series of rotat-ing glass cylinders that produces haunting harmonies when played with a finger wetted in vinegar. Then there is the world's biggest fully chromatic modern flute, 16 feet tall, and the full-size pipe organ. He believes his to be the largest and most varied collection actively played by one person in the world. Paintings and drawings by Lanier have been exhibited in mu-seums and galleries in the United States and Europe.

He is best known, however, for inventing "virtual reality" as a shared ex-perience, and naming it. In his early 20s Lanier founded VPL Research—yes, in a garage in California. It was the first company to provide research labs around the world with the then-almost-magical virtual-reality para-phernalia. When he was 24, his groundbreaking work made the cover of *Scientific American.* Few recent innovations have had such consequences. It is difficult to buy an automobile or fly in an airplane today that wasn't de-signed in virtual reality. The petroleum to fuel them was probably found with Lanier's inventions. City planning, building design, surgery and sci-entific visualization—especially of molecules important to the creation of new drugs and the understanding of proteins and genes—are being rede-fined by virtual-reality imaging. So is the training of police, firefighters, emergency response teams and the military.

In the early 21st century, Lanier was the chief scientist of Advanced Network and Services, the engineering office of Internet2—a coalition of 180 American research universities sharing an experimental next-generation network so powerful that when they fired it up, lights dimmed all over campus, or so the story goes. He led the National Tele-Immersion Initia-tive. It aimed to create alternative worlds in which people at distant sites work together in a shared, simulated environment that makes them feel as if they were in the same room.

"Our social contract with our own tools has brought us to a point where we have to decide fairly soon what it is we humans ought to become, because we are on the brink of having the power of creating any experience we desire," writes Howard Rheingold, an analyst of technology's impact on society. Virtual reality "represents a kind of new contract between humans and computers, an arrangement that could grant us great power, and perhaps change us irrevocably in the process."

Lanier is sufficiently renowned that on the rare occasions when he visits his hometown, people in Mesilla recognize him on the street. That's not so difficult, as he is quite a sight. Lanier has grown up to be a vast bear of a man. A panda bear, actually, is what comes to mind as he pads around in his sandals. His eyes are blue. His skin is so pale as to verge on the albino. His hair naturally falls into sandy brown dreadlocks. They hang below his waist. As he talks, they constantly fly about. No one ever mistakes Lanier for somebody else. Lanier today is far and away the most famous son of Mesilla, New Mexico.

Lanier still isn't entirely sure why his father and mother, Ellery and Lillian, an intellectual and an artist, respectively, chose to move from New York City to Mesilla shortly after the birth of their only child. As recently as Lanier's Vietnam-era childhood, few of Mesilla's streets were paved. There was no television—no signal reached that far. Buildings made out of adobe—walls of dried mud as much as two and a half feet thick—were still common. "The houses were always in a state of being dissolved" during the rare but torrential desert rains, he recalls. "Every single thing was in a constant state of melting back into the elements. It was a constant struggle to rebuild. It was this odd world of constant decay. Strange world. A lot of death play."

When Lanier was growing up, Doña Ana County, in which Mesilla is located, was the second poorest in the United States. "There were diseases that just shouldn't still have existed in America. There were deformities," he says. Cruelty and violence were a part of life. Lanier especially remembers one of his classmates drowning in the school swimming pool. Lanier's recollection of the incident is that it was common knowledge that his classmates had murdered the child and "he deserved it because he was unpopular and I was next."

Lanier's father was a writer for fiction magazines, his mother a concert pianist who also was the family's primary breadwinner. They managed because of her pioneering remote stock trading, buying and selling on Wall Street by phone from the yucca-punctuated desert. "I lived down that ditch," he remarks today, pointing to an irrigation canal. Mesilla is now showing signs of wealth he does not remember from his childhood. People are using metal for fencing, not just woven twigs. The park now has grass.

Lanier had a singular upbringing, which shapes him and his views about the future of human nature. "My childhood is divided into two pieces—when my mom was alive and after she died. When she was alive it was a childhood that was kind of structured in a nice house kept nicely." Then she died, suddenly, in a horrible automobile accident. Lanier was nine. "The year after my mom died I just spent in a hospital trying to die." He suffered a succession of diseases. "I have a lost year there," he says. He recalls little of man's first landing on the moon that year.

Meanwhile, "my father made some disastrous financial decisions. We ended up utterly impoverished. We moved to this cheap piece of land and we lived in tents." Trying to recoup, Lanier's father—always something of a bohemian—decided to erect a homemade home. Astoundingly, he let his strange, brilliant, motherless child design it. The result was spectacularly, predictably unpredictable. "I designed this crazy building," Lanier recalls.

Lanier had read a book by the creator of *The Whole Earth Catalog* lauding the charms of geodesic domes. Many years later he would meet the author, Stewart Brand. "The first time I ever met Stewart I said, 'You know, I grew up in a dome,' and his first words to me were, 'Did it leak?'" Lanier, sputtering, replied, "Of course it leaked; what were you *thinking*?" A rant ensued concerning malpractice. Nonetheless, it was quite a structure. "This house wasn't made only of domes. It had some fantastic crystal forms and weird spires and towers and jutting parts. I mean, it was a really strange house." Much of it has since fallen down.

Architectural follies, of course, hardly began to resolve the central issue of a preteen's life, even if "you couldn't really build a dome without some trigonometry."

"The degree to which I was a social failure is impossible to even state," he says. "It was just extreme beyond . . ." Lanier trails off. It was beyond not having friends. People surrounding him were utterly "mean-spirited, hostile and threatening. If I'd been in any normal place, some kind of serious intervention would have happened. The Spanish and the Mexicans despised each other, the Indians were darkly depressed and just swallowed by hopelessness, and the white culture was redneck—extremely intolerant, boorish, violent and uneducated. They were warring camps. It was just a terribly mean-spirited environment that I happened to be in. I figured out how to connect to people socially much later in life than most people do. In the technical academic world you meet a lot of awkward young men in the math department or whatever. People are always saying, 'Oh, we have a real weirdo this semester,' and there will be some hairy creature in some little dingy basement room who is sort of suckling on a computer."

He laughs. "I mean, they don't even know the *meaning* of awkward young technical guy. They don't even know the *beginning* of the meaning. But I got it eventually. It was just very hard. I remember I would go into a little convenience store and my goal would be to buy something without creating a scene. Without embarrassing myself. I had to practice that."

In this world, any connections were magical. One night when he was 11, the phone system went crazy. Suddenly, "all the phone lines were connected together, so if you picked it up there were hundreds of people. It was kids, and they were all talking to each other. There were all these floating voices. It was just this one night, and what was cool about it, for me, it was like the thing on the Internet where no one knows you're a dog. That was an amazing moment for me."

When he was 14, he found refuge in a summer program for high-schoolers at New Mexico State University in nearby Las Cruces. There, "I think I went through developmental stages that people usually go through pre-puberty—in terms of learning how to have conversations and stuff. I started to learn how to talk to people."

By today's standards, New Mexico State—home of the Aggies, reflecting its rural heritage—was a glorified community college. But because of New Mexico's role in nuclear weapons development it harbored a few math, physics and engineering faculty members with world-class minds. The Manhattan Project scientists had been hidden in Los Alamos up by Santa Fe. The White Sands Missile Range is nearby, as is the Trinity site at Alamogordo, where the world's first atomic bomb was exploded on July 16, 1945. Thus, a neighbor was Clyde Tombaugh, who discovered Pluto. His backyard bristled with telescopes, which he let the young Lanier use. Tombaugh was in charge of designing optical missile trackers at White Sands.

At the end of that summer, Lanier decided okay, I am simply not leaving. He had not graduated from high school, but he was more than clever enough to obfuscate the paperwork. "I just stayed in college instead of going back to high school, and that actually started to work."

The most astonishing thing about New Mexico State was that, thanks to the federal government's interest in nukes, it had a computer facility that Lanier describes as "kick-ass." Some of the pioneers of graphical representation came out of there. This was a very big deal. Graphical representation is what the trash can image on your desktop is all about. It's at the heart of the point-and-click system that you take for granted in

Windows. It's also the foundation of all the games that today pull in more money than do Hollywood box office receipts. As a result of his experience there, Lanier wound up as a programmer in Silicon Valley, joining what commentator Jon Katz calls "an inconspicuous movement, attracting millions of intelligent, technologically aware, community-oriented, self-described outsiders, mesmerized by finding themselves in a club that will not only take them in, but puts them in charge."

Today Lanier lives very near the crest of the Berkeley Hills in a stunning place adventurously created with the help of "an avant-garde seismic engineer." It has a spectacular view of the Golden Gate Bridge. The down payment came from his cashing out of a recent start-up company. Still without a diploma of any kind, Lanier almost absentmindedly collects faculty appointments that keep him on the transcontinental jets he professes to loathe. These postings include the engineering school of Dartmouth, the business school of the University of Pennsylvania, the arts school of New York University, and the computer science department of Columbia. The Micronesian island chain of Palau has issued a postage stamp honoring him. In his spare time he is fundamentally rethinking the historical underpinnings of all computers. He believes the way they work now is whacked. There has got to be a better way.

When faced by the prospect of a sudden transformation of human nature, Ray Kurzweil, Bill Joy and Jaron Lanier each responds from the deep recesses of his soul. Kurzweil worships the power of ideas to resolve all problems; Joy in his lonely fashion engages death; Lanier attributes all his subsequent work to finding "the connection I lost."

The thinking of Kurzweil, Joy and Lanier describes a triangle. Lanier's is not some middle vision between that of Kurzweil and Joy. He is off in an entirely other territory that pokes and prods their technological determinism. Lanier agrees with Kurzweil that it is not tremendously likely that you can stop radical evolution by willing it gone. He agrees with Joy that The Curve could lead to mortal dangers. Yet Lanier would not relinquish transcendence even were that possible. Indeed, he views the prospect of exploring all the ways humans could expand their connections as the greatest adventure on which the species has ever embarked. Lanier's critical difference is that he does not see The Curve yielding some inevitable, preordained result, as in the fashion of the Heaven and Hell Scenarios. "If it turns out Bill or Ray are right, I'll be disappointed mostly because it's such a profoundly dull and unheroic outcome," he says. "It's such a gizmo outcome. There is no depth to it at all."

Lanier believes it is well within the power of the species to transcend to something far beyond the current understanding of human nature. He just views as sterile the prospect of uploading some portion of our brains into computers. Instead, he pictures a rich and tasty brew of opportunities. He can see a vast array of transcendences. He imagines humans making intelligent decisions, exercising creative control. If you were graphing Lanier's idea, it would not be represented by smooth curves, either up or down, as in the first two scenarios. It would doubtless have fits and starts, hiccups and coughs, reverses and loops—not unlike the history we humans always have known. It would be messy and chaotic, like humans themselves. Technology would not be in control. It would not be on rails, inexorably deciding human affairs. At the same time, the outcome would definitely involve radical change.

I call visions like this The Prevail Scenario.

⁓⁓⁓

UNCERTAINTY SUFFUSES The Prevail Scenario. For Lanier, that's not a bug. It's a feature. "The universe doesn't provide us a way to have absolute truth," he says. "I am not fanatical about my ideas. I'm perfectly happy to see where there are holes in them. This idea is something I believe—in the sense that I act on it. But let me tell you the trap I want to avoid falling into." He judges Kurzweil and Joy to be "severe exaggerators and overstaters. Their reasoning is similar to that of a paranoid person in that they find only the little bits that fit into their worldview and build this cage in which they imprison themselves. I'm not willing to be a fanatic and demand that people see that every bit of data supports my view. I want to be given the latitude to present my own thing more softly. I actually perceive it with less of a sense of certainty and bullheadedness. It's just my best guess."

His key point about The Prevail Scenario: "I will argue for perceiving a gradual ramp of increased bridging of the interpersonal gap. I believe that that's demonstrable. I do not perceive it as being an exponential increase. I do not perceive it as something where there is an economy of scale and it's compounding itself and it's heading towards some asymptotic point. I am not saying it's accelerating." The Prevail Scenario, he's saying, is measured by its impact on human society. He is specifically arguing that even if technology is on a curve, its impact is not. This is why he is skeptical about the idea of a Singularity—technology increasing so quickly as to create an imminent and cataclysmic upheaval in human affairs.

In his version of The Prevail Scenario, Lanier is talking about transcendence through an "infinite game." "The future that I'm trying to find is one where people are in the center and there's this ever-expanding game of connecting people that creates a game into the future."

James P. Carse, the emeritus director of religious studies and professor of the history of literature at NYU, in 1986 published a book called *Finite and Infinite Games*. In it, Carse describes the familiar contests of everyday life—games played in business and politics, in the bedroom and on the battlefield. Finite games have winners and losers, a beginning and an end. Finite players try to control the game, predict everything that will happen, and set the bottom line in advance. They are serious and determined about getting that outcome. They try to fix the future based on the past.

Players of infinite games, by contrast, enjoy being surprised. Continuously running into something one didn't know will ensure that the game will go on forever. The meaning of the past changes depending on what happens in the future. "A finite game is played for the purpose of winning, an infinite game for the purpose of continuing the play," Carse says. Infinite games never end, for they are unscripted and unpredictable. Carse sees them as more rewarding, and Lanier vibrates to this chord. Finite players play within the rules. Infinite players play *with* the rules. "Life, liberty, and the pursuit of happiness" is an infinite game, Lanier believes. Infinite games are the real transcendence games. They allow you to transcend your boundaries. They allow you to transcend who you are.

On several levels, Lanier questions what he sees as the finite-game premises of the Heaven and Hell Scenarios. He doesn't doubt that there are exponential processes at work, including Moore's Law and all the rest. But he wonders if they are immutable, and questions the nature of their social impact. After all, by definition you can only measure closed environments with fixed boundaries. A lot of life isn't like that, Lanier points out. Sure, the price of chips is plummeting. But is that making us smarter?

How would it be possible to measure the system known as the U.S. Constitution? Lanier asks. To prevail from horse-drawn days to the present, it has to be a miraculously sophisticated document. Yet we have no units to measure that sophistication, he points out. The Constitution is not a computer operating system but a human operating system. That's the difference between a closed system and an open one. The great irony is that if we can measure something, it can't be all that complex. How can we measure creativity? Human nature is the ultimate example of the immeasurable.

That's why Lanier is far more bent by nerds trying to mold human nature to their closed-system computers than he is concerned about human enhancement. Right now he thinks computers are making us stupider.

How can you say that? I protest. If you call up an airline reservation number, you get an amazingly sophisticated machine that can understand what you're saying and respond in meaningful ways.

"The very nature of oppression has always been to force people to live within the confines of some idea about what a person is," he replies. "That is true whether you're talking about some ancient religious oppressive regime, or a communist regime, or a fascist regime, or one of the big bad industrial-age companies" that reduced people to cogs in their organizational machine. "Or for that matter Freud. There are a lot of people who have this idea about what a human is and expect other people to live within the confines of that theory."

He views the belief that a human is like a computer as the current repression. "In the computer-human loop, the human is the more flexible portion. So whenever you change a piece of computer technology, the chances are that the human users will actually be changing more than the technology itself changed." You quickly learn that there are only certain questions that the airline reservation bot can handle, and only certain words, and you dumb down your activity to deal with its limitations. If you treat a computer like a person, thinking that there might be any real intelligence there, you make yourself stupid.

"I don't think it is particularly dangerous yet. But that's the start of a potential trend that I think could be a big problem." You start by learning how to con your allegedly smart word program. You have to. Otherwise it will automatically fix things from right to wrong, and you won't get what you really meant to type. Then you organize all your finances so that they'll look good to a pathetically simplistic credit-rating computer. When you have machines evaluating people, as in school testing, you have to learn what the machine wants and play its little game in order to establish that you are a satisfactory human being. Keep this up—accepting the notion that we can trust machines to do some of our thinking for us—and you depart from reality, Lanier believes. We model ourselves after our technologies, becoming some sort of anti-Pinocchios. Human spunk begins to evaporate.

"To train ourselves to adapt to a low-grade form in order to get some machine to work is a little bit like asking people to reduce their vocabulary so that language will work better overall," Lanier says. "Or asking people not to play any new musical chords because all the musical instruments are designed for the existing chords or something like that. It shuts down the game."

Lanier's critical concern is connectedness between human beings, not transistors. Suppose that someday bots run convenience stores and dry-cleaning emporia, replacing immigrants. Will a bot ever get to know you well enough to, one spring day, along with your shirts, give you garden seeds for Thai eggplant and melon?

If vapidity is where The Curve is taking us, Lanier wants no part of it. Enhancement, by contrast, doesn't worry him particularly. "If somebody put some brain chip in their head and it's supposed to enhance their memory but instead it makes them weird in some way, as long as they are still part of the game of society in connecting with people, they would probably just be an interesting and eccentric person, as long as they are not homicidal or something like that. I'm not saying there aren't any potential problems. But to me that can be part of an adventure. That doesn't intrinsically scare me as much as a society that voluntarily endures a slow suicide through nerdification, in which they blanch out their own lives of any flavor or meaning. That scares me more."

Lanier is dismissive of what he describes as "the religion of the elite technologists," from Moravec to Minsky, in the halls of "true believers" at Stanford, MIT and Carnegie Mellon. They believe in a key anticipated outcome of The Heaven Scenario: "That computers are becoming autonomous and a successor species."

"My feeling about spiritual questions is that there is a tightrope that I try to stay on, not always successfully. If you fall to the right side, you become an excessive reductionist. You pretend to know more than you do and you become overly rational. If you fall to the left side, you become superstitious and you believe that there are magic tricks of meaning. Staying right on that line is where you're a skeptic but also acknowledge the degree of mystery in our lives. If you can adhere to that, I think that's where truth lies. Sometimes it's lonely and frustrating. For a lot of these questions, I think 'I don't know' is the most dignified and profound answer. A profound 'I don't know' is the result of a lot of work."

Lanier wants to stay open to the possibility that "the world we manipulate here isn't all there is. The world accessible by technologies isn't all there is. I don't want to become a superstitious fool and believe I can say anything about this other world. That's very important. I don't want to start saying, 'Oh, there are these angels here.' The idea of God as an entity that talks and stuff doesn't quite fit for me. It's also not something I'm gonna dismiss." He makes a small joke by pretending to be the systems administrator of all creation: "We have limited privileges in this area."

To describe his version of transcendence in a Prevail Scenario without falling off his tightrope, Lanier likes to talk about octopi. Actually, he also likes to talk about the psychology of early childhood, as well as the day that aliens visit the earth and perceive human nature for the first time. But these to him are all stories about the same thing—a steadily increasing ramp to transcendence that leads to deeper and better ways of bridging the interpersonal gap.

Lanier's Prevail Scenario is the search for a complex, evolving, inventive transcendence. Because it is an infinite game, it never goes into a Singularity, as in the Heaven and Hell Scenarios. Because it's fundamentally imaginative, it doesn't have any such simple measurement. It just expands forever. Human connectedness is "a much more profound kind of ramp," Lanier believes. "The thing about a Singularity hypothesis is that it's profoundly uncreative."

He begins his tale by saying, "We have a certain bag of tricks that was bequeathed to us by our evolutionary past. In many ways we are very lucky in terms of what evolution gave to us. I love the opposable thumb, for instance. It's great. Lots of great things. Don't want to complain. Feel grateful for all of it. But there are a couple of ways in which we are unlucky. The business of being sacks of skin separated by air is one of them.

"Let me start with the Martians, if I could. So one day the Martians are on their own super version of *Star Trek* out exploring the universe and they come to Earth and they are going to send a report back home. I think this is what the Martians would say:

" 'You know, this place is kind of touching, but mostly it's just sad. You know these earthlings—they call themselves people. They are—you're not going to believe this, and you might be grossed out, but here's what they are like. They're separated from each other in these sacks of skin. There are often many feet of air between them when they are communicating. Just atmosphere. Yet they are conscious, they form relationships, they think, they long for one another. But they aren't connected. They can't do mind melds like us. So what do they do—okay, here comes the gross part. Just hold on to your stomach and try not to get too grossed out. So there is an orifice and it's an orifice that they eat with and they breathe with it and they can make these weird sounds out of this same orifice. Now I told you not to get sick, okay? Space exploration is a tough game, and you just have to deal with it, okay? If you can't deal with it, get interested in something else.

" 'Then they have these other orifices, called ears, where sounds go in, and they have this code system and they communicate that way. It's awkward and it's weird and it's disgusting, but it's what they have. I mean, we can only sort of feel for them.' "

End report to the Commandress of the Fleet.

Lanier continues:

"This is how I think we would seem to aliens. I'm presuming these particular aliens have some form of connection where they're entering each other's dreams and so forth. They have much fuller contact between minds. So that sets the tone for this idea about the ramp that I care about—the connectivity ramp.

"You can imagine the ramp starting long ago with the advent of spoken language. You can see it continuing with the advent of reading and writing. You can see it continuing further with the rise of art forms and things like drama."

Hundreds of thousands of years ago, humanoids could talk, even with relatively primitive brains. This means spoken language and the brain co-evolved. They became increasingly complex together. Relatively recently, however, when reading and writing were invented, the brains that were around were the same as ours. "If you had a time machine, you could pluck a baby from back then and raise it now and they'd be the same people," Lanier says.

"So what happened? The only explanation is that the very design of reading or writing was opportunistic. It took advantage of an innate potential in the brain. It was wired in such a way that, given time, one could discover a type of written language that could work in it. So we found that language. I think in a similar way this process can continue. I think actually we're seeing one right now with kids and computers."

He's talking about us transcending by going through something as significant as learning to write—the beginning of civilization, and hence cultural evolution. The biggest thing in 10,000 years. He moves on to another story.

"So this is the one about kids. If kids are little enough, they have trouble distinguishing between the inside and the outside of themselves. They have trouble distinguishing reality from fantasy. Obviously at some point that becomes less so or they don't grow up. For the vast majority of kids, the recognition that the outside and the inside are different really sucks, big time. The reason is that if the whole world is the same thing as fantasy, then you're sort of in this God-like state where you just have to imagine something and it's real. Suppose there was this 200-foot-tall giraffe made of sapphires that could talk like a Power Ranger. It puffs into existence. If you don't know the difference, it's like it's real.

"Now, when you start to realize that there is reality, all of a sudden you enter a different world. It's not so much that it's impossible to do things, but you're just very weak. Everything is really hard. Could you work your whole life and corner the market on sapphires and get city council approval to build this big thing and all that? I mean, yeah, maybe. But it's just like this huge pain in the butt."

Here's one of Lanier's punch lines:

"But there is an interesting thing. The reason you enter reality even though it is a pain in the butt is that it is someplace where you are not alone. It's a place where Mom is real and other people are real and food is real. There are benefits to reality. A very significant one is, you are not alone."

In this view, the fundamental dilemma of childhood—the fundamental choice between two unpalatable alternatives—is this: "You can stay inside your fantasy world and be God-like. But you are terribly lonely. You are also vulnerable in important ways. Or you can enter the real world, where the transition to weakness is not just slight—it's huge. It's going from Lord of the Universe to this pathetic little pink thing that wets itself. There is no bigger gap in status. In order to enter the real world, kids have to lose the largest amount of status that it is possible to lose. They hate it, hate it, hate it. If they can accept that, there is one other little nasty pill to come along, which is mortality. You got those two, you have made it to adulthood. It takes a really long time to get there. I feel like lately in America it takes 40 years to get there. You can think of the connectivity ramp I was talking about as a way of trying to soften the blow of becoming an adult."

The world people are entering today with their computer games is a transcendent step up this connectivity ramp, Lanier believes. "I would argue that when kids respond to online gaming and generate this extraordinary enthusiasm and adeptness at computers—as if out of nowhere—this is what they are really responding to. It's a third way that avoids the dilemma of childhood. If you're in a virtual world with other people, they're real. The virtual world between you and them exists in the same way that the real world exists between you and them. But if you do some weird thing in it, like make a giant sapphire giraffe or whatever, it's real for the other people. So you get the connection and you get rid of the solipsism"—the proposition that nothing exists or is real except one's own self.

"But you keep the imaginativeness. You keep your level of power. You can go in and make the world you imagine but without losing the other people. The virtual world is the first place that's ever been like that—that it gives both things. Games surpassed movies back when games had no production values—when they were just bleeps and bloops. Pac-Man—even Pong. I think this is the explanation. This is what people are really looking for. They sense that there is a kind of reality that has the flexibility of imagination and the potential, at least, for lack of solipsism. I think that's the explanation for kids and computers. Virtual reality is the strongest case."

It's the first major bump up the connectivity ramp in millennia, Lanier feels. It's "some sort of important transcendence involving computers."

"Here is I guess a reasonable place to bring up the cephalopods."

It is? Isn't that kind of a neck-snapping transition?

"It's a very gentle one—you'll see. So cephalopods are our tentacled friends in the sea. They are the fanciest of mollusks. The well-known cephalopods include the octopus and the squid, and there is another one that is similar to each of those called the cuttlefish. That's a favorite of mine.

"They are the most alien creatures of intelligence on this planet. Of the creatures that display intelligence as a survival mechanism, most of the ones that we study are actually not that different from us." Dolphins have different flippers than we do, but they're mammals and vertebrates, like us, like the great apes. They're cousins.

"Cephalopods survive by their wits. Evolution made the trade-off where they lost their body armor and they became soft and vulnerable. Their energy went into being smart instead. They have fantastic eyes and fantastic brains and it all evolved along a separate track, independently. So it's the closest thing to an alien we have to compare ourselves against. We're tremendously lucky to have them on this planet. They give us the best tool we have to gain some insight into what we might be and what we might become. They are the control experiment, the only one we have so far.

"There are a few species that exhibit an incredible behavior that really gives you a picture of a different path toward communication and connection. If you've ever snorkeled and you look at an octopus or a squid in the wild, you might notice that a lot of species can change colors. The way that works is that there are cells in the cephalopods' skin called chromatophores. A chromatophore is a cell with a pigment in it with a particular color and the cell can expand or contract when it's excited. If it expands all the red ones at once, the animal will turn red in the skin. So that's how the trick works. There are a few species—and I'll mention the giant cuttlefish and I'll mention the mimic octopus—where there are individual nerve pathways to each chromatophore. So they have a big map display.

"For instance, in the giant cuttlefish they have an extra lobe in their brain which is a chromatophore lobe. A thought in that brain lobe is immediately projected as animation on the surface of their body. So when you watch them, they can animate their skin—a moving animation. They are comparable in many ways to current laptop screens in their capabilities, except it's their whole skin.

"Now one of the things you might be wondering is, 'Why haven't I heard of this?' The reason you haven't heard of this is that the full range of their capabilities was only documented by camera after computer morphing became popular. When people see it they assume it's a computer graphic. So it's never really had the impact it would have, had they been filmed earlier. If they had been filmed earlier, they would have been the most famous animals for years. I show people, and they simply don't believe it's real. They can change shape. They are morphers. But they can also display things on their skin. Complex patterns. Their camouflage is so good—I have some footage and you will just not believe a computer didn't make it. You can go up to them and it just looks like whatever is there. There will be this thing like a little bit of a rock and a little bit of a plant. It just turns into an octopus and it zips away. You can also watch one settle down and just turn into things. It's amazing. It's wonderful.

"If you watch how a cuttlefish hunts, it goes up to its prey—now remember the cuttlefish, like a person, has a soft body that is vulnerable. So it has to use its wits to hunt. What does it do? It goes up to a giant crab, which it wants to eat. You can see the crab sense that there is a predator. So the crab snaps into this defensive posture, anticipating a fight. The cuttlefish isn't impressed. The cuttlefish turns on a psychedelic light show. It starts morphing and putting up patterns and it really looks a lot like a '60s concert stage light show or something."

You're making this up.

"I'm not making this up! In fact, I'm probably understating it. You can see the crab look at the cuttlefish and the crab just goes, 'Uhhhhh . . .'" The crab is completely confused. The crab is looking at this psychedelic light show, and just as this crab is at its maximum state of confusion the cuttlefish pounces and goes for the equivalent of the jugular. The cuttlefish has a beak, and it just goes in for a kill point on the crab, and the crab doesn't know what hits it.

Wow.

"So it's using art to hunt. But the most interesting behavior is that they animate at each other to communicate. If you look at two of them together, one will make a pattern and then the other one will make a pattern and then they make these patterns and they synchronize and they are animating at each other. There is a cuttlefish animation pattern dictionary in the works. It has over 90 entries now at Woods Hole."

What are some of the things they say?

"Well, the usual stuff. 'Where is the food?' 'Want to mate?' 'Did you hear about Fred? Boy, was he fucked up last night.'" Lanier laughs.

"They have fantastic nervous systems, and the reason they are not running the planet—because I think in quite a few ways they are better set up than we were to evolve. The thing that we have that they don't, is that they don't have childhood. They raise themselves. They are born out of the egg and they live on instinct. If they had childhood and they nurtured their young, then they would have eventually developed culture. Then they would have taken off, started up their own ramps, and I think we'd all be living in the zoos that they are in. They don't have culture. They don't have nurture. They only have nature. If they learn new patterns, they don't pass it on to their young. They start over again with each life. So that's given us the room to get where we've gotten. Otherwise we would have been screwed.

"If you can imagine a version of the cuttlefish with childhoods and they grow up being nurtured and having play and having culture and all this, that would be another path to getting to the same place that I think human children long to get to. They would be born with the ability to turn into what they would want."

That is where Lanier sees The Prevail Scenario intersecting with the power of The Curve in a creative way. In his Prevail Scenario, we use technologies to share means of connecting that in the past were beyond our wildest imagination—like that of the cuttlefish. We use the GRIN technologies—genetics, robotics, information and nano technologies—to devise new realities that are equally inventive. In this way we forge multiple ways of creating success, of rising to transcendence.

THERE ARE UNLIMITED VERSIONS of The Prevail Scenario. Lanier's is merely one of the better-articulated. They all, however, start with these principles:

- Humans have an uncanny history of muddling through—of forging unlikely paths to improbable futures in defiance of historical forces that seem certain and inevitable.

- The wellspring of this muddling through, of this prevailing, is the ability of ordinary people facing overwhelming odds to rise to the occasion because it is the right thing—for example, the British "nation of shopkeepers" that defied the Third Reich.

To these, Lanier starts by adding one more proposition:

- Even if technology is advancing along an exponential curve, that doesn't mean humans cannot creatively shape the impact on human nature and society in largely unpredictable ways.

Thus, Prevail is an odd combination of the marvelously ordinary and the utterly unprecedented. It is so common and so rare—so old and so new—that the history of The Prevail Scenario is less well defined than that of the Heaven or Hell Scenario.

When he complains that he finds the Heaven and Hell Scenarios "un-heroic," for example, Lanier implies that he sees something brave, noble and epic in Prevail. Indeed, "there is good reason that hero stories are our favorite story form. They have survival value," notes the scenarist Brian Mulconry. "When we step out against all logic to save the world, we save ourselves."

Yet many hero myths are not Prevail myths. Jason and the Argonauts, Ulysses, and even the Alamo are not stories about ordinary people. From Hercules to Davy Crockett, their protagonists had already transcended to the glorious, to the larger than life, even before they stepped up to their greatest challenges.

Exodus is closer to Prevail. It is a tale of people so abundantly ordinary that even Yahweh can't take it. At one point he sends an angel to lead them, saying, "Go on to the land where milk and honey flow. I shall not go with you myself—you are a headstrong people—or I might extermi-nate you on the way." They certainly discovered the non-obvious path. Six hundred thousand families, their flocks and herds in immense droves, kvetching and wailing all the way, spend forty years getting to Canaan de-spite considerable odds for a much earlier arrival. The Sinai is not that big a desert. From the Nile Delta, if instead of following a pillar of fire you just walk toward the rising sun, the Promised Land is on your left, hard to miss. An Israeli tank can make the trip in a day. The length of that journey did bind the people, however, scouring the memory of slavery from those who would found Jerusalem—a critical element of their pre-vailing.

Huckleberry Finn may be the archetypal Prevail hero when, in the pivotal moment of "his" novel, he considers struggling no longer against the great forces of civilization and religion arrayed against him. He thinks about how society would shame him if it "would get all around that Huck Finn helped a nigger to get his freedom":

That's just the way: a person does a low-down thing, and then he
don't want to take no consequences of it. Thinks as long as he can
hide, it ain't no disgrace. That was my fix exactly. The more I
studied about this the more my conscience went to grinding me,
and the more wicked and low-down and ornery I got to feeling.
And at last, when it hit me all of a sudden that here was the plain
hand of Providence slapping me in the face and letting me know
my wickedness was being watched all the time from up there in
heaven, whilst I was stealing a poor old woman's nigger that hadn't
ever done me no harm, and now was showing me there's One
that's always on the lookout, and ain't a-going to allow no such
miserable doings to go only just so fur and no further, I most dropped
in my tracks I was so scared. Well, I tried the best I could to kinder
soften it up somehow for myself by saying I was brung up wicked,
and so I warn't so much to blame; but something inside of me kept
saying, "There was the Sunday-school, you could a gone to it; and
if you'd a done it they'd a learnt you there that people that acts as
I'd been acting about that nigger goes to everlasting fire."

Huck decides right then and there to abandon a life of sin, avoid eternal damnation and for once in his life do the right thing by society's
lights. He decides to squeal, to write a letter to Jim's owner telling her
how to recapture her slave.

Then he gets to thinking about human nature:

I felt good and all washed clean of sin for the first time I had ever
felt so in my life, and I knowed I could pray now. But I didn't do
it straight off, but laid the paper down and set there thinking—
thinking how good it was all this happened so, and how near I
come to being lost and going to hell. And went on thinking. And
got to thinking over our trip down the river; and I see Jim before
me, all the time: in the day, and in the night-time, sometimes
moonlight, sometimes storms, and we a-floating along, talking,
and singing, and laughing. But somehow I couldn't seem to strike
no places to harden me against him, but only the other kind. I'd see
him standing my watch on top of his'n, 'stead of calling me, so I
could go on sleeping; and see him how glad he was when I come
back out of the fog; and when I come to him again in the swamp,
up there where the feud was; and such-like times; and would always

call me honey, and pet me, and do everything he could think of for me, and how good he always was; and at last I struck the time I saved him by telling the men we had small-pox aboard, and he was so grateful, and said I was the best friend old Jim ever had in the world, and the *only* one he's got now; and then I happened to look around, and see that paper.

It was a close place. I took it up, and held it in my hand. I was a-trembling, because I'd got to decide, forever, betwixt two things, and I knowed it. I studied a minute, sort of holding my breath, and then says to myself:

"All right, then, I'll *go* to hell"—and tore it up.

It was awful thoughts, and awful words, but they was said. And I let them stay said; and never thought no more about reforming. I shoved the whole thing out of my head; and said I would take up wickedness again, which was in my line, being brung up to it, and the other warn't. And for a starter, I would go to work and steal Jim out of slavery again; and if I could think up anything worse, I would do that, too; because as long as I was in, and in for good, I might as well go the whole hog.

A classic example of The Prevail Scenario is the arguably most perfect film Hollywood ever made, *Casablanca*. Humphrey Bogart as Rick is ensconced in a cozy world of thieves, swindlers, gamblers, drunks, parasites, refugees, soldiers of fortune, genially corrupt French police and terrifying Nazis. Rick's cynicism is his pride; he sticks his neck out for nobody. His only interest is in seeing his Café Américain flourish. And then, of course, of all the gin joints in all the towns in all the world, Ilsa (Ingrid Bergman) walks into his. The rest of the film concerns him betraying his own cauterized heart in service of a higher purpose. As Rick says, "It's still a story without an ending."

Interestingly, the most phenomenally successful film series of the recent era—the *Star Wars, Harry Potter, Matrix* and *Lord of the Rings* movies—are all exemplars of the Prevail myth, from Han Solo's grudging heroism to little people with furry feet vanquishing the combined forces of Darkness. If the ageless way humans process information is by telling stories, what does our hunger for that story say?

All scenarios are to some degree faith-based. They rest upon assumptions that cannot be proven. In fact, that is one of the key points of scenario exercises—discovering what people's hidden assumptions are, in order to hold them up to the light. In a scenario exercise, should you hear someone say, "Oh, that can't happen," that's a surefire sign of an embedded and probably unexamined assumption.

This is not to say embedded assumptions are necessarily wrong. It is simply useful to know what they are, why we believe them to be valid, what the early warning signs would be if messy reality started to challenge them, and what we would do about it if our most cherished assumptions turned out to be flawed.

In both the Heaven and Hell Scenarios, the embedded assumption is that human destiny can be projected reliably if you apply enough logic, rationality and empiricism to the project.

In The Prevail Scenario, by contrast, the embedded assumption is that even if a smooth curve does describe the future of technology, it is not likely to describe the real world of human fortune. The analogy is to the utter failure of the straight-line projections of Malthusians, who believed industrial development would lead to starvation, when in fact the problem turned out to be obesity.

The Prevail Scenario is essentially driven by a faith in human cussedness. It is based on a hunch that you can count on humans to throw The Curve a curve. It is an instinct that human change will bounce strangely in the course of being translated from technological change. It is also a belief that transcendence is unlikely to be part of any simple scheme. Prevail does not, however, assign a path to how this outcome will be achieved. The mean-spirited may say it expects a very large miracle. The more sympathetic may say it expects many millions of small miracles.

It is dangerously wrong to assign probabilities to scenarios and ignore those that strike you as unlikely. History shows that the low-probability, high-impact scenarios are the ones that really shock—Pearl Harbor, for example. It would be unfortunate if, as you lay dying, surrounded by millions of others, your last thought was, *I wish I'd paid more attention to Bill Joy.* It would be unnerving if you woke up one day to find your world unhinged due to the rise of greater-than-human intelligence, and your first thought was, *Didn't Ray Kurzweil say something about this?*

Prevail's trick is that it embraces uncertainty. Even in the face of unprecedented threats, it displays a faith that the ragged human convoy of divergent perceptions, piqued honor, posturing, insecurity and humor will wend its way to glory. It puts a shocking premium on Faulkner's hope that man will prevail "because he has a soul, a spirit capable of compassion and sacrifice and endurance." It assumes that even as change picks up speed, giving us less and less time to react, we will still be able to rely on the impulse that Churchill described when he said, "Americans can always be counted on to do the right thing—after they have exhausted all other possibilities."

To focus his version of Prevail, Lanier adds a fourth proposition:

- The key measure of Prevail's success is an increasing intensity of links between humans, not transistors. If some sort of transcendence is achieved beyond today's understanding of human nature, it will not be through some individual becoming superman. In Lanier's Prevail Scenario, transcendence is social, not solitary. The measure is the extent to which many transform together.

In Lanier's version of Prevail, the *idea* of progress is progressing. Lanier points out that, historically, there have been two measures of the march of human progress. One is technological and economic advance, starting with fire and the wheel and marking points on The Curve up through the steam engine and beyond.

The second ramp is moral improvement. It starts with the Ten Commandments and proceeds through the jury convicting Martha Stewart on all counts. Some find our moral improvement difficult to perceive, pointing to the variety and abundance of 20th century atrocities. It's hard to argue with these people. They may be right. But Lanier thinks those who deny the existence of a moral incline are not in touch with the enthusiasm humans once brought to raping, pillaging and burning. Genghis Khan's Mongols killed nearly as many people as did all of World War II, back when 50 million dead was a significant portion of the entire human race. Their achievement—making the streets of Beijing "greasy with the fat of the slain," for example—is still a marvel given their severely limited technologies of fire and the sword.

Lanier has issues with both of these ramps. Note that Lanier uses the word *ramp* because he does not necessarily believe either is on an exponential Curve. The technological incline is a flawed measure of progress on many levels, Lanier says, most particularly because it suggests that the meaning of humanity can be reduced to zeros and ones. The moral ramp is a problem because, taken to its logical outcome, it requires more energy than humans have, and also can lead to holy wars. So his version of Prevail rests on the proposition that a third ramp exists and that it is the important one. That is the ramp of increased connection between people.

Many of the flaws Lanier sees in using technology as a measure of progress begin with his experience as a software scientist. Lanier seriously questions whether information technology will work well enough anytime soon to produce either Heaven or Hell. He completely believes that the moment nanobots are poised to eat humanity, for example, they will be felled by a Windows crash. "I'm *serious* about that—no joke," he says. "Legacy code and bugs all get worse when code gets giant. If code is at all similar in the coming century to what it is now, super-smart nanobots will run for nanoseconds between crashes. The fact that software doesn't follow Moore's Law is the most important factor in the future of technology." DARPA has similar concerns. It is fundamentally rethinking how computers work. As Col. Tim Gibson, a program manager for DARPA's Advanced Technology Office, put it, "You go to Wal-Mart and buy a telephone for less than $10 and you expect it to work. We don't expect computers to work; we expect them to have a problem. If a commander expects a system to have a problem, then how could he rely upon it?" In fact, Lanier sees full global employment as the main virtue of increasingly crappy giant software. The only solution will be "the planet of the help desks." Everybody on earth will have to be employed taking phone calls giving advice on how to make the stuff work.

There are a host of reminders of the limits of technological prognostication. In 1950, in the article "Miracles You'll See in the Next 50 Years," *Popular Mechanics* claimed that in the year 2000, eight-room houses with all the furnishings, completely "synthetic in the best chemical sense of the term," would cost $5,000. To clean it, the housewife would simply hose everything down. Food "out of the reach of any Roman emperor" would be made from sawdust and wood pulp. Discarded rayon underwear would be made into candy. Spreading oil on the ocean and igniting it would divert hurricanes. The flu and the common cold would be easily cured. The rooftop family helicopter would accomplish voyages of over 20 miles, including much commuting. For short trips, the answer would be one's teardrop-shaped, alcohol-burning car. *Popular Mechanics* didn't get it all wrong, of course. They noted that the telegraph companies were hitting hard times in the year 2000, because of the fax machine.

Nonetheless, there are some distinct categories into which bad predictions fall:

- *The enterprise turned out to be a lot more complicated than it sounded.* This is why we don't have robotic maids, or electricity from nuclear fusion, or an explanation for what causes cancer.

- *The cost/benefit ratio never worked out.* This is why we don't have vacation hotels in orbit.

- *The future was overtaken by new technologies.* This is why automotive standard equipment does not include CB radios.

- *Bad experience inoculated us against the plan.* This is why there are so few new nuclear fission power plants.

And most important:

- *Inventors fundamentally misunderstood human behavior.* This is why we have so few paperless offices.

"We should start from the point of view that it is best to make the assumption that we know less than we think we do about reality," Lanier says. "It's hard to know for sure—it's a guess—but probably there's a lot more to reality than we think. If you think a human is just like a naturally occurring technology that's almost understood—. If you think that a human is something that you just have to figure out a few little things about, but basically, the underlying theory about it is coming together and all you have to do is trace those genes and proteomics and a little bit of stuff about how neural networks work and in about another 20 or 30 years, basically, you'll have it nailed—. If you believe that that would be a complete description of what a human is—. There is this danger that you might have missed something and you have reduced what a human is."

Needless to say, his peers pillory Lanier for his heresy. He is the one guilty of linear thinking, they believe. In the near future we will not so much write software, laborious line by laborious line, as grow it—reverse engineering the techniques we find in nature and adapting them to our software needs. "If we are still plunking around with software in 2012 or 2015, that would be a really bad sign for people who expect a real-soon-now Singularity," says Vernor Vinge. If, however, you start seeing large networks reliably coordinating difficult tasks such as air traffic control by learning from their experience, or parallel processors behaving like biological cells, The Curve will be on track to change society, he says. Kurzweil is hurt that anyone would compare his elaborate methodology to harebrained rabbit-out-of-the-hat predictions from the past. He scoffs at the notion that software is not improving. He points out that his company's voice-recognition software in 1985 cost $5,000 for a 1,000-word vocabulary. By the turn of the century it cost $50 for 100,000 words, and the newer software was much more accurate and easier to use. He acknowledges that software is not advancing as fast as hardware, but he estimates its value doubles every six years—still an exponential increase. He also accuses Lanier of "engineer's pessimism." That suggests Lanier is simply displaying the melancholy naysaying of someone who can't face another programming deadline. It is also a subtle slur. Calling Lanier an engineer suggests that he is not so much a scientist, much less a visionary, as he is a cubicle-inhabiting code monkey. Lanier replies, hey, these guys talk a great game and wave their arms. I actually *do* this stuff. If they can write the software, they can prove I'm wrong. That retort suggests some people with hefty credentials are dilettantes and poseurs. This altercation goes around and around.

Lanier sees more virtue in measuring progress through the second ramp—moral improvement. "I have to admit that I want to believe in one particular large-scale, smooth, ascending curve as a governor of mankind's history," he says. "Specifically, I want to believe that moral progress has been real, and continues today." Should you start with the revolutionary proposition that "all men are created equal" in 1776, Lanier suggests, you can then plot the graph of increasing dignity and autonomy through the abolition of slavery in the United States with the Civil War, women gaining the right to vote, the abolition of legal racial discrimination with the civil rights struggle, American empathy with those with whom we were supposed to be at war in Vietnam, the widespread acceptance of the sexes being treated equally, the breaking down of legal barriers against gays, and now an increased insistence that animals are not machines but feeling beings who should not be made to suffer gratuitously at the hand of humans. "You could plot all these on a graph and see an exponential rate of expansion of the 'circle of empathy,' " Lanier says.

This empathy notion is that people draw a mental circle around themselves. Inside the circle is everyone we care about and for whom we have deep compassion and understanding. Outside are the ones for whom we don't. "Most people, when they're young and idealistic, tend to want to draw the circle pretty large," Lanier observes. "Indeed, it would be lovely to draw it really, really large, to be able to live life in such a way that one caused no harm at all." The problem is, if you draw the circle too large, you starve. If you try to kill no living creature, what about those bacteria? If you say, "Okay, not bacteria, but I'll try not to kill insects," well, what about those bugs you might find in your flour? The point is that you have to set some limits. Otherwise, universal empathy "takes so much energy that you can't do very much else," he says. There is also, of course, the opposite hazard of drawing the circle so small that you cut off people who are important to making you who you are.

Lanier ultimately finds the circle of empathy troublesome as a measure of The Prevail Scenario. For one thing, the technological elite is trying to co-opt it. Those who worship the idea that computers are becoming sufficiently smart to be a successor species to humans would have you believe that soon we will be morally obligated to bring silicon beings inside our circle of empathy. Lanier thinks that is perilous hogwash. He thinks it cheapens the standing of humans inside that circle.

He also thinks that focusing on a process of increased morality is dangerously narcissistic. That's "the tragedy of religion and the tragedy of most utopias," he says. "If your utopia is based on everybody adhering to some ideal of what is good, then what you're saying is, 'I know what is good, and all of you will love the same goodness that I love.' So it's really ultimately about you." Others will be good your way or be tied to a stake surrounded by kindling.

This is how Lanier gets to his ultimate measure of the success of The Prevail Scenario. It is the third ramp of progress—the ramp of increased interpersonal connections. That ramp, historically, starts with the invention of language and then moves to writing, drama, literature, printing, film, the telephone, radio, television, the Internet and so forth. What you are measuring is an increase in the quantity, quality, variety and complexity of ways in which humans can connect to each other. Not ways in which they become identical, but ways that they become closer. It's the increased solution to the problem the Martians felt sad about when they encountered those sacks of skin surrounded by air.

The connectedness ramp is not measured by inventions. The test is interesting group behavior. Lanier doesn't care, for example, that millions of people are now participating together in online games. These he mostly finds tedious. Progress is in the emergence of interesting human societies. "This is where I see the action right now," he says.

For example, at the University of St. Andrews in Scotland, where a certain royal hottie studied art history in the early 21st century, Prince William couldn't even go out for drinks with friends without being tracked electronically by a pack of wired women. "A quite sophisticated text messaging network has sprung up," an "insider" told the *Scottish Daily Record*. "If William is spotted anywhere in the town then messages are sent out" on his admirers' cell phones. "It starts off quite small. The first messages are then forwarded to more girls and so on. It just has a snowball effect. Informing 100 girls of his movements takes just seconds." At one bar, the prince had to be moved to a safe location when more than a hundred "lusty ladies," so alerted, suddenly mobbed the place like cats responding to the sound of a can opener.

This is the sort of interesting growth in human social connection Lanier has in mind. It's called "swarming," a behavior that is transforming social, work, military and even political lives worldwide, especially among the young. It is the unintended consequence of people, cell phones in hand, learning that they can coordinate instantly and leaderlessly.

"It's the search for peak experience, something that's really going to be special," says Adam Eidinger, a Washington political organizer. "It happened to me just last week. There was a concert." His cell rang, and the call was from Bernardo, "one of the biggest swarmer cell-phone people I know. 'Where are you? There are all these people here!' And he wasn't just calling us. He called 25 people. Pretty soon everybody he knew was sitting on the grass, and none of them knew they were going to be there that morning."

This is the precise opposite of a 1962-style *American Graffiti* world. Then you had to go to a place—the strip, the malt shop—to find out what was going on. In the early 21st century, you found out by cell phone what was going on, and then you went to the place where it was happening.

Swarming is a classic example of how once-isolated individuals discover a new way to organize order out of chaos. It is a tick on The Curve of the connectivity ramp. The whole point is to bring people together for face-to-face contact. Swarming is also leading to such wondrous social developments as "time-softening," "cell dancing," "life skittering," "posse pinging," "drunk dialing" and "smart mobs."

Howard Rheingold is an apostle of swarming. A colorful character who tastefully paints his black dress shoes with moons, stars, planets and flames, Rheingold has for a generation examined society's unintended and imaginative uses of new technology.

He helped pioneer virtual communities (a phrase he invented and wrote a book about) before most people had even heard of e-mail or seen a cell phone. He began this work so far in the dim and murky past— 1988—that pundits then saw as preposterous the idea of human relationships being created simply by typing into the ether. This was before *flesh-met* entered the lexicon of the early adopters, as in: "Oh yeah, we know each other real well—although I don't think we've ever flesh-met."

As a cell phone increasingly becomes something that a teenager gets with her driver's license and it shrinks from a tool you carried to a fashion item you wear, Rheingold sees a profound shift in society. "They amplify human talents for cooperation," he says.

This is by no means all fun and games. The gear is used by "some of its earliest adopters to support democracy and by others to coordinate terrorist attacks," says Rheingold, author of *Smart Mobs: The Next Social Revolution.* Smart mobs are a serious realignment of human affairs, in which leaders may determine an overall goal, but participants at the lowest possible level—who are constantly innovating—create the actual execution on the fly. They respond to changing situations without requesting or requiring permission. In some cases, even the goal is determined collaboratively and nonhierarchically. It is the warp-speed embodiment of the French revolutionary's maxim "There go the people; I must follow them, for I am their leader."

The key to why mobiles are an uptick in the ramp of human connectivity is that they move communications out to wherever and whenever humans roam. Especially as e-mail is piped to your mobile, one sees behavior like that in the Philippines in 2000 and 2001. There, former president Joseph Estrada, accused of massive corruption, was driven out of power by smart mobs who—alerted by their cell phones—swarmed to demonstrations, gathering in no time. "It's like pizza delivery," said Alex Magno, a political science professor at the University of the Philippines. "You can get a rally in 30 minutes—delivered to you."

Cell phones driving political change is part of a ramp of political connectivity with mythically Prevail overtones. These include fax machines enabling Tiananmen Square, photocopiers fueling the Polish Solidarity uprising, cassette recordings firing the Iranian revolution and shortwave radios aiding the French Resistance. The difference with cell phones was the amazing speed with which people could swarm. It created not only a new kind of protest but also a new kind of protester. "It's a great way to get people who are in offices involved," Christina Bautisto, who works in Manila's financial district, said of her fellow professionals. "They don't have to spend all day protesting. They just get a message telling them when it's starting, and then they take the elevator down to the street. They can be seen, scream a little and then go back to work." In Washington, mobile-mediated swarms are regular highlights of the World Bank and International Monetary Fund protests. "I don't want to give away all our tricks," says Eidinger, the political activist. "But wireless plays a huge role." That includes everything from little "Family Pack" communicators from Radio Shack on up to sophisticated channel-skipping radios that are not easily monitored, all of which are used by "flying squads" to respond quickly to unanticipated opportunities. Cell phones are in constant use by lawyers seeking court orders designed to complicate the lives of the authorities as the protest is still evolving.

The U.S. military has been one of the earliest institutions to both fear and see the possibilities in swarming. John Arquilla co-authored *Swarming and the Future of Conflict* for the Office of the Secretary of Defense. He sees swarming—"a deliberately structured, coordinated, strategic way to strike from all directions"—as spearheading a revolution in military affairs. "In future campaigns," Arquilla says, leaders might benefit by simply "drawing up a list of targets, fixed and mobile, and attaching point values to them. Then units in the field, in the air and at sea could simply pick whatever hadn't yet been taken. The commander would review periodic progress, adjust point values if needed from time to time, and basically stay the hell out of the way of the swarm."

Despite these sober implications, in the early 21st century social swarms were easily the most interesting examples of the ramp of connection. Social swarming involves sharing your breath with others in real time. It means pulsing to the rhythm of life with your posse. It means a nonstop emotional connection to your clan.

It's Saturday night in Washington, and between the art-show openings of twilight and the after-hours clubs near dawn, the tribe that swarms touches down at Gazuza. Single, in their twenties and thirties, and wired, the members of the hive flit into the stylish Dupont Circle club as they hear that at this instant, the action is here. Bill Luza, 35, an architectural designer dressed all in white, is old enough to regale the crowd with tales of days so ancient that his first cell phone was the size of a loaf of bread. It came with its own shoulder bag. Today, of course, to be young is to be cognitively welded to a mobile. "You always want it near you," somebody says. "You take the phone out of your purse and leave your purse behind. You take your phone even when you don't take your purse or your keys. It's like a little person." Luza raises his head from a call. "That one was from Argentina," he remarks casually.

All right, Mr. Lanier, you say your measure is interesting group behavior? The swarmers laugh when "cell dancing" comes up. This is the choreography of two people who are vaguely in the same area but can't find each other. "It's a locator service," says Anna Boyarsky, 21, an intern at *National Geographic*. "My younger brother was in town. We were going to meet up for lunch. 'I'm at M and something,' he said." She had him start walking down the street, calling out landmarks. Suddenly, she crowed, "I see you, I'm at the other corner."

"Drunk dialing" brings blushes of recognition. "Saying things that you shouldn't be saying because the cell phone's in your pocket and you're drunk," someone acknowledges knowingly. "Stupid things," says Angie Hacker, an intern with the U.S. Public Interest Research Group. "My best friend at home, she broke up with this guy she went out with for two years. She calls him and like, 'I know you're not over me. I know you feel that way. You're just going out with that other girl because she's around.' And then she hung up."

"Ohhhhhh. I have a friend," says Corinne Fralick, 21, an intern at the Center for Policy Research on Women. "Every weekend, one o'clock in the morning, she calls me. She's totally trashed, and in California—three-hour time difference—to tell me how much she loves me, how much fun she's having, how much fun I'm having. Talking about everything. 'The boy I kissed earlier.' No point to the conversation. The cell phone companies must love it."

More seriously, everyone acknowledges that being constantly in touch with the rest of the swarm is changing their sense of time, place, obligations and presence—indeed, the texture of their lives. Which would be Lanier's point.

The very fabric of their time has softened. Remember arranging to meet at a specific time, like 8, at a specific location? Forget it. The new hallmark of squishy lives involves vaguely agreeing to meet after work and then hashing out the details on the fly. A time-softened meeting starts with a call that says, "I'm 15 minutes away." It's no longer unforgivable to be late, as long as you're in contact. "If you didn't have the cell phone, you'd make more of an effort to be on time," says Kaine Kornegay, 21, an intern in the Senate. "It's more socially acceptable to be late, because you've given notice that you would be."

"With that, the problem is resolved because the information was transmitted, although not his physical body," chimes in Ky Nguyen, 30, a freelance writer. "There's a level-of-service agreement. You expect people with cell phones to be available all the time. If they don't call back quickly, that's interpreted as a snub, and it causes anger. It would not be the same calling a land line because you might be out, so taking a day to get back could seem perfectly reasonable. You get mad at each other when those expectations vary from actuality. Sometimes it's because of a failure to perform on the part of a person. But at others, it's just a failure to communicate the level of expectation—what one person is expected to provide versus what another person expects to receive."

The expectations for connectedness can be astonishingly high. In an earlier conversation, Shirleece Roberts, 21, a senior at Rutgers who likes to use text messaging, had said of her swarm, "Everything is based around the cell phone. Where we're going to meet. Where we're going. Whether we're lost. Where we're at. How to get there. Everything."

Roberts is constantly pinging her posse. "When I get off work, going to the gym, I tell them, 'Meet me there.' If I'm going to the store or to the movies or out to eat, I'll tell them. If we're at parties or clubs and get split up, we'll send a message that says 'Meet me outside.' You talk to all your friends, all day, every day. Before you come to work, when you get off work, during work, before going to bed. See what we're doing. Going to sleep or going out." The last thing Roberts does at the end of the day is send a text message that says, "Good night."

There can be a dark side to all this. Swarmers run the risk of skittering like water bugs on the surface of life. By being quickly and constantly connected, they can avoid deep contact in time-consuming and meaningful ways. "If I've shown up and not found the love of my life, not had a love-at-first-sight experience," at one location, says Bernardo Issel, a writer, "then I have the opportunity to find out if there are other events going on where that might happen. You're flitting from one place to another. You're more likely to pursue superficial engagements rather than deep pursuits. It contributes to this certain MTV approach to life where you engage in something for a few minutes and then there's a commercial." Boyarsky agrees, though she adds, "You have to have a grip on reality. Unless you know what is real—what is a real friendship and relationship—neither can have an effect on you. If you know what is real, then you know that the cell phone is not a real relationship. It's a connection, but not a person. It allows you to connect to other people, but it's not them, and not you."

"It's a sign of commitment when you turn off the phone," Boyarsky says. "When somebody turns off their cell phone for you, it's true love."

Lanier has a word he uses to describe this kind of intensity: *flavor*. "To my mind *flavor* is simply the word for whatever it is that defines the circle of empathy that I don't know how to describe scientifically or technologically but that I think I can see. It's some kind of meaning beyond the thing itself." To him *flavor* captures whatever it is about the human enterprise that is ineffable and marvelous.

This is why Lanier is not at all panicked by the prospect of an increasingly enhanced humanity. As long as it narrows the gaps between people and has flavor, he is content. In fact, he can't wait for the ramp of connectivity to give him the communication powers of his beloved cuttlefish. Sure, he says, some people will use their enhancements to advance their flavorless conformist careers, with everybody looking the same because they've all got the same tall, blond ideals. But that's going to get real boring, not to mention that genetically engineered lawyers are going to be a variation on The Hell Scenario.

At some point, a device will appear that will be to biological enhancement what the Kurzweil 250 synthesizer was to music. It will finally put the powers of creation in the hands of people with some real imagination. When that day occurs, Lanier would like wings, please. Wings that have the presentation capabilities of the skin of his beloved cuttlefish, to be precise. He would like wings he could unfurl on which he might display whatever he imagined. He would have not just words for things as he tried to connect to others, but pictures—moving pictures, from his artistically original mind. He thinks it would be pretty hot on a date.

If I were a Natural, I ask, and I came across an Enhanced with cuttlefish display wings, would I be able to connect with such a creature?

"I actually think this is a yes," he says. He admits that there may be a little initial recoiling in horror, but by that decade, he figures, there will be enough Enhanced with unusual attributes that good old-fashioned human curiosity will take over. "Whether you can connect to this person is the responsibility of both parties. I'd want to be somewhat surprised by somebody who went to that much trouble, to have wings."

For The Prevail Scenario to work, he believes, you will have to have a world in which you have both differences between people and opportunities for intense connectedness. The measure of success would be the extent to which you could communicate more deeply and completely with others in a flavorful way.

So Lanier thinks the answer to whether the Natural could have an intense connection with such an Enhanced would be based entirely on the content of her spirit and on what that person revealed of herself on her wings.

"If somebody has display wings integrated into their body," Lanier says, "and all she can do is show *Gilligan's Island* reruns, I mean, I don't want to know that person," he says.

"That wouldn't even be funny once."

———

TRYING TO ARRANGE a journey to wherever Kurzweil, Joy and Lanier might happen to be tomorrow is like playing pool on a table where the balls jump like bullfrogs. Of them all, Jaron Lanier is the hardest to nail down. So when he abruptly announces that if I can get myself to the Denver airport, he'll pick me up for a road trip down the Front Range on his way to visit his ailing dad in New Mexico, I hop to it.

Wrestling his way out of the snowy elevations, Lanier shows up at the airport in a Ford Windstar, a minivan at the unfashionable end of the mommy-mobile spectrum. It was the only vehicle he could find in Colorado, he explains, for which the rental company wouldn't charge extra for him dumping it far, far away—doubtless because they were relieved to see it depart. So there we soon are, tooling through the mountains and deserts west of the 100th meridian with tape recorders, batteries, legal pads, pens, laptops, shoulder packs and a backseat rapidly filling with snack food wrappers. Two desperadoes headed for the border, looking like we've just ditched the middle-school soccer team. *Oh yeah. Out of our way. We bad.*

The high point of the trip, of course, is Lanier showing off Mesilla. Its old Spanish plaza remains remarkably genuine. A wedding is going on at the *iglesia,* the church at the north end of the square. It would not be out of place a thousand miles south. Some businesses have begun to make a pass at gentrification. There's a bed-and-breakfast and a bookstore. The ground floor of the old courthouse now features the Billy the Kid Gift Shop. But even amid the predictable supply of badges in the shape of a six-pointed star saying "Sheriff" at the top and "New Mexico" at the bottom, with a wide selection of boys' and girls' first names embossed in the middle, one can find certain blooms of authenticity. Lanier spies a collection of Indian flutes for sale. "Oh, I know the guy who makes those," he says. "Those are cedar flutes. They're really good." He picks one up and suddenly the tourist trap is filled with amazing trills as he triple-tongues this flute and flutters his fingers over the holes. Everything stops. All turn to listen to the music Lanier is coaxing out of these humble instruments.

Leaving the plaza, we pass the Palacio Bar. Lanier has vivid childhood memories of watching people carrying their shotguns into this bar. He remembers the sound of gunfire. He's never been inside, he muses.

"Great!" I say. "Let's have a beer!"

Lanier looks and thinks the way he does completely without benefit of drink or drugs. He is a lifelong teetotaler. His imagination is completely fueled by his own onboard supply of neurochemicals—one of his more remarkable achievements. So there we are, bellying up to the bar of the Palacio, and Lanier orders a Diet Coke. I look at him in horror and quickly order a Cuervo Gold and a draft, just to avoid trouble. The Palacio is still pretty unadulterated. But it's not the way Lanier imagined it. It's more of a family place now. Yet over the bar there's a list of deadbeats who have welshed on their tab, labeled "Bad Eggs." On top of the thermostat behind the bar there lies a sapper—one of those whippy metal things with weights at either end covered with leather that cops used to carry in their hip pockets, calling it "a persuader." The bartender says he's never had to use it. But it remains prominently displayed.

Lanier may have no use for mind-bending substances, but he does take a fine and studied interest in his home state's regional cuisine. This is how we end up peeling off the Interstate for breakfast at the town of Hatch. Hatch, which is even less prepossessing than Mesilla, prides itself on being the chili capital of New Mexico. No small boast. The universal solvents of every meal in the state are green chili salsa and red chili salsa. A local sternly lectures me for not recognizing the value of spreading chili jam on one's breakfast toast, and stirring it into one's yogurt.

New Mexico's chilis are intriguing. They're not the Texas instant-hit, blow-the-back-of-your-head-off kind. New Mexico's chilis have layers of flavor. They enter into a conversation with your food and your taste buds in a complex way.

On our way out of town, after a vast platter of huevos and salsa at La Palma Cafe, we espy this little cinder-block hole-in-the-wall place called Flores Farms, on Hall Street, Hatch's main drag. Out front hangs row after row of *ristras*. These are fat decorative braids of dried chilis, some as many as six feet long. We definitely need some of these, so we pull in, and behind the counter is Felipe Mendoza. Mendoza has the sort of terrific, sun-darkened, wizened Mexican-American visage over which the photographer Ansel Adams, who roamed these parts, would have swooned. We get into a very deep conversation about the various kinds of chilis he has, when to roast them, how to crush them, the correct proportion of cilantro to add and so forth. This fills the better part of an hour. Wandering around the tiny shop, examining its wares in the course of all this, we discover two big maps on the wall, one of the United States and one of the world. On close examination both turn out to be riddled with pinholes. When a customer comes in, Mendoza will give him a pushpin to stick in the maps to show how far away from Hatch, New Mexico, he's from. At the end of the year the shopkeepers remove all the pushpins and the process starts all over again.

This is only the fifth month of the year, but already the maps are full of pushpins. Not only from Miami to Seattle, but including Russia, India, Iraq and some little island off the coast of western Australia nearer to the Antarctic than anything else. That's the one that really impresses Mendoza.

Much later it occurs to me that Felipe Mendoza's world is a metaphor for Prevail. It is this intensely local yet vastly global arrangement that's very complex and very authentic whose pivot literally is flavor. Mendoza is no poster child for going back to some static nature, Lanier observes. Mendoza talks about all the varieties of chilis they are experimenting with to see how far they can push their business. His livelihood depends upon people coming to Hatch and saying Hatch is special. He is both a man of the world and is grounded in that place. He is clearly somebody who has the flavor of the valley. He is the essence of being connected while relishing differences.

Lanier thinks it important that we carefully pick which ramp on which to focus as we ride The Curve of exponential change. It is impossible to tell humans to spurn further evolution.

Lanier has sympathy for those who ask why we can't just be satisfied with all we have. It's hardly an irrational point of view, he says, but it is an impractical one. That's not our nature. "You can't have a clever species sitting around on the planet with nothing to do," he says. "Trouble *will* ensue. Something bad *will* happen. So it's essential to have long-term goals. These ramps are not just for fun. They're actually for our survival because of our nature. We want to choose our ramp wisely, and I think the one I've outlined, I think it works." He laughs. "I'm thinking of it as a technologist, like designing a ramp for mankind. This is a good ramp. It's respectful of others. It doesn't say, 'I know the right way.' It assumes differences. It's psychologically extremely challenging. And it is based on a real measurable achievement—whether people have understood each other—instead of some fantasy."

Who knows whether Lanier's version of Prevail will prevail. But Hatch provides evidence that his scenario could be credible. For by the side of the road, in a secluded part of New Mexico, there stands a small monument, decorated with *ristras*, pointing the way to an infinite game in which people are at the center, flavor is not blanched out and the goal is connections as complex as those musicians like to make. It might be a long shot. But if we can evolve in that direction, we might even manage to find an ambitious and persistent way to expand the ineffable and the marvelous and the resilient in the human spirit.

The Prevail Scenario

There are two key elements to any version of The Prevail Scenario:

- Humans have an uncanny history of muddling through—of forging unlikely paths to improbable futures in defiance of historical forces that seem certain and inevitable.

- The wellspring of this muddling through, of this prevailing, is the ability of ordinary people facing overwhelming odds to rise to the occasion because it is the right thing—for example, the British "nation of shopkeepers" that defied the Third Reich.

To these, Jaron Lanier shapes his version of The Prevail Scenario by adding:

- Even if technology is advancing along an exponential curve, that doesn't mean humans cannot creatively shape the impact on human nature and society in largely unpredictable ways. Technology does not have to determine history.

- The key measure of Prevail's success is an increasing intensity of links between humans, not transistors. If some sort of transcendence is achieved beyond today's understanding of human nature, it will not be through some individual becoming a superman. In Lanier's Prevail Scenario, transcendence is social, not solitary. The measure is the extent to which many transform together.

Predetermined elements:

- Few. The Prevail Scenario views uncertainty about the specifics of the future of human nature as one of its more plausible features.

Critical uncertainties:

- Are The Curves of exponential change smoothly accelerating, or are they susceptible to unexpected slowdowns, reversals or stops?

- Will The Curves of technological change produce a smooth curve of change in human culture?

Big differences between The Prevail Scenario and both the Heaven and Hell Scenarios:

- In Prevail, humans are picking and choosing their futures in an effective manner. They are actually succeeding in practical ways to slow change that is seen as negative or accelerate change that is seen as positive. This is not to be confused with mere rhetoric that has little functional outcome. Nor is it to be confused with protests that result in little actual change, or change that merely alters the outcome by moving the relentless Curve from one part of the globe to another—from North America to Asia, for example.

Early warning signs that we are entering The Prevail Scenario:

- Resistance to The Curves of change is actually having an effect worldwide.
- Certain technologies that affect human development and enhancement are globally seen as worth slowing down or stopping, in the way that the use of nuclear weapons was effectively prevented for the second half of the 20th century.
- Technologies that were seen as inevitable turn out to take much longer to develop than anticipated. Predictions common in the early 21st century begin to sound as silly as those of the middle of the 20th century, such as the paperless office, hotels on Mars and self-cleaning houses.
- Researchers voluntarily stop working on topics they view as too dangerous.
- Researchers decline funding for certain topics that they view as too fraught with human peril, putting their ethics ahead of their promotions, tenure, graduate students and intellectual curiosity.
- Researchers decline funding from organizations they view as too laden with problems, such as corporations and the military.
- Moore's Law, which projects the swift repeated doubling of computer power, is discovered no longer to be a reliable guide, because it has hit fundamental physical limits.
- Computational power is no longer seen as achieving exponential growth because of the inability of software to keep up the pace of innovation.
- There is little correlation between any exponential change in technology and the development of human society.

CHAPTER SEVEN

Transcend

Transcend: To go beyond in some respect, quality or attribute; to rise above, surpass, excel, exceed.

—Oxford English Dictionary

Transcendence is the belief that you can win, break even, or get out of the game.

—Clay Shirky

Genetic engineering is wicked, the crowd in the charm-free auditorium at Yale University is being told. They fidget as bioethicist George Annas scolds them. Look at all those rosy promises made when we were being sold nuclear power and manned space flight, he says. Tinkering with our gene pool—altering human nature as we have understood it for millennia—is unspeakable.

Annas is as compact, pudgy and generally affable as a stuffed toy with a comb-over. But right now, he's on a tear. Nobody should change humanity without consulting society, says the chair of the Department of Health Law, Bioethics and Human Rights at the Boston University School of Public Health. The burden of proof should be shifted to researchers. Make them prove harmless any species-altering or species-endangering experiments. The world is too dangerous already.

That's the old precautionary principle, responds Gregory Stock, who is sitting to his left. Stock, the anchorman-smooth author of *Redesigning Humans* who sees biotechnology leading to The Heaven Scenario, is the director of the Program on Medicine, Technology, and Society at the UCLA School of Medicine. Annas' suggestion is just another way of saying that we should never do anything for the first time, Stock goes on. How could anyone have established there was no harm in creating the first laser? The precautionary principle is just a way to block research without admitting that that is the goal. Science works by making mistakes. We shouldn't torture ourselves with hypotheticals. Minimize the actual problems that arise. Allow people individual choice; let them advise those who follow about the paths not to be taken.

Our mistakes are called children, Annas seethes. We should try this on twenty generations of primates first.

It is unlikely that parents who treasure their offspring would rush out to use any treatments, much less genetic ones, that they didn't believe were fully validated and safe, says Stock. We're going to go on this adventure. We should not just accept but embrace the new technologies, because they're filled with promise. And because we can.

It's hard to be against the future, says Annas. It's not transhumanism we should worry about; it's dehumanization.

Someone from the audience asks Annas how he is going to stop this technology. What makes you think government is up to the task?

World government is a pipe dream, but a good pipe dream. We should think and talk about it, Annas replies.

The debate rages for more than an hour. Stock closes his portion by predicting that future humans will look back at this glorious moment, when all ways to alter humanity were being developed, and marvel. It's an enormous privilege to be alive at this time, he says, finishing his delivery of The Heaven Scenario.

Annas ends his presentation of The Hell Scenario with a warning: We are not very good at preventing harm to the environment, the pain of poverty, the horror of genocide. He hopes we use the precautionary principle as our guide to prevent future calamities.

The debate organizer, James "Jay" Hughes, a health policy professor at Trinity College, sympathizes with Stock. He says that arguing in favor of the transformation of human nature is "like arguing in favor of the plow. You know some people are going to argue against it, but you also know it's going to exist."

"Detailed regulation is not possible and probably not desirable," Australian high court justice Michael Kirby later tells the crowd. Kirby serves as a bioethics adviser to the United Nations High Commissioner for Human Rights. "This is not defeatism or resignation. It is realism.

"All one has to do is read the science journals to know these issues are on the table today," Kirby says. "One thing I can say with certainty from my experience is that the wheels of law, of the legislative process, grind very slowly within nations and slower still internationally. The progress of science, on the other hand, is ever accelerating. If anything, we've been surprised at how quickly technology has progressed. It's worth taking on these issues intellectually now, rather than in crisis later."

~w~

I N AN ORNATE ROOM in a Romanesque building on the Yale campus in New Haven, there is a shrine to the relationship between knowledge and wisdom. Over the elaborately carved fireplace in one corner of Linsly-Chittenden Hall, formerly the university library, is a proverb: "Through wisdom is a house builded and by understanding it is established and by knowledge shall the chambers be filled with all precious and pleasant riches." Along one wall is a stained glass mural created in 1890 by Louis Comfort Tiffany, stretching dramatically for 30 feet. The colors are striking—creamy pearl, opalescent pinks, brilliant peacock blues, deep golds and muted antique rose in marbleized tones. In the center of the stained glass, in long flowing gowns, three angels represent Light, Love and Life. A dozen more beautiful and remarkably modern-looking young women portray the muses of Religion, Music, and Art. Near the center, Science is attended by romantic figures representing the virtues that undergird it, Research and Intuition. Behind them are a bevy of other qualities on which Science depends: Devotion, Labor, Truth, Perception and Analysis.

One fine June New England weekend, these goddesses of knowledge and wisdom preside over some intense conversation about the transcendence of human nature. The gathering, dubbed "The Adaptable Human Body: Transhumanism and Bioethics in the 21st Century," is co-sponsored by the estimable Yale Interdisciplinary Bioethics Program Working Research Group on Technology and Ethics and a young organization called the World Transhumanist Association.

Transhumanism is a loosely defined movement that started in the 1970s but is gaining heightened attention as the genetic, robotic, information and nano technologies—the GRIN technologies—make the transhumanists' interest in engineered evolution increasingly credible. Those like Annas who do not wish to see human nature altered are taking transhumanism seriously enough to attack it.

Transhumanists are keen on the enhancement of human intellectual, physical and emotional capabilities, the elimination of disease and unnecessary suffering, and the dramatic extension of life span. What this network has in common is a belief in the engineered evolution of "posthumans," defined as beings "whose basic capacities so radically exceed those of present humans as to no longer be unambiguously human by our current standards." "Transhuman" is their description of those who are in the process of becoming posthuman—the metamorphosis they believe, not without good reason, some of us are entering right now.

Transhumanists view human nature as "a work-in-progress: a half-baked beginning that can be remolded in desirable ways through intelligent use of enhancement technologies." In Web sites, publications and meetings, they advocate enabling those "who so wish, to live much longer and healthier lives, to enhance their memory and other intellectual faculties, to refine their emotional experiences and subjective sense of well-being, and generally to achieve a greater degree of control over their own lives." Transhumanists say that for them, this positive goal has "replaced customary injunctions against 'playing God' or 'messing with Nature' or 'tampering with our human essence' or other manifestations of 'punishable hubris.'" They believe it naïve to think the human condition and human nature will remain pretty much the same for much longer. Instead, they believe the GRIN technologies are fundamentally changing the rules of the game.

Hardly knowing what to expect from my first encounter with a couple of hundred transhumanists, I've come up to Yale ready for almost anything. And indeed, there is a smattering of sweet but very strange people. Natasha Vita-More, the artist formerly known as Nancie Clark, brings up the possibility of a person someday soon issuing a mental command to make her skin milky white in the morning, tawny brown in the afternoon and a midnight blue to match her gown in the evening. An occasional actress, dedicated bodybuilder (her motto: "Flex my mind, flex my body") and president of the libertarian Extropy Institute, Vita-More's interests include future body design. She offers a conceptual model of an optimized human who has built-in sonar, a fiber-optic cable down the spine and a head full of nanotech data storage. Vita-More's biography suggests she was born in the first half of the 1950s. She has a Left Coast–slender figure, brunette hair that falls below her collarbones, musky perfume, boots, tight jeans, a large mouth painted red, publicity materials that feature glamour portraits, and a face that seems to be on a little too tight. She expertly tosses around the jargon of the genetic, robotic, information and nanotechnology trades with a Santa Monica sense of wonder at all the future possibilities the cosmos might bear. Those of her male listeners who look like they could benefit from less pizza and more time in the sun respond as if they have encountered that celebrated deity, the geek goddess. This is clearly not the first time Vita-More has run into this reaction. She does not seem to hate it. Those who view people like Vita-More as flakes dismiss by association many of the ideas they embrace. Critics of The Singularity, for example, love to brush aside that idea as "the Rapture of the nerds."

Far more typical of those in attendance, however, are heavy hitters such as Annas and Stock and representatives of formidable organizations such as the American Medical Association and the National Science Foundation. This is a gathering of people who are thinking deeply about what transcendence might mean. "All futurism is black and white. All reality is gray," says Hughes. Indeed, the gathering will be written up in the sober and scholarly *Wilson Quarterly*, in addition to lengthy pieces across the political gamut from *Reason* magazine to the *Village Voice*. All discuss what would happen if humans engineered themselves into something so very far beyond our current nature.

One of the interesting things that already can be said about this young century: it is the time when the idea of transcending human nature has become controversial. On the face of it, this is striking—first, that earnest and intense people are taking the idea seriously; second, that some equally clearheaded people are against it.

Of course, people who object to fiddling with human nature do not view that program as progress. They ponder what might be lost. They fear that a search for transcendence might instead prove damning. They know what happened when Eve chose to eat from the Tree of Knowledge. They may be right.

Yet it is still a remarkable point in history at which we have arrived. One wonders if there was a comparable debate the last time such a change loomed—when the species for better or for worse faced the possibility of moving out past the capabilities of the chimpanzee, past the Neanderthal, past the Cro-Magnon. Who knows. Maybe there was considerable wailing around the campfire about the new people stealing the spirits of the beasts by painting them on cave walls.

The Transcend proposition rests on three premises:

- The undeniable competitive advantages that the genetic, robotic, information and nano technologies convey on those who embrace them for economic, medical, educational, military, or artistic reasons suggest that these methods will continue to advance at an ever-increasing rate.

- So many of these technologies—"designer babies," augmented cognition, metabolic makeovers, anti-aging medicine and all the rest—can alter basics of the human condition. If they can modify our minds, memories, metabolisms, progeny and personalities, it seems reasonable to think that these procedures might well have an impact on what it means to be human.

- The history of technologies as disruptive as these suggests that there will be unintended consequences. We will be surprised by many of the outcomes.

If you accept these three propositions as reasonable bets, what you're looking at is that rare bird, the high-probability, high-impact scenario. Transcend builds on, expands and gives measure to The Prevail Scenario, in which technology does not control us, but we control technology.

If you find it practical to think that the genies of these technologies are unlikely all to be bottled up, you have several choices. One is to be convinced that the future is bleak, in the fashion of The Hell Scenarios of Bill Joy, Susan Greenfield, Martin Rees and others who argue that technology is controlling us and the outcome likely will be catastrophe.

Or you can agree with R. Buckminster Fuller that "we are called to be architects of the future, not its victims," and take an active hand in shaping human destiny. There are important ethical questions about the desire to engineer better children, for example. What happens if people get to pick the sex of their offspring and the outcome is a socially disruptive and even warlike surplus of frustrated young males, as is the case already in South Korea, Pakistan, India's Delhi, China, Cuba, Azerbaijan, Armenia and nearby Georgia? If a child is created with genes selected to endow him with the marvelous traits of a Yale valedictorian or an Olympic sprinter, will he face even more oppressive parental expectations than in the past because now he is supposed to live up to his design? Suppose our desire for children who are well adjusted, well behaved, sociable, attentive, high-performing and academically adept is fulfilled by drugs? Will that get in the way of them developing character? These are all good questions raised in *Beyond Therapy: Biotechnology and the Pursuit of Happiness,* the remarkably literate and thoughtful 2003 Report of the President's Council on Bioethics, of which Leon R. Kass is chair. Kass is a medical doctor, a professor in the Committee on Social Thought and the College at the University of Chicago, and a fellow of the conservative American Enterprise Institute. These questions go to issues of what we want childhood to be.

"Life is not just behaving, performing, achieving," the report says. "It is also about being, beholding, savoring. It is not only about preparing for future success. It is also about enjoying present blessings. It would be paradoxical, not to say perverse, if the desire to produce 'better children,' armed with the best that biotechnology has to offer, were to succeed in its goal by pulling down the curtain on the 'childishness' of childhood."

Even assuming that in our role as architects of the future we pick and choose the elements that we implement, I have yet to encounter a persuasive argument that the advantages conveyed by the GRIN technologies are likely to be stopped worldwide—short of a cataclysm. In fact, the more problems we face, the more rapidly we probably will reach for an ingenious and seemingly miraculous fix. Nor have I seen a case made that convinces me I'd like to live in a world in which human imagination were so entirely blocked.

I do not wish to be cast as an opponent or a debunker of the social critics of technology. I hope I have presented them and their scenarios fairly. Readers should examine their arguments carefully. They offer important reasoning regarding the cautions we should consider. I wish we'd had such an informed discussion before we embraced nuclear power. It could well have benefited everybody—including the electricity industry.

In the absence of an attractive alternative, however, I elect to light out for the Territory in the words of Huckleberry Finn. I choose to examine the possibility that human nature might continue to evolve and be improvable, and to consider what transformation might actually look like and what it might mean. "What is a man? A seed? An acorn unafraid to destroy itself in growing into a tree?" asks David Zindell in *The Broken God*.

Exploring the Transcend hypothesis adds specificity, measurements and means to the goal of controlling our evolution in the fashion of The Prevail Scenario. At the very least it casts light on our current age by causing us to wonder about our present definitions of human nature and evolution and the meaning of transcendence.

THE CENTRAL ARGUMENT about the future of human nature is whether it is fixed and immutable, once and forever, or whether it can continue to evolve.

In an e-mail exchange, I asked Francis Fukuyama about his view. He is the author of *Our Posthuman Future* who sees biotechnology altering the underpinnings of human nature as a Hell Scenario.

Take a human of 10,000 years or so ago, I wrote to Fukuyama. Genetically identical to us, but living before The Curve of change kicked into a new gear with cultural evolution. Before people create writing, reading, and arithmetic, formal logic, world religions, cities, global trade and the whole deal. Before people could go from the first powered flight to a moon landing in 66 years. Your basic subsistence-level hunter-gatherer. Before civilization. Would we regard him as fully human? Why? Do his genes alone do the job? Has the steep ramp of cultural evolution had no effect on what makes us human today? Does civilization have nothing to do with defining what we now consider to be human nature?

"In one sense the answer to this question has to be yes," he replied. "That's what we would say about, say, an uneducated member of our own society if we posed the question—does this person deserve to enjoy human rights? The faculty itself is what makes us human, and not what we put into it. On the other hand, part of our nature is to develop those faculties, so the actual human is both the faculty and the faculty's content."

I'm not talking so much about rights, I e-mailed back. I don't expect anyone would recommend treating our 10,000-year-old man like a chimpanzee. What I'm struggling with is a definition of human nature. Does just the container—the physical human vessel—bound it? Or do the contents help define us? Isn't the container now filled with the distilled wisdom of our billions of forebears as has been captured by civilization? Isn't part of human nature now man-made? Doesn't the rise of rigorous logic and increased empathy count? If so, then one might say we have evolved over the past 10,000 years. Therefore, we might continue to do so. If not, then cultural evolution has no importance. Correct? Or am I missing something?

"But the rights are important," he responded. "If we didn't feel the uneducated primitive human was really human, we would not give him/her rights. But nobody would be willing to assert that we lose something of our essential humanness by being uneducated."

Agreed, I wrote back. The rights are important. But here's the crucial question—is our ancestor's version of human nature identical to ours? That gets to the nut of it, for me. Let us agree and assume that all versions of human nature are sacred and worthy of reverence, and let us set that aside for a moment. Has civilization produced a more evolved human being or not? Has the new contents in the old container made a difference?

Fukuyama did not respond. But if you agree with Thomas Hobbes that the life of our 10,000-year-old man in nature is nasty, brutish and short—the product of unfettered selfishness—then I think one has to conclude that all these millennia of billions of humans storing and sharing and cooperatively building on each other's wisdom—the content—has to count as part of evolution. It is inheritable, has variation and contributes to reproductive fitness. Even the least educated among us is not raised by wolves, feral and wild. He grows up shaped by contemporary humans who own televisions, who have been shaped by modern society.

Felipe Fernández-Armesto is no libertarian apologist for technology. The University of London historian believes embryos should be recognized as humans and abortion banned. Nonetheless, in *So You Think You're Human? A Brief History of Humankind,* he writes: "We have to face challenges to our concept of humankind and, where they are valid, confront the consequences for some of our dearest concepts: human rights, human dignity, human life—ideas for which, in the recent past, we've begun to re-shape states, re-fashion laws, and fight wars."

He likes the dynamic view offered by the German sociologist Justin Stagl, who says human nature is a shifting combination, changeable over time. This blend includes a "biological heritage" of ill-developed instincts and reflexes. But it also involves our legacy of transformation "from a biologically determined to a socio-culturally determined being" for whom culture has become natural. Above all, it embraces our "utopian potential"—our vocation to transcend our failures and defects, "to strive to attain superhuman goals and avoid the inhuman." Our human nature may be grounded in our animal nature, but our ability and eagerness to develop our "better nature" are unique.

Fernández-Armesto writes:

There is still no agreement about what "human nature" is—what, beyond trivial or temporary features of our physiologies or our cultures that happen to have been thrown up by history or evolution, is common to and exclusive to the creatures we recognize as human. Human nature, if it is proper to speak of such a thing, is not fixed: it has changed in the past and could change again. Its continuity with the natures of other animals is part of its fluidity. . . . How much our nature has to change before our descendants cease to be human is a question we are not yet ready to answer. In this respect it resembles the question about when, in the course of evolution, our ancestors became human—which is also unanswerable at the present stage of our thinking and knowledge.

That humans are uniquely rational, intellectual, spiritual, self-aware, creative, conscientious, moral, or godlike seems to be a myth—an article of faith to which we cling in defiance of the evi-

> dence. But we need myths to make our irresoluble dilemmas bearable. And our claims for our nature are more: not mere myths but also aspirations, still waiting to become true. . . . For now, if we want to go on believing we are human, and justify the special status we accord ourselves—if, indeed, we want to stay human through the changes we face—we had better not discard the myth, but start trying to live up to it.

That's the most satisfying description of human nature I've found. But we probably never will firm up a detailed definition of human nature upon which we can widely agree until we first encounter creatures with a genuinely different kind of intelligence to whom we might compare ourselves. Human nature is certainly not like *those* guys, we'll then say, and we'll start making lists of how that might be so, because making lists of such differences is part of human nature.

Certainly, humans are astonishingly assorted in their behavior and norms. Read the tabloid and supermarket press if you doubt this; the bizarre variety of arrangements humans can come up with is what keeps the penny press in business. In fact, what anthropologists have traditionally done for a living is collect the vast array of variations in our contents. Some bristle like cats confronting a Rottweiler at the very idea something called "human nature" even exists. Mary Catherine Bateson is the distinguished author of *Composing a Life* and the daughter of Margaret Mead. Mead, author of *Coming of Age in Samoa,* achieved fame starting in 1928 for advancing the notion that nurture is far more important a determinant of behavior than nature. To this day, to seriously suggest in her daughter's presence that the phrase "human nature" is meaningful and important is to gain a life experience long to be treasured in memory.

Elaborate global ideologies have been founded on varying definitions of human nature. The huge edifice of Christianity is based on the fundamental principle that an all-powerful God creates human nature in his image and likeness. Then in the late 1800s along comes Karl Marx, who says, baloney. He explicitly rejects the notion that there exists some God who has anything to do with our human nature. For Marx, human nature is the sum total of all our social interactions. He proceeds to construct this elaborate, world-moving ideology, saying that if you can alter the social relationships upon which human nature rests, you can change human nature. The historical results are spectacular. Millions die at the hands of Stalin, Mao and Pol Pot before it becomes clear that this can't be how human nature works. In the mid-20th century arrives Jean-Paul Sartre, who says of the social-relationship proposition, baloney. That's not what human nature is. He and the other existentialists say human nature is not dependent on anything or anyone. Individuals are condemned to be free, asserts Sartre in his famous phrase. Each is free to shape who he or she is going to be; this is indeed the great human challenge. That notion keeps undergraduates busy in late-night bull sessions for two generations. Then, in the late 20th century, sociobiologists come along and say, baloney. Edward O. Wilson in his book *On Human Nature* scores Sartre's model as not bloody likely, saying human beings are a product of evolution; sure, you can have tremendous variety, but human nature is fundamentally bounded by our own genetically determined, species-specific patterns of behavior.

Our DNA shows that the human species as we know it is indeed quite new—100,000 years or so old at most—and therefore remarkably unified. We share 98 percent of our genome with chimpanzees—not much less than we share with each other. The genetic difference that makes us human is barely significant, statistically. Humans respond with shared tendencies and behaviors to whatever challenges and opportunities we are offered by geology, geography and the global distribution of wild plants and animals. Few researchers can afford to ignore any longer the thousands of human universals. They include such not particularly deep examples as art, athletics, bodily adornment, calendar, cleanliness training, community organizing, cooperative labor, courtship, dancing, division of labor, education, ethics, etiquette, family, feasting, fire making, folklore, food taboos, funeral rites, games, gestures, gift giving, government, greetings, grief, hairstyles, hospitality, housing, hygiene, incest taboos, inheritance rules, joking, language, law, luck superstition, marriage, mealtimes, medicine, modesty, mourning, music, mythology, numerals, personal names, property rights, puberty customs, religion, sexual restrictions, soul concepts, status differentiation, surgery, tool making, trade, visiting and weather-control rituals.

With the rise of the genetic, robotic, information and nano technologies, however, we're talking about remaking this container. That is why it is so important to figure out what's up with human nature now. As Alvin Toffler wrote in *Future Shock* in 1970, "The future always comes too fast and in the wrong order." The Curve is handing us something profoundly new and overwhelming in its power. Look at how dramatically the first few technologies aimed at modifying our metabolisms and minds have changed our behavior. The birth control pill sparked an acceptance of casual sex at the same time it led to native-born population levels actually dropping in most of the developed world. Antidepressants such as Prozac ignited concerns about humans turning into zombies at the same time they saved the lives of a lot of miserable people. Those were primitive and halting first steps.

The deeper question is whether these GRIN technologies can alter basics of the human condition. Can we imagine them changing the way we shape truth, beauty, love or happiness? Can we imagine altering the seven deadly sins—pride, envy, gluttony, lust, anger, greed and sloth? Or the virtues of faith, hope, charity, fortitude, justice, temperance and prudence? Some transhumanists think—or at least hope—the answer to some of those questions may be yes.

Nick Bostrom seems a good man to ask about transcendence. Not only does he write extensively and learnedly about posthumanity, but he cares less about the gear than the philosophy. Bostrom looks the classic Scandinavian—tall, trim, blond, blue-eyed, square-jawed. Born in 1973, he has the slightest trace of a lilting Swedish accent, wire-rimmed glasses, close-cropped hair, and pants a good four inches too long, puddling around his heels. The co-founder of the World Transhumanist Association, Bostrom has impressed donnish committees at places such as Oxford, where he is a British Academy research fellow with the faculty of philosophy; Yale, where he taught; and the London School of Economics, where he earned his PhD.

So tell me, I ask, how did you get into the transhumanist business?

He was born in Helsingborg, he replies, a long way from anything important. Helsingborg is 300 miles south of the bright lights of Sweden's big city, Stockholm. It is on the shores of the Öresund Strait that barely separates Sweden from Denmark. A bridge has finally been built, after a hundred years of discussion, linking the Swedish peninsula directly to Western Europe. But at the time Bostrom was growing up, Helsingborg was "a hundred thousand people by the sea," not at all touristy. It was simply the best place to catch a ferry on the way to someplace else. He remembers it for its beech forests. "In the spring when all the leaves come out there is this very sort of bright green light and the ground is simultaneously covered with white flowers. Um, what are they called in English? Wood anemones? *Vitsippor* in Swedish. It's very beautiful."

His parents were prudent members of the conventionally nonreligious solid middle class. His mother was an administrative assistant for a fiberglass company. His father was a banker and noted stock market analyst.

Dad was averse to taking risks with his own money, however, so the family was never wealthy.

Bostrom recalls an excruciating adolescence. He loathed school as a prison in which he never encountered anything or anyone challenging. An only child, he felt acutely isolated and different. One day in the library at the age of 15 changed his life. "I picked up a book and went on home and started reading and this whole new world opened up," he says. It was *Thus Spake Zarathustra,* by Friedrich Nietzsche.

I laugh incredulously. At the age of *15?* I ask. *Zarathustra*—a literary masterpiece from the early 1880s—is a provocative, passionate, dark and legendary work of philosophy. It is written in a combination of prose and poetry that, the author would have you believe, reflects the voice of the Persian prophet Zoroaster, who, after years of meditation, has come down from the mountain to offer his wisdom to the world. It is the source of the phrase "God is dead." Bostrom read *Zarathustra* first in Swedish and then in the original German. Nietzsche led him to such titanic what-does-it-all-mean philosophers as Schopenhauer and then Goethe—the author of *Faust*—and then back to the works of the ancient Greeks. You know, your typical teenage coming-of-age story.

"It really felt like I had been asleep for the last 15 years and only just woken up. It was really fascinating," Bostrom says. "It opened up a whole new world of learning and poetry and literature and art. I felt this sense that I had wasted a lot of time. If I wanted to amount to anything, I would have to not lose another day and really get cracking." Bostrom started a crash program of self-directed home schooling—there weren't a lot of other people in Helsingborg who shared his new passions. That allowed him to pass exams for his high school diploma early and eventually wind up at Stockholm University, where he received a dual master's in physics and philosophy in 1992. This led to his continued rise through academe, including getting his doctorate in the philosophy of science from the London School of Economics in 2000 and on to Yale and then Oxford, which probably has the largest philosophy department in the world. While in London, he discovered on the Web other people who were thinking about the future of human nature. "So then it seemed to me that if you actually wanted to do something good in the world, the most important thing to do seemed to be to try to encourage more understanding or study of the potential implications of technologies," he says. That's how he came to co-found the World Transhumanist Association in 1998. Oh yes, in the middle of this, he worked as a stand-up comedian.

When we're talking about transhumanism, we're talking about transcending human nature, I say to Bostrom. In 1948, T. S. Eliot received word that he had won the Nobel prize as he was writing the play in which he coined the term *transhuman*. One notion of transcendence is that you touch the face of God. Another version of transcendence is that you become God. Does the word *transcendence* mean anything to you? And if so, what?

"I have this conviction that life can be very, very good and wonderful," Bostrom replies. "I remember especially when I was 16 and sitting in this beech forest with white flowers and reading poetry. There were moments that you just realized how wonderful things could be. Then you go back to school and you forget it. It fades from consciousness. You get stuck in the gray routine. It's almost like you are walking in your sleep." Bostrom decided not to sleepwalk through the rest of his life. "I guess I said to myself I don't want to forget just how good things can be."

He is determined to break through to higher ground. "We are biological organisms. The difference between the best times in life and the worst times is ultimately a difference in the way our atoms are arranged. In principle that's amenable to technological intervention. This simple point is very important, because it shows that there is no fundamental impossibility in enabling all of us to attain the good modes of being, and it's very probable that we can discover far better and more wonderful modes of being than anybody has yet experienced. This is the basic goal of transhumanism. Technological progress makes it harder for people to ignore the fact that we might actually change the human nature."

Bostrom imagines that there might be pleasures whose blissfulness vastly exceeds what any human has yet experienced. He can imagine much cleverer philosophers than us. He can imagine new and different kinds of artworks being created that would strike us as fantastic masterpieces. He can imagine a love that is stronger and purer than any of us has ever felt—including preserving romantic attachment to one's partner undiminished by time. Our thinking about what is possible for humans to attain is likely constrained by our narrow experience, he believes. We should leave room for the possibility that as we develop greater capacities, we will discover values that will strike us as more profound than those we can realize now, including higher levels of moral excellence.

What we now consider natural is not necessarily desirable or morally good, he points out. Cancer, malaria, dementia, aging and starvation are all ills to be fixed, in his view. So is our susceptibility to disease, murder, rape, genocide, cheating, torture and racism. "If Mother Nature had been a real parent, she would have been in jail for child abuse and murder," he says.

He hardly denigrates what can be achieved by education, philosophical contemplation or moral self-scrutiny—the methods for human improvement suggested by classical philosophers such as Plato, Aristotle and Nietzsche. He strongly supports the goal of creating a fairer and better society. Transhumanism developed directly from the centuries-old secular humanist tradition, notes Jay Hughes, the Trinity professor who is executive director of the World Transhumanist Association. That includes the liberal democratic idea that all people should be equal before the law. But Bostrom doesn't see that as preventing us from expanding our biological capacities.

If you don't buy the idea that human nature is static and exclusively determined by the kind of biology you could find in the people who crossed the Bering Strait into North America thousands of years ago, Bostrom says, then it's possible to take a new look at the bigger picture, the long-term fate of humankind, and to welcome the prospect of using technology to change the rules of the game. Currently, he believes, the game is human, but it is not humane.

The inventory of projects to be worked on is long. Gregory E. Pence's list of things we don't need includes pain, infirmity, mental illness, overpopulation, involuntary death, stupidity, cowardice, biological cravings no longer good for us such as those for burgers and fries, diseases that kill children and progressive diseases such as Alzheimer's that destroy minds. Pence, who presented his case at the Yale gathering, is the author of the text *Classic Cases in Medical Ethics: Accounts of Cases That Have Shaped Medical Ethics.* He professes in the philosophy department of the University of Alabama.

What we could use, Pence says, is more memory, better immune systems, cells that do not age, stronger skeletons with more muscle mass, more talent in the visual and performing arts as well as better jokes, an increased ability to process vast amounts of information quickly, an increased ability to do advanced math, an ability to speak many languages, an absence of genetic disease and a greater sense of wonder and curiosity. All of these we can soon achieve with the GRIN technologies, Pence believes.

Nor is that the end of it. The looks of Aubrey David Nicholas Jasper de Grey of Britain's University of Cambridge are almost as striking as his ambitions. His chestnut hair is swept back into a ponytail. His russet beard falls to his sternum. His mustache—as long as a hand—would have been the envy of Salvador Dalí. His research area is called "strategies for engineering negligible senescence," or SENS. It means curing aging. The well-named de Grey thinks—as do some researchers at the National Institute on Aging of the National Institutes of Health, near Washington—that the first person who will gracefully make it to the age of 150 is already alive today. He thinks scientists soon will triple the remaining life span of late-middle-age mice. The day this announcement is made, he believes, the news will hit people like a brick as they realize that their cells could be next. As the prospect looms of exceedingly long life—on the order of 5,000 years—he speculates that people will start abandoning risky jobs, such as being police officers, or soldiers. He thinks people will start putting more of a premium on health than wealth. Twirling the ends of his mustache back behind his ears, he says slyly, "So many women, so much time."

Pence's big caveat is that the adoption of these changes has to be voluntary, not state-mandated. Government control of these technologies, in his view, is at the core of eugenics—that reviled early-20th-century movement devoted to improving the human species through the control of hereditary factors in mating. It led to Hitler exterminating Jews, Gypsies, the handicapped and others he saw as polluting the "master race." Individuals voluntarily embracing these changes, by contrast, are not engaged in eugenics, Pence says. Not that this comforts the critics of transformation. Some fear that The Enhanced will see those at the bottom of society not as disadvantaged and worthy of our support but simply as candidates for repair. But if it's voluntary, others may respond, what is the problem with uplifting the least able among us?

Life is unfair, noted Lee M. Silver once. There is no such thing as genetic justice right now. Natural athletes have more red blood cells than your kid, he points out. Parents will want to know why they can't give their kids advantages that others have. Silver is a professor at Princeton University in both the Department of Molecular Biology and the Woodrow Wilson School of Public and International Affairs. He is the author of *Remaking Eden: How Genetic Engineering and Cloning Will Transform the American Family*.

It is human nature to desire to give advantage to our children, says Silver, who himself is short. The market will respond. There are only three ways advantages are distributed: chance (or God, if you are persuaded the deity gets personally involved), the state and parents. If you are convinced that a total ban on the creation of such advantages is unlikely, and you view government distribution as parlously close to eugenics, that leaves parents as the appropriate decision makers, Silver says. Let them choose what is desirable. (Maybe such children will not be called The Enhanced. Maybe they will be called The Chosen.)

Can we screw this up? Can we, by our well-meaning attempts to reduce suffering and increase opportunity, reduce human character? Without a doubt. Just ask Leon Kass, the bioconservative responsible for the *Beyond Therapy* report. He even makes the case for continuing to experience anguish, decrepitude and death. "A flourishing human life is not a life lived with an ageless body or an untroubled soul," he writes, "but rather a life lived in rhythmed time, mindful of time's limits, appreciative of each season and filled first of all with those intimate human relations that are ours only because we are born, age, replace ourselves, decline, and die—and know it."

Bostrom does not embrace such limitations. "We humans lack the capacity to form a realistic intuitive understanding of what it would be like to be posthuman," he says. "Chimpanzees can't imagine the ambitions we humans have, our philosophies, the complexities of human society, or the depth of the relationships we can have with one another." Just so, Bostrom feels, our present modes of being are but a narrow slice of what is permitted by the laws of physics and biology.

What about humans dividing up into The Enhanced, The Naturals, and The Rest? I ask. That could become a nightmare of slaughter. Whenever two species compete for the same ecological niche, it usually ends badly for one of them. An increase in business for Pestilence, War, Famine and Death—the Four Horsemen of the Apocalypse—would hardly seem a likely measure of transcendence.

"You might have a very broad continuum" in a generation or two, Bostrom says. "There are people in Africa who have nothing and starve, and yet, you know, there are wealthy Americans. That spread could be slightly larger if the American millionaires are not only rich but in addition have memory enhancement implants and expect to live 150 years. There could be herds of almost posthumans, and then slightly less transhumans and then sort of augmented humans. The reason we don't have tall people conspiring against little people, or vice versa, is that there is no obvious cutoff point, and it's just one continuum living in the same world. I guess it depends partly on whether enhancement technology should result in totally separated groups with radically different levels and nothing in between or whether it's more like a continuum. The latter might make it easier to avoid some of the worst forms of rivalry and repression. But even so, obviously it's a big concern."

Christine L. Peterson, founder and president of the Foresight Institute, a California nonprofit formed to help prepare society for the effects of nanotechnology, agrees. "The goal is peaceful coexistence among traditional humans, augmented humans and machine-based intelligences," she says. "The analogy is to entities more powerful than humans, like governments and corporations. We come up with checks and balances. We always protect weaker members of society against those who want to push them around." Jay Hughes, who was once a Buddhist monk, has as his big aim for the future the creation of "societies not trying to kill each other or prevent each other from reproducing." He sees that as a far higher and more moral choice than trying to keep everyone the same. "Levelers would not have allowed the pyramids to be built, or Chartres cathedral," he says.

At the same time, Bostrom points out, "technology could work either way. It might turn out that some of the greatest contributors to inequality—for example, severe disabilities—might be the first and easiest to correct. So you could actually get a decrease in inequality." Yet he acknowledges that "even if people have the same rights under the law there could still be something disturbing about this society where people are sort of destined from their birth not to have any chance of reaching the high levels just because they lack the enhancements. It's a bit like that today, but it could be more so." Bostrom hopes for a leveling up, not a leveling down. He sees as the noblest goal a society where everybody who wants to could become Enhanced.

It's telling that Nietzsche, the philosopher who transformed Bostrom's young life, was the one who created the idea of the *Übermensch*. Literally, it means "overman," but is more commonly translated in comic-book fashion as "superman." The overman, says Nietzsche, is the entity that will follow humans on the evolutionary ladder. "Man is a rope, tied between beast and Overman—a rope over an abyss," Nietzsche writes. "What is great in man is that he is a bridge and not a goal." He might as well have called the *Übermensch* the posthuman. In him Nietzsche sees strength, courage, nobility, style and refinement. He is devoid of human timidity, continually aspiring to greatness and living life as a creative adventure. In Nietzsche's view, we should aim at becoming such an admirable posthuman, even if we who are "all too human" find this ideal so idealistic as to be unfulfillable.

Remember the opening scene in Stanley Kubrick's *2001: A Space Odyssey* in which the protohumans are flinging bones about and one of the bones soars into the sky, only to morph into the spaceship of advanced humankind? Remember the swelling theme music behind it, the five clear and haunting horn notes followed by the kettledrums and the full triumphant rush of the Vienna Philharmonic Orchestra? That's Richard Strauss' *Also Sprach Zarathustra*, embodying the ascension of man into spheres reserved for the gods—a deliberate choice. The movie ends with an astronaut catapulted into mysterious posthumanity.

The hard question is this: What if Kass is right to worry, and yet, as seems likely, our evolution continues? Right now, the argument is usually cast rather fruitlessly between the proponents of the Hell and Heaven Scenarios. One side sees the dangers and wants everything stopped. The other side sees the promise and serves as cheerleaders. They talk past each other.

For The Prevail Scenario to prevail—for us to be the masters of change and not its pawns—we have to recognize the dangers at the same time as we accept that transformation is coming, and figure out how our solutions will accelerate at the same pace as our challenges. Figuring out how to expedite the response of our culture and values also helps us learn what these tests of our humanity are telling us about human nature.

For example:

- Slobodan Milosevic, the former Serbian leader, kills a hundred thousand people and is accused of crimes against humanity. Does his behavior make him a monster to be removed from the protection of the law, to be forever cast outside the bounds of humanity? Or does he remain all too recognizably one of us?

- In 2004, a new technology for the first time allowed women great flexibility in timing motherhood. Women are able to freeze pieces of their ovary when they are young and fertile, enabling them to have full careers, perhaps culminating in becoming CEO of a big corporation, without fretting about the biological clock. They can then thaw these eggs and start having children in their 60s, 70s or 80s, if they wish. Is this human? I ask a neighbor celebrating her 62nd birthday. She doesn't see "retirement age" as a time to start winding down. She sees it as the opening of brand-new chapters. You've got 40 exciting years to go, I tease her. Yes I know, she replies, looking me straight in the eye. You can do anything you want in all that time, she says. Anything at all. That's why I threw out my husband.

• In the early 21st century we have been having an interesting societal conversation about human nature, signaled by the decision of the Vatican to weigh in on the debate. An issue once far on the fringes of public discussion has moved swiftly to center ring in North America and Western Europe. Suppose two people of the same sex want state recognition of their committed union—gay marriage. Is that just another wrinkle in evolution? Or is it an abomination? Is gay marriage an early warning of significant legal and political controversies to come over what constitutes genuinely human behavior? If bonds between people of the same gender are beyond the pale, how will we feel when people start having relationships with their digital companions?

• What happens when your kid comes home crying after failing once again to compete against bigger, faster, stronger, smarter, more talented, cuter, better-behaved kids whose rich parents have given them big tweaks? What will be your gut reaction to their ability to succeed in ways that your kid can't? Do you say, "Don't worry, dear, we love you just the way you are and, besides, just because other parents are willing to take risks with their kids' bodies and minds doesn't mean we have to"? Do you remortgage the house to try to catch up? Or do you try to get the tweaked kids exiled from your school?

• What happens when a girl shows up who is so intent on connecting to you that she wants to show you the contents of her mind on her cuttlefish wings? Do you banish her from the tribe?

This stuff is going to get pretty hairy, pretty quick.
If transcendence is inevitable, how will we manage it?

————ɯɯ————

IF YOU BELIEVE that matter and energy are all that make up the universe, and you think the GRIN technologies will finally enable you to manipulate just about any matter you can imagine, then you have a stirring proposition on your hands. It's sort of a reverse Pascal's Wager.

Blaise Pascal was the brilliant French scientist, mathematician and physicist of the early 1600s who is credited with inventing everything from an amazingly advanced calculator to the roulette wheel and the wristwatch. He helped create the calculus of probabilities and was a pioneer of decision theory. He argued that God cannot be proven logically to be real, yet each of us must decide whether or not to believe. If you choose to believe in God and the deity turns out not to exist, he reasoned, you haven't lost much. If you choose to conduct your life devoutly and God does turn out to exist, the rewards are infinite. Therefore, he concluded, it pays to play the odds.

A reverse Pascal's Wager, then, would be one in which you decide that since it is possible that God does not exist, it might be reasonable, if you have the opportunity, to attempt to do the job yourself.

Some sputter. The very idea of aspiring to godlike powers to them is blasphemous. "Genetic engineering," writes Michael J. Sandel, a professor of political philosophy at Harvard, is "the ultimate expression of our resolve to see ourselves astride the world, the masters of our nature. But the promise of mastery is flawed. It threatens to banish our appreciation of life as a gift, and to leave us with nothing to affirm or behold outside our own will."

"When genetics moves out of the realm of disease and into the study of human traits, especially when an intent to alter traits is implied or openly stated, the discomfort level appropriately rises, and questions about 'playing God' are often raised," writes Ted Peters, professor of systematic theology at the Graduate Theological Union in Berkeley. "Perhaps if we could be confident that humans would play God as God does—with infinite love and compassion—the concern would be lessened."

Yet suppose you accept as a compact definition of human nature, "the inclination to steal fire every chance we get." The odds are good somebody's going to reach for these awesome powers, somewhere in the world, soon. So the big question is where we're going to get the wisdom to use such strength properly. That takes us to the question of whether we can manage our evolution successfully.

Co-evolution—problems prodding humans to solutions that offer new opportunities, which in turn create new challenges, in an endless spiral—has an extremely long and distinguished history. The Little League fastball is a prime example. Only humans are capable of it. Apes can heave things around, but they can't pitch. They don't have the brains.

The first hominids capable of walking erect—usefully freeing up their hands to seek food—couldn't run fast enough to catch many animals. Therefore they were stuck with eating a lot of fruits and nuts unless they stumbled across an animal that had been crippled by a more formidable predator. This limited them to living in tropical areas where such a diet

was feasible. One day, however, a desperately hungry protohuman aston-
ished herself by pegging a rock one-handed at a rabbit or a bird and hitting it on the first try. This was an inflection point in history. Suddenly,
she had an overwhelming advantage. She had become a predator capable
of acting at a distance—the first significant one on the planet, notes
William Calvin, the University of Washington theoretical neurobiologist
and best-selling author on the evolution of the brain. She and her children suddenly had vastly more protein available to them than everybody
else. A population explosion of their offspring ensued. That's what always
happens when you open up a new evolutionary niche.

Nailing a rabbit with a fastball is difficult. It is an extremely brain-
intensive activity. Targeting the prey, figuring out where it will be in a
few seconds, coiling up your body, explosively using every major muscle
group in sequence to pitch and at the last thousandth of a second releasing your finger so precisely that the rock will go where the animal will
be—and doing all this in a few heartbeats—takes a great deal of processing power. We have trouble creating a machine that can do it, as the history of the anti-missile missile program demonstrates.

The only way protohumans managed was by stumbling onto brain
specialization. Evolution rewarded with better eating those whose left
brain concentrated on developing throwing ability. Quickly—as these
things go—the brain size of their talented offspring tripled in size. (The
two halves of apes' brains still don't coordinate quickly. The best apes can
do is a two-handed over-the-head throw if they wish to crack a skull with
a boulder. This is the *only* reason they don't have fat contracts with the
Yankees.) Those who could specialize suddenly discovered they could
come out of the trees and into the plains, living handsomely off the prey
they found there, kicking off the process by which humans ended up oc-
cupying every portion of the globe. As it happens, the areas of the brain
that offer immediate nutritional payoff by allowing this complex behavior
are the ones now colonized by language. Bigger brains enabled hunting
parties to create strategy. It also is the start of cultural evolution—handing
down to the next generation, for example, ways of creating sharp spear
points that work much better than baseball-sized rocks. Flaking hard
rocks also throws sparks, which is a handy way of creating fire on demand. The rest is history.

Is it possible that we might continue to see that sort of co-evolution, in
which the use of new tools enables more specialized and capable human
understanding? Will we in turn fundamentally change some of our prem-

ises about technology as we continue to evolve? Are humans already changing as a result of the Information Age?

That's exactly what we've been doing since the 1940s, says Don Kash.

Don E. Kash is long and lean, with the kind of hearty laugh that sounds like he just heard a joke he shouldn't repeat. He has been studying innovation for more than 40 years. He's developed corporate case studies from Silicon Valley to India. He teaches at the schools of public policy at George Mason University in Fairfax, Virginia, as well as at Tsinghua University in Beijing—"the MIT of China."

"There's always been innovation," Kash says. "But we learned something in World War II that has become permanently embedded in our minds. And that is, you can build organization systems that can do almost anything you can think of, and they will do that without anyone understanding how it's done. No one person knows precisely how the organization accomplished it."

World War II was a hinge in co-evolution history, Kash points out. The war was won with devices that did not exist when the war started. Radar, code-breaking computers and the atomic bomb are conspicuous examples. Its lessons resound to this day.

Kash recalls his case study of Intel. Why is Intel so successful? Well, he recalls being told, it's because we use minimum information. Huh? What do you mean minimum information? Well, we progress by solving problems one at a time. How do we solve problems? Well, we decide on a solution and we try it. And if it works, we go right on. If it doesn't work, "we try sumptin' else." In this fashion, the new is routinely created not by individual geniuses, as mythology might have it, but by faceless teams of ordinary people. Science no longer paves the way for engineering; usually it's the other way around. Intel figures out a way to make wires only a few molecules thick, and why that might work is at best of passing interest—as long as it does. Science can take years if not decades to catch up with an adequate explanation of the device's quantum mechanics. It is the final triumph of Edison over Einstein.

This is a profound change in human circumstances, Kash notes. "You go to executive management classes and they tell you, 'If you don't innovate, you die. The path to success is to make what you're doing obsolete.' My children get up every morning thinking there will be new capabilities. Why do they think that? Because of some Platonic model? Hell, no," he says, pausing dramatically for emphasis. "It's because—That's. What. They. Experience. Every. Day."

The problem is that our ways of thinking about the world have not caught up with these new realities, Kash says. For example, "we have a conceptual model that takes physical finiteness"—shortages—"as its core starting point assumption. Throughout most of human history, this was the experience. Economics is all bound up in this. We now live in a world, however, where our experience is not shortage. Our experience is in fact trying to get people to consume things, because we have too many of them." We still look at the world, however, using eons-old assumptions that haven't held true in generations. We still worry about scant supply when "the great management issue in the world is not scarcity, it's surplus. The main public policy issue across the board is what do you do with surplus capacity? Food is the classic example." If people starve in Africa, it's not because of any lack of food in the world. Absent some climatic meltdown, rich nations produce so much food that they can ship it halfway around the world and still charge so little as to force small local farmers out of business, driving them off their tiny plots and into the sprawling fetid cities. For decades, famines have invariably been caused not by any global lack of food but by the food not getting to the right people—frequently because corrupt regimes withhold it to control their own populations. This is not to say hunger is not an enormous challenge. Rather, the core problem is different from the one we have been trained to consider. The disruptive factor is the opposite of our myths.

In terms of physical, material objects now, anything imaginable is possible. "Our capabilities are now so great that nature is not something you study in order to understand how things work," Kash notes. "Nature is something you study to figure out how to engineer and change it. Including human nature. Galileo got into deep shit because he said there wasn't any difference between the human mind and the divine. There wasn't anything that God could understand that we couldn't understand. Well, I think that's not true. But there isn't anything that God can do that we can't do. That's a different story. It really is sort of dazzling." Yet "what we have is still a set of ideas about how the world operates that we hang on to," Kash says. "And we have a social system that behaves totally different than our ideas.

"If you are dealing with The Singularity, the definition is that none of the rules work anymore," says Kash. "If none of the rules work anymore, then you either say, 'Well, what the hell, just ride it out and see what happens,' or you try to formulate some new rules. You start with a rule that says that in managing society, you have to do it by trial and error. We have a rhetoric now in which politicians and corporate leaders have to talk as if they were in a position to control and manage," despite the abundant evidence that this is not remotely the case. The central rule of society today is run-and-gun adaptation.

The critical issue in co-evolution is time. Can we speed up the societal understanding response such that co-evolution has a chance? It's child's play to focus on the rate of increase in the problems. The big question is whether we are also seeing an accelerated increase in the rate of solutions. Are we seeing a rise in adaptability? In time-to-market work-arounds and muddle-throughs? In empathy? In beauty and love?

In a cynical era, it is easy to dismiss those questions out of hand with a snort and a "Hell, no." Culture and values always change more slowly than technology. But evidence exists that increasingly interconnected billions of individuals are coming up with answers that are "good enough" to deal with their new local realities.

Look at history. The printing press allowed the rise of such coordinated and cooperative action as democracy, science and global trade, notes Howard Rheingold, the fellow who wrote *Smart Mobs*. The ensuing spread of self-government, rationalism and complex webs of enterprise is an example of millions of individuals acting collectively to produce outcomes vastly beyond the power or even understanding of any individual or even any nation. None of these was feasible before there was a means for these individuals easily to link up, share information and act on it through the printed word. Democracy, science and vast markets each presented transformational approaches to problems that once seemed insolubly complex.

The question is whether other institutions of co-evolution are emerging in our generation. Consider, for example, the "gift economy." The Net from first creation was not built by greed. Those who built it openly offered wisdom and information for free. Their legacy is all around us. Look at the results to any question you ask of your search engine. Most of the answers that pop up clearly were put there by people with no hope of immediate financial gain—pictures of the beach on the island of Bequia, hymns in German to antique Harley-Davidsons, the warning signs of frostbite, the botany of mistletoe, short stories by Anton Chekhov. It's endless. Now we have begun to freely share the very guts of our own computers—both the content and the processing power—individual to individual, peer to peer. Music sharing—which already involves more people per year than vote in American presidential elections—is just the beginning. It points toward a gift economy that is altering the basics of the marketplace.

Everybody knows the best kind of birthday present. It's the one that shows how well you understand the other person. "It's the stuff you secretly care about—that honors your spirit," says Lewis Hyde, author of *The Gift*, the modern classic on the subject. "There's a sign of connection in the way the gift is given." It's this creation of a human bond.

What's new is the notion that you can build an economy around gifts.

"What we idealists often like to call a gift economy—the selfless offering of value without expectations of direct returns—is a rather modern concept," says Jim Mason, a cultural anthropologist from Stanford who has conducted extensive fieldwork in Papua New Guinea. "It is an open engagement with a community. The question of how this interacts with a market economy is one we clearly have no answer to."

"It's work-as-gift rather than work-as-commodity," says Richard Barbrook, of the Hypermedia Research Center at the University of Westminster in England. "At no time since the invention of money have gifts ruled like they do now," Hyde says. A gift economy is indeed an economy—you can rationally expect that if you tender a gift, sooner or later you will receive some kind of return. But the return is indirect. The Bible is full of examples. "Cast thy bread upon the waters," the Book of Ecclesiastes commands. Not only do five loaves and two fishes offered freely by Jesus and his disciples feed the multitude, according to the book of Matthew, but the leftovers fill 12 baskets.

The deep structure of the Internet has an undeniably utopian cast. "E-mail is the oddest thing," says Hyde, a 1991 recipient of the MacArthur Foundation "genius" grant. "All the computers and servers and connections—it has a gift economy feel to it. Unlike the telephone, there's not somebody there charging me when I'm using it. And I don't know the anonymous benefactors who have made it work."

"The Net is haunted by the disappointed hopes of the sixties," says Barbrook. From the earliest days, with the battle cry "Information wants to be free," the Internet was seen as a space where people could find ways to collaborate without the need for either governments or markets to mediate social bonds. And indeed, it has worked out that way. To this day, it is notoriously difficult to charge for anything on the Net, because there is so much of value there for free. Look at the endless self-help sites, such as those for widows and widowers. They aren't "run" by anybody. People there cluster spontaneously around their needs and desires. Untold numbers of doctors who specialize in exotic diseases participate as volunteers in support groups. Many are the grief groups that sport their own psychologists. Tim Berners-Lee, the inventor of the Web, gave away his discovery. "One thing that was clear to him was that if it was going to be successful, he would have to give it away," Hyde says. "The irony is that to become a commercially viable medium, it had to begin with a public domain action."

Is this co-evolution? In just the first 2,000 days after the Web was born, the world built some 3 billion Web pages readily available to the public. That's 1.5 million a day, one for every two humans on Earth, and the total is growing exponentially. This is a construction feat that would impress the pharaohs. If some government had tried to order this vast project into existence, what would it have cost? asks Kevin Kelly, author of *New Rules for the New Economy*.

"There is not enough money in the world to do this," Kelly says. "This is an impossible thing we've done. It's a remarkable human achievement of Renaissance proportions in 2,000 days. It's unbelievable. Americans already send 600 billion e-mails a year. This is transformative, the scale and speed with which we have made this. That is the gift economy. It's an act of faith. A holy act. This gift exchange is socializing us to a degree not seen before. The typical person today is engaged in more relationships with more people in more dimensions than ever before. It's amazing given the number of people involved. I'm not so naïve as to think the gift economy is going to replace the market economy. But I do think the gift economy is an essential underpinning of the market economy."

The most influential thinker of the 20th century seeking to unify the truths of science and religion was the French Jesuit scientist Pierre Teilhard de Chardin. In his 1940 magnum opus *The Phenomenon of Man*, Teilhard argued that someday our technology would allow us to create a web of thought and action that would make the world more complex, diverse and alive, moving humankind toward an ultimate evolution. He called it The Omega Point. He propounded the notion that the earth might be one big single living organism, with all the elements of it—from the people to the birds—connected like cells in a body. The goal of evolution, he suggested, is to link up individual human minds, bringing an explosion of intelligence and even global consciousness to this mammoth being. The attention this notion received, especially in the sixties, was of an airy, hand-waving, late-night-dorm-session sort. It was hard for serious people to imagine how such a global consciousness would ever be wired up in any practical way, and even harder to observe any concrete evidence of its existence. With the rise of the World Wide Web, however, some scientists, such as Murray Gell-Mann, winner of the 1969 Nobel prize in physics and a pioneer in the study of complex systems at the Santa Fe Institute, began to think they might be looking at the first evidence that maybe Teilhard was right. "The Internet has accelerated a phenomenon of people finding one another with all sorts of consequences, some wonderful and some terrifying," Gell-Mann said.

"What we're hoping for is a global increase in the collective intelligence of the human species, without which we cannot survive on this planet," says Ralph H. Abraham, one of the progenitors of complex systems theory and a professor of mathematics at the University of California at Santa Cruz.

"The Web is mediating a collective thought process that has feedback effects," says Robert Wright, author of *Nonzero: The Logic of Human Destiny*, which argues that evolution is aimed in a positive direction—perhaps toward a Heaven Scenario. It is reminiscent, Wright says, of Teilhard's idea that technology would connect minds into a "brain for the biosphere as the human species consciously assumes stewardship of the planet." It explains "why serious people take Teilhard seriously," he says.

Is this evidence of co-evolution continuing? Do "the use of new tools enable more specialized and capable human understanding?" asks William Calvin, the neurobiologist. "Will we in turn fundamentally change some of our premises about technology as we continue to evolve? Are humans already changing as a result of the Information Age? Yes, and I suspect computer use in preschool years will eventually create some softwiring that will make for more capable adults in other areas than just mousing and typing."

Look at the open-source movement, which allows thousands of people collaboratively to find solutions to difficult problems over the Net, without any of them claiming ownership of the result. Look at eBay, which brings individuals together to trade in the planet's largest flea market. Look at the way wireless computers allow people to bring cities to life as they sit out on benches or inside coffeehouses, speeding up the pace of response of the metropolis—which has always been what cities have been about. Look at how all of these create new kinds of work. The central question of co-evolution is not what the computer will become but what kind of people we are becoming. Can human understanding about human understanding increase? Can we learn what actually makes teams work? Can we truly understand cognition? Do we have a moral obligation to use enhancement technology to make ourselves beings who are more compassionate, moral and wise? Is it our only chance for survival? The planet comes with an expiration date. If a mutated virus doesn't get us, there's a stray asteroid out there with our planet's name on it, and if we dodge that rock, sooner or later, the sun is scheduled to explode. Is our only way out to continue the march from prehuman to early human to human to transhuman to posthuman, in order to tame the forces of the universe?

The significance of these questions is encapsulated in the words of the scenarist Arie de Geus: "The ability to learn faster than your competition may be the only sustainable competitive advantage."

"Adaptation means you're making quick adjustments until you can formulate some kind of grand theory down the line," Kash says. "I wish that were not the case." He would prefer that leaders expounding high-minded principles rule society. But Kash is forced to recognize that the great doctrines of our old stories very well may no longer remotely fit the facts. So the only available answer is to adjust as necessary. Quickly. Until you can come up with a new and grand story that can hold society together.

HUMANS FIND AN ABSENCE of explanations for how the world works profoundly unsettling. That's why the search for this new grand story becomes important. Yet when you start talking to professionals who are thinking about what this narrative might be like, you find it to be an almost entirely secular group—the subject of God rarely comes up. I am not particularly religious myself, but the American people overwhelmingly are. So it occurred to me to wonder what transcendence might have meant historically to the worldwide range of the devout. You'd think over the last three thousand years or so, we might find a few hints in their work as to how to think about this fix.

Karen Armstrong is a witty, self-deprecating pixie of a woman with a quality British accent. Among the most eminent authors about God and religion writing in English today, she has produced refreshing, lucid and compelling biographies of Muhammad and Buddha and books relating their beliefs to Christianity and Judaism, including her best-selling and fearlessly titled *A History of God*. In 1969, at the age of 24, she left the Roman Catholic convent she had entered as a teenager. She returned to an unrecognizable world—Vietnam, the Beatles, Vatican II, feminism, the sexual revolution. But she did not abandon her search for meaning. She is now resolutely secular—her tailored suits are accented by the subtly cheerful geometric colors of designer Missoni scarves. But she remains fascinated by what the German philosopher of history Karl Jaspers referred to as the Axial Age—a period of unique and fundamental focus on transcendence that is "the beginning of humanity as we now know it," she says. All over the world, humans simultaneously began to wake up to a burning need to grapple with deep and cosmic questions. All the major religious beliefs are rooted in this period. "The search for spiritual breakthrough was no less intense and urgent than the pursuit of technological advance is in our own," she says. "That's quite endorsing, actually. Instead of seeing your own tradition as an idiosyncratic, lonely quest, it becomes part of what human beings do, part of a universal search for meaning and value. This is the kind of scenario that the human mind goes through in its search for ultimate meaning."

"If there is an axis in history, we must find it empirically," Jaspers wrote.

The spiritual process which took place between 800 and 200 B.C. seems to constitute such an axis. It was then that the man with whom we live today came into being. Let us designate this period as the "axial age." Extraordinary events are crowded into this period. In China lived Confucius and Lao Tse, all the trends in Chinese philosophy arose, it was the era of Mo Tse, Chuang Tse and countless others. In India it was the age of the Upanishads and of Buddha; as in China, all philosophical trends, including skepticism and materialism, sophistry and nihilism, were developed. In Iran Zarathustra put forward his challenging conception of the cosmic process as a struggle between good and evil; in Palestine prophets arose: Elijah, Isiah, Jeremiah, Deutero-Isaiah; Greece produced Homer, the philosophers Parmenides, Heraclitus, Plato, the tragic poets, Thucydides, Archimedes. All the vast development of which these names are a mere intimation took place in these few centuries, independently and almost simultaneously in China, India, and the West.

The new element in this age is that man everywhere became aware of being as a whole, of himself and his limits. He experienced the horror of the world and his own helplessness. He raised radical questions, approached the abyss in his drive for liberation and redemption. And in consciously apprehending his limits he set himself the highest aims. He experienced the absolute in the depth of selfhood and in the clarity of transcendence.

Armstrong is fascinated by the human universals operating amid the tumultuous upheaval of that cultural revolution. What caused dispersed civilizations simultaneously to develop these broad, transcendent ideas? There is no human culture that does not incorporate some notion of religion. Even nonbelievers develop systems such as Marxism that sport all the trappings of religion. This evidence causes Armstrong to believe that religion is an essential human need, as unlikely to be outgrown as our need for art. She sees religion as a universal search for meaning and values. She believes it is hardwired.

"Human beings cannot endure emptiness and desolation," Armstrong writes. "They will fill the vacuum by creating a new focus of meaning." Think of the constellations in the night sky. Humans eagerly connect dots and come up with the most elaborate—even poetic—tales, adorning them with heroes and myths, rather than tolerate randomness. The desire to believe goes way back in evolutionary history. "At the start of the 20th century, sociologists said religiosity would decline because of public education and rise of science; instead, it got bigger," notes Michael Shermer, a leading scourge of superstition and bad science who is editor-in-chief of *Skeptic* magazine and author of *Why People Believe Weird Things: Pseudoscience, Superstition, and Other Confusions of Our Time.* "All of this stuff is linked to the desire for there to be Something else with a capital S. A force or a power. It's the basis of mythology—all that Joseph Campbell *Power of Myth* stuff. We love all that. That's why *Star Wars* and the Force were so popular." After all, rationality is hardly a secret. "It's pretty widely publicized," says Shermer. "There's a lot of popular science writing and TV shows. It's not a mystery."

Maybe this tells us something about human nature. That we are pattern-seeking, storytelling animals. If one sees belief as reflecting a hardwired need for meaning and values, then perhaps in the Axial Age we filled the emptiness of our emerging consciousness with the highest aspirations for human nature we could possibly imagine.

This raises the interesting question of whether we are due for a new Axial Age. If our narratives of how the world works are not matching the facts, are we seeking a new era of sense, intelligibility, clarity, continuity and unity? If profound restatements of how the world works arose all over the planet the last time we had a transition on the scale of that from biological evolution to cultural evolution, will it happen again as we move from cultural evolution to technological evolution?

Betty Sue Flowers, of the University of Texas at Austin, notes it's been a while since "the Enlightenment and the Renaissance gave us a sense of coherence. There was a benevolent God that invented the universe, even if it were a clockwork frame. That framework has been up for grabs—it has fallen away. For a long time it didn't bother us. But now we are facing strong questions. Should we indeed ban the cloning of humans? For that you need a larger frame. We do not have that agreed-upon larger frame. This is a spiritual crisis. It's not about science."

Flowers was the editor of *The Power of Myth*, the book by Joseph Campbell with Bill Moyers that accompanied the hugely popular and honored PBS series of the same name. A poet, editor, educator and scenarist, she was the Kelleher Professor of English and director of creative writing at UT before becoming director of the Lyndon Baines Johnson Presidential Library.

Flowers sees coming a cultural revolution that seeks to address today's confusions about how things really do fit together. She says: "What's emerging is an interesting amalgam. It comes from our economic myths of globalization, that everything fits together. And that overlays our environmental work about the way things fit together. Even if it's a remote snail, it has intrinsic value. There is an interconnectedness of things. There is a value somehow in the way things are connected—the web of life. That's the next Enlightenment."

The importance of creating such a commonly held framework, Flowers believes, is "it synchronizes human activity. It distinguishes what 'outside the box' is. It gives you a way to move forward together."

If we're talking about managing transcendence—of coming up with specific ways to Prevail—how would we measure success? How would we know whether or not we were progressing in the program to shape our next humans? When asked what we want for our children, we usually say "happiness." So one of the more obvious places to start might be whether or not we were seeing an increase in happiness.

There are three levels of happiness, Martin E. P. Seligman points out. They involve the pleasant life, the good life, and the meaningful life. Seligman is president of the American Psychological Association, Fox Leadership Professor of Psychology at the University of Pennsylvania, and the author of *Authentic Happiness*. Each of these levels of happiness could be influenced by a transformation of human nature, and each might be a good measure of whether we are managing transcendence effectively.

The pleasant life is the easy one. It consists of having as many positive emotions as you can. "That's the Hollywood view of happiness, the Debbie Reynolds, smiley giggly view of happiness," Seligman says. It's about base pleasures, raw feelings, thrills, orgasms. That one's going to be a snap to enhance—the drugs alone.

More interesting is the good life. It means the fulfillment of potential. That is what Thomas Jefferson had in mind when he talked about the pursuit of happiness. He was picking up on Aristotle's idea of *eudaemonia*, defined as "the exercise of vital powers along lines of excellence in a life affording them scope." Seligman says, "Aristotle talks about the pleasures of contemplation and the pleasures of good conversation. When one is in eudaemonia, time stops. You feel completely at home. Self-consciousness is blocked. You're one with the music. There are six virtues we find endorsed across cultures. They are nonarbitrary—first, a wisdom and knowledge cluster; second, a courage cluster; third, virtues like love and humanity; fourth, a justice cluster; fifth a temperance and moderation cluster; and sixth a spirituality and transcendence cluster. We sent people up to northern Greenland, and down to the Masai, and are involved in a 70-nation study in which we look at the ubiquity of these. Indeed, we're beginning to have the view that those six virtues are just as much a part of human nature as walking on two feet is."

This second form of happiness involves the full exercise of your vital powers. If our vital powers *and* our scope are dramatically, almost unimaginably transformed, it's hard to see how this pursuit of happiness will not be enhanced in a measurable way.

The third form of happiness that is inevitably sought by humans is the pursuit of a meaningful life. "There is one thing we know about meaning," says Seligman, "that meaning consists in attachment to something bigger than you are. The larger the thing that you can credibly attach yourself to, the more meaning you get out of life. Aristotle said the two noblest professions are teaching and politics, and I believe that as well. Raising children, and projecting a positive human future through your children, is a meaningful form of life. Saving the whales is a meaningful form of life. Fighting in Iraq is a meaningful form of life. Being an Arab terrorist is a meaningful form of life. Notice this isn't a distinction between good and evil. That's not part of this. This isn't a theory of everything. This is a theory of meaning, and the theory says, joining and serving in things larger than you that you believe in while using your highest strengths is a recipe for meaning. One of the things people don't like about my theory is that suicide bombers and the firemen who saved lives and lost their lives both had meaningful lives. I would condemn one as evil and the other as good, but not on the grounds of meaning."

It's impossible that there will be a drug for meaning, Seligman says. But if meaning suggests deploying your greatest strengths in the service of something you believe is larger than you are—pursuing the infinite game—that would seem to go to the heart of the measure of The Prevail Scenario: increased human connections. "Religion isn't about believing things," Armstrong says. "It's ethical alchemy. It's about behaving in a way that changes you, that gives you intimations of holiness and sacredness. It doesn't really matter what you believe as long as it leads you to practical compassion. If your belief in a traditional God makes you come out imbued with a desire to feel with your fellow human beings, to make a place for them in your heart, to work to end suffering in the world, then it's good."

Introducing compassion into the equation is at the core of meaning. "Without more kindliness in the world, technological power would mainly serve to increase men's capacity to inflict harm on one another," Bertrand Russell once wrote. Compassion may thus be at the core of successfully managing transcendence—of coming up with a practical way to Prevail over the blind forces of change.

"Evolution moves toward greater complexity, greater elegance, greater knowledge, greater intelligence, greater beauty, greater creativity, and more of other abstract and subtle attributes, such as love," observes Ray Kurzweil. "And God has been called all these things, only without any limitation: infinite knowledge, infinite intelligence, infinite beauty and so on. Of course, even the accelerating growth of evolution never achieves an infinite level, but as it explodes exponentially it moves rapidly in that direction. So evolution moves inexorably toward our conception of God, albeit never quite reaching this ideal. Thus the freeing of our thinking from the severe limitations of its biological form may be regarded as an essential spiritual quest."

"Someday after mastering winds, waves, tides and gravity, we shall harness the energies of love," writes Pierre Teilhard de Chardin. "And then, for the second time in the history of the world, man will discover fire."

———⁓∾⁓———

ONCE UPON A TIME, drifting through the eternal magic lands of New Mexico, Jaron Lanier got to musing about meaning, ceremony and ritual.

He reflected on the time he was speaking at a conference and the subject of the use of embryonic stem cells came up. He recalls a man of the cloth getting up and ripping into the panelists. "Even if it's just some little speck on a petri dish, if it's human, it deserves dignity and you guys are taking away our dignity—you're just a bunch of boys with technological toys. You have no knowledge of life—you are a disgrace," Lanier remembers him saying.

This denunciation started Lanier thinking.

"I turned around," he recalled, "and I said, 'What kind of dignity do we care about? Do we care about dignity that is just granted? Or do we care about dignity that is earned?'

"So I'm Jewish. If there is one thing in life that is not dignified, it's entering adolescence. So we have this thing called a bar mitzvah. The bar mitzvah is a ritual that is kind of a nuisance. It's kind of expensive. It requires a lot of people to participate. And you know what? It creates a little bit of dignity. Not always enough. But it does have a function. It creates some awareness, some community involvement and some responsibility and a little bit of pride, and there you get dignity.

"Dignity is something people have to create. So I said, 'You religious people, instead of sitting on your duffs and watching us and then critiquing, you should be the ones figuring out where the dignity comes from for all this. Why is it that when medicine changed and we started to have operations, that the religious people weren't there with new rituals to try to lend some sense of comfort, dignity, meaning and community connection when somebody has an operation? Why is it still only the same old birth, marriage and death? Why are you guys sitting on the sidelines? So I challenge you. I don't want to be living in a world in 20 years where this is a nonritualistic way to do stem cell research. There is a way to turn this around, to bust through, and that's my third way. My way through these things is, instead of sitting back and assessing, you have to actively create new culture.'" The most important thing is not to leave it to the scientists.

I like Lanier's notion. It resonates with my sense of human nature. As Ray Kurzweil says, "The essence of being human is being creative." If culture and values are ever going to shape our technological evolution, we will need to mark our transcendence at every point, to show that we're treating it seriously and taking responsibility. The nice thing about ceremony and ritual is that while churches can and should get involved—with their vestments and their sanctuaries—it can start from the bottom up. It can start with individuals and small groups—including those who describe themselves as spiritual if not religious—taking ownership of their future and inviting others to stand and witness. At these rituals we can deliberately seek patterns and tell stories—stories that perhaps can begin to contribute to the master narrative of what is happening to us. Our new Axial Age, if it is to come to pass, has got to start somewhere. Might as well be around our campfires.

George Bernard Shaw, who in his time was called the British Nietzsche, was so fascinated by the transcendent that he wrote a play called *Man and Superman*, in which he argues that the reason to never stop attempting to Prevail is to avoid the disease that most dangerously afflicts humankind:

> DON JUAN: . . . My brain is the organ by which nature strives to understand itself. . . .
>
> DEVIL: What is the use of knowing?
>
> DON JUAN: . . . To be able to choose the line of greatest advantage instead of yielding in the direction of least resistance. Does a ship sail to its destination no better than a log drifts nowhither? . . . And there you have our difference: To be in Hell is to drift; to be in Heaven is to steer. . . . At least I shall not be bored.

Perhaps it is with our devotions that we can start choosing to steer. Right now the stories we tell do not match the facts. You can see it in the way we handle our first primitive enhancements—our face-lifts, our Botox injections, our Viagra prescriptions, even our knee replacements and pacemaker implants. We still seem to be a little embarrassed about them, even while the number of procedures soars every year at double-digit rates. Will we forever keep mum about our obviously intense desire to break the bonds of mortality? Or should we lift the taboo and start dealing with it?

Shall we be bashful about these lines we are crossing because we do not have a way to make them meaningful? Or should we start marking these rites of passage as an important part of the future of human nature? Think about what happens when the first-grader whose hand you are holding is old enough to take her SATs. By then there long will have been several means on the market to improve her score by 200 points or more. They no longer seem remarkable. "Those pharmaceuticals she takes? They simply help her express her natural abilities," you say. "Like vitamins. They're no different from the memory pills the boomers gobble up to banish their 'senior moments.' Her attachments and implants? So now she is always connected to Google. Big deal. It's just simpler this way. Without her laptop, her enormous backpack bends her over that much less. She was hell bent and determined to have herself pierced anyway. Might as well have those damn things do something useful, like help her think faster. Hey, maybe these will help her get into Yale. Stranger things have happened. It worked for that babysitter she used to have, and he was thick as a brick. Now if they could just invent a new way to pay the tuition. . . ."

Can we picture devotions marking the great significance of a young person getting her first cognition piercing—awakening her mind directly to the Web of all meaning? What about a rite of maturity in which someone is formally recognized as knowing enough worth keeping that the larger society marks the occasion of his well-deserved first memory upgrade? Should we have a liturgy of life everlasting as a person receives her first cellular age-reversal workup?

These rituals could have important content, important aspects of story. They could say, "Never forget who you were; always respect what you've become. You are a part of us, no matter how far you roam." They could include a formal admonition to use these new powers only for good. They could include the observation that we may be playing for the highest stakes. We cannot detect any other intelligence in the universe. Maybe that's because every other species in the cosmos has flunked this transcendence test horribly, leaving no trace behind. The playwright and former president of the Czech Republic Václav Havel sees "transcendence as the only real alternative to extinction." This is serious. This may be the ultimate final exam.

Will these rituals do any good?

I don't know. Do baptisms, marriages and funerals—sanctifying birth, copulation and death—do any good? My experience says yes. At the very least they are celebrations of transformation where people cross barriers—barriers of class, gender, region, race and religion. They bring us together by officially marking and embracing critical moments. On these occasions, human connections that are rarely achieved elsewhere routinely occur.

If we are embarking on a path in which we stand to transform ourselves more than at any brief period in our species' time on earth, we are creating new critical moments. Perhaps we might start formally marking the occasions.

If we did, inviting those we know from all walks of life and all levels of ability to these ceremonies, it would continue to knit together the fabrics of all the different kinds of human natures to come.

It would be about creating the third happiness, the happiness of being part of something much larger than us.

It would be about continuing to march up the ramp of human connectedness.

That, after all, might just possibly be the ultimate transcendence.

It might be the point of this final exam.

CHAPTER EIGHT

Epilogue

Que sera, sera. Whatever will be, will be.
The future is not ours to see. Que sera, sera.

—Doris Day, 1956

We can only see a short distance ahead, but
we can see plenty there that needs to be done.

—Alan Turing, 1950

ACCORDING TO ancient Washington lore, a flap occurs when somebody inadvertently tells the truth.

In the early 21st century, there were several flaps involving DARPA. Admiral John Poindexter had created no end of trouble, distress and woe by pursuing a plan to mine databases that would enable the government to collect vast information on the citizenry in the name of fighting terrorism. Just when that was dying down, a DARPA plan surfaced that would have mapped information about expected futures. It would have allowed savvy people to make financial bets revealing their anticipations of events in the Middle East, such as prospective assassinations of various countries' leaders. The balloon went up again. For his previous involvement in the Iran-contra scandal of the Reagan administration, Poindexter had always been a lightning rod. But for these flaps, Poindexter finally was canned.

All of this had nothing to do with human enhancement, but the Defense Sciences Office took the fall. Powerful staffers of the Senate Appropriations Committee went looking for a way to rap DARPA's knuckles. (You thought people elected to Congress made our laws? How quaint.) The staffers went down the list of DARPA's projects, found the ones with titles that sounded frighteningly as though they involved the creation of a master race of superhumans, and zeroed out their budgets from the defense appropriations bill. There is scant evidence they knew much, if anything, about these projects. But we will probably never know the details, because significant people are determined that the whole affair be forever shrouded in mystery. The levels of secrecy were remarkable even for DARPA; they were astounding by the standards of the notoriously leaky Senate. Even insiders said it was hard to get a feel for what the facts really were. It took months of reporting and questioning, poking and prodding, even to get a formal "no comment" either from the leadership of the Senate Appropriations Committee or from Anthony J. Tether, the director of DARPA.

A careful study of DARPA's programs a year later, however, showed little change. Considerable creative budgetary maneuvering ensued. The peas of quite a few programs now reside under new, and much better camouflaged, shells. "They're saying, 'Okay, this is the second strike. Do we have to go three strikes?'" one manager said. "It doesn't stop anything. We'll be smarter about how we position things." Meanwhile, he said, new human enhancement programs are in the pipeline, "as bold or bolder" than the ones that preceded them. The slap by the Senate staffers did get DARPA's attention, however. Tether reportedly is "covering his tracks. He wants paper. It's like now he's insisting on documentation. Before we move ahead, he wants somebody in the Army to say, 'If you do this, we will use it.' Which makes it somewhat more difficult. But you can always find somebody with stars who is willing to say, 'Yes, if you can demonstrate that this is safe and effective, we'd use it in a heartbeat.'"

In the course of all this, the four-year hitch of Michael Goldblatt, the head of the Defense Sciences Office, came to its scheduled end. Goldblatt is the fellow whose enthusiasm for spending millions on human enhancement—including the creation of the celebrated telekinetic monkey—is fueled by the plight of his daughter, Gina, who is wheelchair-bound by her cerebral palsy.

The timing of his departure was convenient. "I've given Tony the ammunition," Goldblatt says. "What I suggested he do is to tell the Senate staffers when the next briefing season starts soon, 'We got rid of that crazy Goldblatt. He was out of control, a cowboy, looking for a way to kill us all, pump us full of drugs.'" Referring to the two men succeeding him, Goldblatt says he told Tether to add, "'We've got in his place not only a consummately loyal U.S. Air Force officer, Steven Wax, but a world-famous pediatrician and critical-care specialist who ran one of the biggest and most prestigious hospitals in the country, Brett Giroir, chief medical officer for the Children's Medical Center of Dallas. He will ad-

dress all of the issues of ethics and everything else. He is a man who has
dedicated his life to not only doing no harm but to doing no harm in the
most innocent population—children. He'll be the watchdog and the fil-
ter and et cetera, et cetera.'"

Relaxing in his Virginia living room, I chide Goldblatt for the incredible naïveté he and the Defense Sciences Office displayed in not thinking
its plans to enhance humans would arouse controversy. If it were my
agenda to carve you guys a new one, I told him, I could do it with one
word: *Frankenstein*. Didn't it occur to anybody that you were playing
with fire?

"Only in jest. We are a community that talks to ourselves. It lives in a
world of technically competent and astute people talking to other technically competent and astute people talking about potential and possibility."

He rises and gets a copy of a small poster. On the left is a quote from
The New York Times dated October 9, 1903. It says, "The flying machine
which will really fly might be evolved by the combined and continuous
efforts of mathematicians in from one million to ten million years." On
the right is a quote from Orville Wright's diary, dated October 9, 1903.
"We started assembly today," it says. The caption reads "DSO vs. others."

"The very nature of the people who work in DARPA—they are not
good at political gamesmanship," Goldblatt says. "For the most part, people live in their own worlds. They start the day with the scientific journals, not the newspapers. To define them as nerds would be wrong. They
are very bright, very broad. Maybe *innocent* would be a better word. The
average program manager at DARPA will tell you, 'I'm gonna make the
world a better place based on my knowledge. I'm gonna give people capabilities they never thought they could get. I'm gonna push technology
along.' And when you do that, then you get the Bill Joys of the world saying, 'Do you understand the consequences of this technology?'

"I think for a lot of people, technology is a race. It's a competitive race.
'If I don't do it, somebody else will do it. If I don't do it, the world will
never hear how great my ideas were. I'll never be as famous as I want to
be. I'll never get the recognition I want.' People who want to be champi-
ons or something—at DSO, we've got them in spades."

Despite these reveries, Goldblatt is far more excited about his future
this sultry August afternoon than he is about reviewing his past. The next
Monday he is becoming chief executive officer of a company in the legendary Interstate 270 biotechnology corridor of Montgomery County,
Maryland, not far from the National Institutes of Health. The company is

called Functional Genetics. It has a revolutionary plan to fight disease. He is so energized, in fact, that he jumps up and suggests we go for a walk.

The founder of Functional Genetics is Stanley N. Cohen, he explains as we stroll through Goldblatt's leafy McLean neighborhood. Cohen, in collaboration with Herbert Boyer, is credited with launching the genetic engineering industry in 1973. They showed a way to stably reproduce DNA. This fundamental transformation of molecular medicine—putting the DNA of one organism into the cell of another—is the underpinning of every biotech company in existence. Cohen has received the National Medal of Science, the National Medal of Technology and a raft of other honors.

Functional Genetics has developed a proprietary way of rapidly isolating cells in humans that have traits relevant to many diseases, and of figuring out ways to shut off genes that play a role in the spread of the disease. To complete their infectious life cycle, viruses are wholly dependent on the cell machinery of the individual they attack. For example, there is a protein in human cells that is required for viruses to emerge or "bud" from infected cells. If there is no protein, there is no way for the virus to spread. Nail the protein that the human's body produces, and it would appear you have a way to stop a virus dead in a way the virus could never overcome. A virus like AIDS. Maybe all viruses.

Functional Genetics also knows which protein forms the plaques between nerve cells that is a hallmark of the onset and progression of Alzheimer's. The company also is dissecting the biochemical pathways that mark the life cycle of tumors of the breast, kidney and lung, looking for ways to deny these cancers what they need from the human body.

They call the products they're working on "anti-infectives." What's revolutionary about them is that all efforts to combat pestilence up until now have focused, one way or another, on attacking the disease organism. But "what if you kill the organism by denying it the metabolic machinery of the host?" Goldblatt asks. "You let the organism invade and you just put up a roadblock so it can't hijack the host." What if you could tune up the human, making her invulnerable to AIDS, Alzheimer's, cancer or any other disease you chose?

There are several ways to achieve this. One is to develop drugs that, for example, nail the target protein. But the approach that really excites Goldblatt is the one they're working on for pigs. "Pigs are susceptible to something called African swine fever. The only cure for African swine fever is that you kill the infected pig and a hundred thousand of its nearest neighbors. That's the only cure. It's wiped out many pigs in Portugal, in many parts of Africa. So what we're going to test is whether or not we can create a breed pig that is uninfectible by African swine fever."

And you would do that how?

"Well, knocking down the gene that allows a small protein, which is absolutely essential to the life cycle of the virus."

Now wait a minute. You're talking about genetically engineering an invulnerable pig. Is this somatic or germ-line? Are you talking about fixing pigs one by one? Or are you talking about permanently knocking down a gene in a pig such that none of its offspring would ever have this gene, making them forever invulnerable?

"Well, it could be either," Goldblatt replies.

Isn't that where you find the path to enhanced humans? I ask.

"That's right, it's very similar," Goldblatt replies.

Humans have been attacking disease, using magic or medicine, since the dawn of time. But now, it would appear, there might be a better way. Forget about attacking the pestilence.

How much more elegant it would be, the denizens of Functional Genetics believe, if we could just create better humans.

Who could argue with that?

Acknowledgments

I'm out on the border, I'm walkin' the line
Don't you tell me 'bout your law and order
I'm try'n' to change this water to wine.

— Eagles, "On the Border"

Don't ever forget that
You just may wind up
In my song.

— Jimmy Buffett, "Mañana"

M Y PERSONAL VILLAGE, to which I always seem to stay rooted, no matter how far I stray, continues to be the newsroom of *The Washington Post*. For allowing me to do the newspaper reporting that led up to this volume, I am indebted to Gene Robinson, Deborah Heard, Steve Coll, Len Downie and Don Graham. They did not flinch, much, when I started filing reports of Borgs, nomads, myths, madness, globally conscious flocks, gift economies, Wexelblat Disasters, redheads, social swarms and all the rest. What a remarkable sense of forbearance this tribe displays, allowing me to linger at the edges of their campfires, listening to the tales being told.

All errors of fact, interpretation or emphasis in this book are entirely my own. Experience, alas, demonstrates the unlikelihood that perfection has been achieved. Suggested elaborations from readers for later editions are welcome through www.garreau.com.

Nonetheless:

American journalism's premier discoverer of cultural meaning is my longtime sparring partner, inspiration and friend, the Pulitzer prize–winning Henry Allen. Henry and I are the R2D2 and C3PO of our trade. When I stubbornly head into forbidding terrain, there he is, waving his hands and telling me what a fool's errand I'm on this time. Yet he is always *right* behind me. In this instance, I am even more indebted to Henry than usual, for he was my main editor during the newspapering that led up to this volume and, as is frequently the case with an editor who can read your mind, he was sometimes damn near my un-bylined co-author. Luckily, his cognitive functions are now sufficiently random that he no longer recognizes as his own the entire exquisitely crafted paragraphs that he stuck into my copy.

Linton Weeks and Paul Richard were devoted comrades asking questions for which I did not have answers and challenging assertions I had not thought through.

The Greek chorus throughout much of this, in the shadowy realm of unauthorized local virtual reality, included my colleagues Frank Ahrens, Libby Copeland, Paul Farhi, Marc Fisher, Ann Gerhart, Jerry Knight, Lynn Medford, Linda Perlstein, Ken Ringle, Roxanne Roberts, Sandy Rovner, Kathy Sawyer, John Schwartz, Cassandra Stern, Chris Stern, Desson Thompson, Linton Weeks, and the late but never forgotten Richard Pearson and Bob Williams.

Other colleagues who played a larger role than they may realize, frequently with their skepticism, often with their information, but sometimes just by listening at important times, include Peter Carlson, Mary Hadar, Steve Hunter, Mark Leibovich, Michael Lutzky, Phil McCombs, John Pancake, Steve Reiss, Ian Shapira, Matt Slovick, Hank Stuever, Robert Thomason, Rick Weiss, Mary Lou White and Teresa Wiltz.

At the School of Public Policy at George Mason University, Roger Stough and Kingsley Haynes lured me into being a senior fellow. Crucial to this book were the years of faculty brown-bag lunches. They evolved from a colloquium on the future of universities, to the future of human networks, to the future of cultural revolutions, to the meaning of security, to transcendence. The drop-ins there were special, ranging from Frank Fukuyama to Seymour Martin Lipset to Hal Morowitz to Joshua Epstein of Sugarscape renown. But the real secret was the diversity of the regulars. George Johnson is university president emeritus and a recovering English professor. Don Kash investigates chaos, complexity and innovation. Jack High is an international economist. Mark Addleson's area is organizational learning, but his biggest contributions probably came from his willingness to examine alternative ways of knowing. The best part of that time, however, was working with my teaching partner, mentor and friend, George Cook. Cook has a remarkable ability to clear the air. Unlike the rest of us, he has done practical things like run a large corporation and win local elections.

Sometimes as I stared out into the middle distance, reflecting on my conversations with software pioneers who think of squid as alternative intelligences, or Navy commanders intent on regenerating lost limbs, a question would occur to me: How do I get myself into these situations? It's striking how often the indirect answer was Stewart Brand. One of the most important things that happened to me after the publication of my last book was getting a call from that National Book Award winner and Internet pioneer, inviting me to join the scenario-planning organization known as Global Business Network. This introduced me to "the tribe that lives in the future." It is a collection of remarkable people united by the belief that the best way to anticipate the future is to invent it yourself. I am particularly indebted to GBN's founders—Brand, Napier Collyns, Jay Ogilvy, Peter Schwartz and Lawrence Wilkinson. Danica Remy gently but firmly guided me onto the Internet not only before most people knew it existed but before the term *dot-com* was coined in the press. I would be sorely remiss if I did not mention the companionship and blazingly original thinking of Kim Allen, Nancy Bambic, John Perry Barlow, Steve Barnett, Mary Catherine Bateson, Raimondo Boggia, Nicole Boyer, William Calvin, Andrew Campion, Denise Caruso, Lynn Carruthers, Manuel Castells, Doug Coupland, Don Derosby, Eric Drexler, Esther Dyson, Chris Ertel, Tina Estes, Oliver Freeman, Frank Fukuyama, Katherine Fulton, Graham Galer, J. C. Herz, Danny Hillis, Chuck House, Bill Joy, Adam Kahane, Eamonn Kelly, Kevin Kelly, Art Kleiner, Jaron Lanier, Jaap Leemhuis, Amory Lovins, Thomas Malone, Dan McGrath, the late Don Michael, Brian Mulconrey, Michael Murphy, Nancy Murphy, Richard O'Brien, Laura Panica, Walter Parkes, John L. Petersen, Anu Ponnamma, Paul Saffo, Lee Schipper, Clay Shirky, Alex Singer, Erik Smith, John Stanning, Alex Steffen, Karen Stephenson, Bruce Sterling, Susan Stickley, Hardin Tibbs, Kees van der Heijden, Vernor Vinge, Peter Warshall, Steve Weber and John Wilson. I shall never forget the water fight among the rafts on New Mexico's Rio Chalma starring the regal Pamela McCorduck. Or the time back in 1998 when Viagra was new. I presented a tab to William Gibson, arguably America's greatest living storyteller about the future. He displayed his incisive connection to the zeitgeist by responding: "It does *what*?"

The best strategic plans are always created in hindsight, a DARPA executive once told me. True enough. Yet as a reporter, I find it disconcerting how often a pile of one's own stories resolves itself into a larger pattern only retrospectively, and then only with the help of others. Among those who early on were kind enough to explain to me the overarching themes in my work were Betty Sue Flowers of the University of Texas at Austin, Sherry Turkle of the Massachusetts Institute of Technology and Peter Leyden, former managing editor of *Wired,* now with GBN. In interviews that started in July 2001 and continued through April 2004, Leyden drew out of me the unifying threads in my project. He also exhaustively reviewed the manuscript as it was produced, balanced by the insightful and penetrating Deborah Heard, that consummate professional who also read it chapter by chapter, correctly loathing any writing that did not relate to real people.

The denizens of the Defense Sciences Office of the Defense Advanced Research Projects Agency, especially Michael Goldblatt, took me into their midst after reading my previous work chronicling culture and values, and on my promise that if allowed to understand their operation in depth, I would do my level best to be accurate and fair. I hope they feel I have acquitted myself honorably. This manuscript was not reviewed by the agency in any way. I did not ask for or receive any security clearances. All interviews were on the record.

In addition to the DARPA program managers I talked to in Virginia, I would like to thank the many principal investigators around the country not otherwise named in the text who spent time getting me in the picture, especially Alun Davies of Rinat Neuroscience Corp.; Bob Fitzsimmons of The Technical Basis LLC; Robert S. Full, who studies geckos and other inspirational creatures at the Poly-PEDAL Lab of the University of California at Berkeley; Wayne Jonas of the Samueli Institute; Hami Kazerooni, master of the exoskeleton project at the University of California at Berkeley; and Bill Rojas of MindTel LLC.

At the National Science Foundation, I was aided by William Sims Bainbridge, Robert Eisenstein, Mihail C. Roco, Philip Rubin and most especially Ruzena Bajcsy and Curt Suplee.

In another one of those turns of events that demonstrate there is a deity and She has a sense of humor, Orville Schell, Cynthia Gorney and Brad Inman providentially stepped in to subsidize my California interviews by making me a senior fellow at the Graduate School of Journalism of the University of California at Berkeley during the writing of this book.

During my California stays, I was particularly indebted to Denise Caruso, who has thought deeply about how to make our future turn out right; Jamais Cascio, who has been thinking about posthumanism for years and was particularly incisive in challenging the scenario logics that appear in this volume; Erik Smith, for the conversations and the couch; and Janice Robertson, for the pallet on her floor and her dazzling smile.

John Brockman, my agent, understood this book before most people. Almost as important, he runs Edge, available at www.edge.org Its mission: "To arrive at the edge of the world's knowledge, seek out the most complex and sophisticated minds, put them in a room together, and have them ask each other the questions they are asking themselves." Like GBN, Edge was an invaluable network of networks.

My editor at Doubleday was Roger Scholl. The indexer was Charlee Trantino.

Others who made conspicuous contributions include Gary Anderson, USMC; Philip Anton of RAND; Roger Brent of the Molecular Sciences Institute; Geoff Cohen; Charles DeLisi and Kenneth Lewes, who organized the pioneering conference on the future of human nature at the Frederick S. Pardee Center for the Study of the Longer-Range Future, Boston University; Kenneth M. Ford of the Institute for Human and Machine Cognition at the University of West Florida; Thomas A. Furness III of the Human Interface Technology Lab; Robin Hanson; Jenny H. Holbert, USMC; David Isenberg; Shaun Jones, USN; Michael Marien of Future Survey; Christine Peterson of the Foresight Institute; Jens G. Pohl; Chris Phoenix and Mike Treder of the Center for Responsible Nanotechnology; Howard Rheingold; Anders Sandberg of the Royal Institute of Technology in Stockholm; Bernardo Seissel; Brigadier General Robert E. "Rooster" Schmidle, USMC, who has got to be the most nonlinear thinker ever to achieve that rank; John Sheehan, SJ; Wick Sloane; John Smart of the Institute for Accelerating Change; Steven Spielberg, who was kind enough to include me in the think tank he convened to anticipate the future of human nature portrayed in the film *Minority Report*; Ilkka Tuomi; Caroline Wagner of RAND; the ever clueful David Weinberger; and Barry Wellman.

Olwen Price, Jeannie Tracy and Chris Kelleher transcribed my interviews. Arlene Brothers organized the suggested readings. I particularly admired the note Price, in her properly Welsh fashion, appended to one of my DARPA interviews: "I hope there won't be too much talk about monkeys. I have been a member of antivivisectionist organizations for 30 years."

For the human contact that kept me ticking through the long slog, hunched over my keyboard in the foothills of the Blue Ridge Mountains of Virginia, I am especially indebted to Erika and Rob Payne, Dennis "Gomer" Pyles, and Kim Sieber.

Most important was my family. Adrienne Cook Garreau, the novelist, was my most trusted editor. Simone Chenette Garreau and Evangeline Reed Garreau were a constant inspiration, keeping me informed of the social impact of accelerating change upon the young, and reminding me with their conversations how unmistakably they were my target audience. These women have shaped my life. This book is aimed at them.

Suggested Readings

This is a highly selective, sporadically annotated list of intriguing opportunities for further reading. It is definitely not a collection of prime sources, nor is it aimed at specialists. Some of the works important to them are exotic indeed—not to be inflicted casually on any sentient creature. If your taste runs to the detailed, many more citations can be found in The Notes.

I hope this is a balanced collection. Because of their importance to various debates, I have included many works that I approach with healthy skepticism, and some with which I flat-out disagree. I have, however, spared the reader most of the works that strike me as egregious hokum.

This is also an opportunity for me to bow deeply and from the waist to the giants in so many fields who have influenced my thinking. Many of them contributed interviews to this work. Much got edited out of this book in the service of focus and pace. I bleed nonetheless.

Since the subjects addressed in this book are evolving, I plan to post further information over time at www.garreau.com. I also can be reached through that site, or through The Garreau Group—the network of my best sources—at 6045 Pilgrim's Rest Road, Broad Run, VA 20137 USA, 540-347-1414, team@garreau.com.

The Curve

Beniger, James. *The Control Revolution: Technological and Economic Origins of the Information Society.* Cambridge, MA: Harvard University Press, 1986. ASIN: 0-674-16985-9.

How the ever-increasing pace of the Industrial Age generated the need for information that eventually became the center of our economy.

Christensen, Clayton M. *The Innovator's Dilemma: When New Technologies Cause Great Firms to Fail.* Cambridge, MA: Harvard Business School Press (Management of Innovation and Change Series), 1997. ISBN: 0-875-84585-1.

> Why well-run companies—specifically the well-run ones—are particularly susceptible to being destroyed by upstart outfits hitching their wagon to new and disruptive technologies. The message is to make your own best products obsolete with innovative ones before the competition can.

Drucker, Peter. *Innovation and Entrepreneurship.* New York: Harper & Row, 1985. ISBN: 0-060-15428-4.

> America's most respected management guru preaches that innovation consists of the purposeful and organized search for change and the opportunities such change might offer economically and socially.

"FUTUREdition": http://www.arlingtoninstitute.org/products_services/futuredition.html or sign up for a free subscription at: http://www.arlingtoninstitute.org/futuredition/index.html#SUBSCRIBE

> A free electronic newsletter edited by John L. Petersen of The Arlington Institute that offers a very useful scan of readings on the frontiers of science, technology, media, geopolitics, the environment, and social perspectives.

"Future Survey: A Monthly Abstract of Books, Articles, and Reports Concerning Forecasts, Trends, and Ideas About the Future, a World Future Society Publication," Michael Marien, editor. Monthly newsletter from: World Future Society, 7910 Woodmont Ave., Suite 450, Bethesda, MD 20814, USA. www.wfs.org/fs

> A spectacularly useful overview of all the good new books and articles on topics addressed in this book and many others. A monumental achievement. I don't know how Marien does it.

Gleick, James. *Faster: The Acceleration of Just About Everything.* New York: Pantheon, 1999. ISBN: 0-679-40837-1.

Hanson, Robin. "Economic Growth Given Machine Intelligence." http://hanson.gmu.edu/aigrow.pdf

———. "Is a Singularity Just Around the Corner?" In *Journal of Transhumanism,* April 10, 1998. http://www.transhumanist.com/volume2/singularity.htm

———. Hanson's Web site: http://hanson.gmu.edu/home.html

> Hanson is that unlikely combination, an economist who thinks about The Singularity. As he says, "I am addicted to 'viewquakes,' insights which dramatically change my worldview."

Ilkka, Tuomi. "The Lives and Death of Moore's Law." In *First Monday*,
Vol. 7, Number 11, November 2002. http://www.firstmonday.dk/
issues/issue7_11/tuomi/

————. "Kurzweil, Moore, and Accelerating Change." Joint Research
Centre, Institute for Prospective Technological Studies, working paper,
August 27, 2003. http://www.jrc.es/~tuomiil/articles/Kurzweil.pdf

> Tuomi contends that Moore's Law has been subject to both cultural
> overstatements and bad data. He proposes that processor innovation
> is not supply driven but results from users of information technol-
> ogy being able to innovate new social uses for semiconductors faster
> than engineers have been able to develop improved technology.
> Tuomi sees the potential for stunning productivity increases
> through the intelligent use of technology, but argues that the future
> of semiconductors finally is determined by social innovation.

"Interesting-People," aka "Farber's List." Archives at http://www.inter-
esting-people.org/archives/interesting-people/

> An e-mail list run by David J. Farber, Distinguished Career Professor
> of Computer Science and Public Policy at Carnegie Mellon Univer-
> sity, that pushes to you a vast quantity of whatever Farber thinks is in-
> teresting on any subject on his mind. If you can handle the sheer
> volume, you will find addressed many subjects related to The Curve
> and all sorts of other things. I don't know many of the digerati who
> don't belong to this list. To subscribe, send a note to dave@farber.net
> telling him who you are and why you want to be included.

Kash, Don E. *Perpetual Innovation: The New World of Competition.* New
York: Basic Books, 1989. ISBN: 0-465-05533-8.

> Particularly useful is its account of the World War II origins of the
> most dramatic rise of The Curve.

"KurzweilAI.net": http://www.kurzweilai.net/

> This is an extremely useful and professionally maintained site. Its
> most valuable feature is at the bottom of the splash page where it says
> "Enter your address to subscribe to our news." If you do, every
> weekday morning at around 6:15 A.M. U.S. Eastern, you get a well-
> edited digest of important, useful and balanced reports from highly
> respectable sources, which over time will cause you to be hard-
> pressed to think that The Curve isn't real.

Rogers, Everett M. *Diffusion of Innovations.* New York: Free Press, fourth
edition, 1995. ISBN: 0-02-874074-2.

> The classic 1962 work, including research dating to before World
> War II, on how innovation enters the mainstream.

Toffler, Alvin. *Future Shock*. First published 1970. New York: Bantam Books, 1984. ISBN: 0-553-27737-5.

————. *The Third Wave*. New York: William Morrow, 1980. ISBN: 0-688-03597-3.

Two of those rare books about the future that look better the older they get.

Vinge, Vernor. Address to NASA Vision-21 Symposium, March 30–31, 1993. Downloadable from http://www-rohan.sdsu.edu/faculty/vinge/misc/singularity.html or http://www.frc.ri.cmu.edu/~hpm/book98/com.ch1/vinge.singularity.html

Vinge's original paper on The Singularity.

The GRIN Technologies: Genetic

Davies, Kevin. *Cracking the Genome: Inside the Race to Unlock Human DNA*. Baltimore: Johns Hopkins University Press, 2002. ISBN: 0-801-87140-9.

Dawkins, Richard. *The Selfish Gene*. New York: Oxford University Press, 1976. ISBN: 0-198-57519-X.

————. *The Blind Watchmaker*. New York: W. W. Norton, 1986. ISBN: 0-393-02216-1.

Elliott, Carl. *Better Than Well: American Medicine Meets the American Dream*. New York: Norton, 2003. ISBN: 0-393-05201-X.

Entine, Jon. "The coming of the über-athlete." In Salon.com, March 21, 2002. http://www.salon.com/news/sports/2002/03/21/genes/index_np.html

Disruptive change usually can be most clearly seen wherever there is the greatest competition. In this case, sports. See also Sokolove, below.

Hall, Stephen S. *Merchants of Immortality: Chasing the Dream of Human Life Extension*. Boston: Houghton Mifflin, 2003. ISBN: 0-618-09524-1.

Heinberg, Richard. *Cloning the Buddha: The Moral Impact of Biotechnology*. Wheaton, IL: Quest Books, 1999. ISBN: 0-8356-0772-0.

Judson, Horace Freeland. *The Eighth Day of Creation: Makers of the Revolution in Biology*. Plainview, NY: Cold Spring Harbor Laboratory Press, expanded edition, 1996. ISBN: 0879694785.

Lightman, Alan, ed., et al., *Living with the Genie: Essays on Technology and the Quest for Human Mastery*. Washington: Island Press, 2003. ISBN: 1-55963-419-7.

McGee, Glenn. *Beyond Genetics: Putting the Power of DNA to Work in Your Life*. New York: William Morrow, 2003. ISBN: 0-060-00800-8.

Pollack, Robert. *Signs of Life: The Language and Meanings of DNA*. New York: Mariner Books, reprint edition, 1995. ASIN: 0-395-73530-0.

From a protégé of James Watson.

Ridley, Matt. *Genome: The Autobiography of a Species in 23 Chapters*. New York: HarperCollins, 1999. ISBN: 0-060-19497-9.

———. *The Red Queen: Sex and the Evolution of Human Nature*. New York: Perennial, 2003. ISBN: 0-060-55657-9.

Shreeve, James. *The Genome War: How Craig Venter Tried to Capture the Code of Life and Save the World*. New York: Knopf, 2004. ISBN: 0-375-40629-8.

Silver, Lee M. *Remaking Eden: Cloning and Beyond in a Brave New World*. New York: Avon Books, 1997. ISBN: 0-380-97494-0.

Sokolove, Michael. "In Pursuit of Doped Excellence: The Lab Animal." In *The New York Times Magazine*, January 18, 2004. http://www.nytimes.com/2004/01/18/magazine/18SPORTS.html?pagewanted=print&position=

Stock, Gregory, and John Campbell, eds. *Engineering the Human Germline: An Exploration of the Science and Ethics of Altering the Genes We Pass to Our Children*. New York: Oxford University Press, 2000. ISBN: 0-195-13302-1.

Watson, James. *The Double Helix: A Personal Account of the Discovery of the Structure of DNA*. New York: Atheneum, 1968; New York: Touchstone, 2001. ISBN: 074321630X.

The GRIN Technologies: Robotic

Asimov, Isaac. *I, Robot*. New York: Gnome Press, 1950; New York: Spectra, 1991. ISBN: 0-553-29438-5.

The origin of "The Three Laws of Robotics," which anticipate built creatures capable of moral reasoning.

Brooks, Rodney A. "The Relationship Between Matter and Life." In *Nature*, January 8, 2001, 409:409–11. http://www.ai.mit.edu/projects/lbr/lm/2001/nature.pdf

———. *Flesh and Machines: How Robots Will Change Us*. New York: Pantheon, 2002. ISBN: 0-375-42079-7.

From the director of the MIT Artificial Intelligence Laboratory, who is also chairman and chief technical officer of iRobot, the company that brought to market Roomba—America's first cheap, practical, sweeping and vacuuming robot.

Clark, Andy. *Being There: Putting Brain, Body and World Together Again.* Cambridge, MA: MIT Press, 1997. ISBN: 0-262-03240-6.

———. *Natural-Born Cyborgs: Minds, Technologies, and the Future of Human Intelligence.* New York: Oxford University Press, 2003. ISBN: 0-195-14866-5.

The argument that we have already crossed the line to being a blend of mind and machine. Resistance is futile, as the saying goes.

Cohen, John. *Human Robots in Myth and Science.* London: Allen & Unwin, 1966; New York: A.S. Barnes, 1967. ASIN: B0007OOGSM.

Dennett, Daniel C. *Brainchildren: Essays on Designing Minds (Representation and Mind).* Cambridge, MA: MIT Press, 1998. ISBN: 0-262-04166-9.

———. *Consciousness Explained.* Boston: Little, Brown and Co., 1991. ISBN: 0-316-18065-3.

———. *Kinds of Minds: Toward an Understanding of Consciousness.* New York: Basic Books, 1996. ISBN: 0-465-07350-6.

———. *Darwin's Dangerous Idea: Evolution and the Meanings of Life.* New York: Simon & Schuster, 1995. ISBN: 0-684-80290-2.

Dennett is a provocative, not always successful, but highly ambitious and much-debated thinker about consciousness who argues it arises from means echoed by information technology, suggesting that consciousness could be created.

McCorduck, Pamela. *Machines Who Think: A Personal Inquiry into the History and Prospects of Artificial Intelligence.* Wellesley, MA: AK Peters Ltd., second revised edition, 2004. ISBN: 1-568-81205-1.

Minsky, Marvin. *The Society of Mind.* New York: Simon & Schuster, 1986. ISBN: 0-671-60740-5.

A highly original and controversial view of the nature of intelligence from a godfather of artificial intelligence.

———. "Will Robots Inherit the Earth?" In *Scientific American,* October 1994, 271:86–91. http://web.media.mit.edu/~minsky/papers/sciam.inherit.html

Moravec, Hans. *Mind Children: The Future of Robot and Human Intelligence.* Cambridge, MA: Harvard University Press, 1988. ISBN: 0-674-57616-0.

———. *Robot: Mere Machine to Transcendent Mind.* New York: Oxford University Press, 1999. ISBN: 0-195-11630-5.

———. His Web site for "Mere Machine": http://www.frc.ri.cmu.edu/~hpm/book98/

From the grand old man of robotics.

Penrose, Roger. *The Emperor's New Mind: Concerning Computers, Minds, and the Laws of Physics.* New York: Oxford University Press, 1989. ISBN: 0-198-51973-7; Oxford University Press, 2002. ISBN: 0-192-86198-0.

————. *Shadows of the Mind: A Search for the Missing Science of Consciousness.* New York: Oxford University Press, 1994. ISBN: 0-198-53978-9.

Those interested in creating machines that think inevitably wind up interested in how we might create machines that at least appear conscious—capable of self-awareness and personhood. Penrose's argument is that the problem with trying to create such machines is that in humans, such consciousness is a quantum artifact, difficult and maybe impossible to replicate in our creations. His position is hugely controversial and deeply intriguing.

Searle, John R. *The Mystery of Consciousness.* New York: New York Review of Books, 1997. ISBN: 0-940-32206-4.

————. *The Rediscovery of the Mind.* Cambridge, MA: MIT Press, 1992. ISBN: 0-262-19321-3.

————. "Minds, Brains and Programs." In *Behavioral and Brain Sciences*, 1980, 3(3):417–57. http://members.aol.com/NeoNoetics/Minds-BrainsPrograms.html

————. "The Myth of the Computer." In *New York Review of Books*, April 29, 1982. http://www.nybooks.com/authors/369

Searle is an eloquent and outspoken critic of the notion that all a computer needs to achieve consciousness is sufficiently enormous processing power.

Walter, William Grey. "An Imitation of Life." In *Scientific American*, May 1950, 182(5):42–45.

————. "A Machine That Learns." In *Scientific American*, August 1951, 185(5):60–63.

————. *The Living Brain.* London: Gerald Duckworth, 1953. ASIN: B0006DDBTS.

Wood, Gaby. *Living Dolls: A Magical History of the Quest for Mechanical Life.* London: Faber & Faber, 2002. ISBN: 0-571-17879-0.

The GRIN Technologies: Information

Barlow, John Perry. "A Declaration of the Independence of Cyberspace." http://www.eff.org/~barlow/Declaration-Final.html

Berners-Lee, Timothy. *Weaving the Web.* San Francisco: HarperSanFrancisco, 1999. ISBN: 0-062-51586-1.

Tim Berners-Lee is the sainted creator of the World Wide Web.

Brand, Stewart. *II Cybernetic Frontiers.* New York: Random House, 1974. ISBN: 0-394-49283-8.

————. *The Media Lab: Inventing the Future at MIT.* New York: Viking Penguin, 1987. ISBN: 0-140-09701-5.

Brate, Adam. *Technomanifestos: Visions from the Information Revolutionaries.* New York: Texere, 2002. ISBN: 1-587-99103-9.

You would think such a collection of musty writings could not be engrossing reading, but you would be wrong. The creators of our world, who were not only technologists but humanists, reveal what they thought they were doing at the time.

Brockman, John. *Digerati: Encounters with the Cyber Elite.* San Francisco: HardWired, 1996. ISBN: 1-888-86904-6.

Bush, Vannevar. "As We May Think." In *Endless Horizons,* Washington: Public Affairs Press, 1975. First published in *The Atlantic Monthly,* July 1945. Available online at http://www.theatlantic.com/unbound/flashbks/computer/bushf.htm or http://www.cs.sfu.ca/CC/365/mark/material/notes/Chap1/VBushArticle/

In this landmark work, the wartime Director of the Office of Scientific Research and Development proffers the stunning idea that in the peace, science should seek to amplify man's mind, the way a trip-hammer amplifies his fist.

Castells, Manuel. *The Information Age,* vol. 1, *The Rise of the Network Society.* Malden, MA: Blackwell, 1996. ISBN: 1-557-86616-3.

————. *The Information Age,* vol. 2, *The Power of Identity.* Malden, MA: Blackwell, 1997. ISBN: 1-557-86873-5.

————. *The Information Age,* vol. 3, *End of Millennium.* Malden, MA: Blackwell, 1998. ISBN: 1-557-86871-9.

These three volumes are towering, magisterial works on the intersection of information technology and society.

Gershenfeld, Neil. *When Things Start to Think.* New York: Henry Holt and Company, 1999. ISBN: 0-8050-5874-5.

Gibson, William. *Neuromancer: Remembering Tomorrow.* New York: Ace Books, 1984. ISBN: 0-441-56959-5.

Gibson is arguably his generation's foremost future writer. This, his most revered work, is the one that gave birth to the phrase "cyberspace." I am also fond of *Count Zero, Mona Lisa Overdrive, Virtual*

Light, in which a terrible villain who somehow sports my name makes an exceedingly brief appearance, *Idoru,* and especially *Pattern Recognition.*

Hafner, Katie, and Matthew Lyon. *Where Wizards Stay Up Late: The Origins of the Internet.* New York: Simon & Schuster, 1996. ISBN: 0-684-81201-0.

Hillis, W. Daniel. *Pattern on the Stone.* New York: Perseus Books Group, 1999. ISBN: 0-465-02596-X.

A remarkably engaging and short explanation of computers by the pioneer of the massively parallel ones.

Hiltzik, Michael. *Dealers of Lightning: Xerox PARC and the Dawn of the Computer Age.* New York: HarperBusiness, 1999. ISBN: 0-887-30891-0.

Licklider, J.C.R., and Robert Taylor. "Computer as a Communications Device." Archives, Massachusetts Institute of Technology, Cambridge, MA, 1968.

———. *Libraries of the Future.* Cambridge, MA: MIT Press, 1965. ISBN: 026212016X.

———. "The Truly SAGE System, or, Toward a Man-Machine System for Thinking." NAS-ARDC Special Study, Archives, Massachusetts Institute of Technology, Cambridge, MA, 1957.

———. "Man-Computer Symbiosis." In *IRE Transactions on Human Factors in Electronics,* HFE-1, (March 1960): 4–11. http://memexorg/licklider.html.

———. "Man-Computer Symbiosis: Part of the Oral Report of the 1958 NAS-ARDC Special Study, presented on behalf of the Committee on the Roles of Men in Future Air Force Systems," November 20–21, 1958.

Founding documents of the merger of mind and machine, from the director of command-and-control research for the Pentagon's Advanced Research Projects Agency, DARPA's ancestral organization.

Torvalds, Linus, and David Diamond. *Just for Fun: The Story of an Accidental Revolutionary.* New York: HarperBusiness, 2001. ISBN: 0-066-62072-4.

From the inventor of the Linux operating system and the open-source movement, the latter of which may indeed be revolutionary if it changes how we create biology.

Turing, Alan M. "Computing Machinery and Intelligence." In *Mind,* October 1950, 59:433–60. The article also appeared in G.F. Luger, ed., *Computation and Intelligence: Collected Readings.* Cambridge, MA: MIT

Press, 1995. Also available at: http://www.loebner.net/Prizef/Turing Article.html

> Can machines think? The origins of "The Turing Test" to determine if a machine is displaying intelligence.

Ullman, Ellen. *Close to the Machine: Technophilia and Its Discontents.* San Francisco: City Lights Books, 1997. ISBN: 0-872-86332-8.

Waldrop, M. Mitchell. *The Dream Machine: J.C.R. Licklider and the Revolution That Made Computing Personal.* New York: Viking, 2001. ISBN: 0-670-89976-3.

> My favorite history of how we got to the point where we are surrounded by personal information devices. The path was abundantly non-obvious.

The GRIN Technologies: Nano

Center for Responsible Nanotechnology newsletter: http://responsible-nanotechnology.org/contact.htm

> Good free newsletter from the Eric Drexler wing of the church.

Drexler, K. Eric. *Engines of Creation: The Coming Era of Nanotechnology.* New York: Anchor, 1986. ISBN: 0-385-19972-4.

————, and Chris Peterson with Gayle Pergamit. *Unbounding the Future: The Nanotechnology Revolution.* New York: William Morrow, 1991. ISBN: 0-688-09124-5.

> Eric Drexler is a founding visionary of nanotechnology. *Engines of Creation* was his seminal work.

"E-drexler.com." http://e-drexler.com/

Feynman, Richard P. "There's Plenty of Room at the Bottom: An Invitation to Enter a New Field of Physics." In The California Institute of Technology's *Engineering and Science*, February 1960. http://www.zyvex.com/nanotech/feynman.html

> The paper that began it all.

Foresight Institute: http://www.foresight.org

> The Foresight Institute, home of the thoughtful and serious Eric Drexler and Christine Peterson, is attempting to prepare society for the overwhelming effects of the "strong" nanotechnology—involving molecular assemblers—that they believe is on the horizon.

Freitas, Robert A., Jr. *Nanomedicine, Volume I: Basic Capabilities.* Austin, TX: Landes Bioscience, 1999. ISBN: 1-570-59645-X.

————. *Nanomedicine, Vol. IIA: Biocompatibility*. Austin, TX: Landes Bio-science, 2003. ISBN: 1-570-59700-6.

————. Nanomedicine Web site, http://www.foresight.org/Nanomedicine/index.html

Interesting and very early work at the intersection of biology, nanotechnology and robotics.

Merkle, Ralph. "Nanotechnology: What Will It Mean?" In IEEE Spectrum Online, September 5, 2004. http://www.spectrum.ieee.org/WEBONLY/resource/speakm.html

————. Merkle's home page: http://www.merkle.com/

The co-inventor of public key cryptography is a pioneering thinker about and cheerleader for nanotechnology. He won the 1998 Feynman Prize in Nanotechnology for theory, and is a director of Alcor, the cryonics firm that freezes dead people in the belief that someday, when technology has moved sufficiently far along, they can be reanimated and cured of whatever killed them.

Mulhall, Douglas. *Our Molecular Future: How Nanotechnology, Robotics, Genetics and Artificial Intelligence Will Transform Our World*. Amherst, NY: Prometheus Books, 2002. ISBN: 1-573-92992-1.

Mulhall refers to these as the GRAIN technologies.

"National Nanotechnology Initiative": http://www.nano.gov/

The home page of the outfit spending your taxpayer dollars to improve "human health, economic well being and national security" through nanotechnology.

Roco, Mihail C., and William Sims Bainbridge. *Societal Implications of Nanoscience and Nanotechnology: NSET Workshop Report*. Arlington, VA: National Science Foundation; National Science and Technology Council (NSTC), Subcommittee on Nanoscale Science, Engineering, and Technology (NSET), March, 2001; and New York: Kluwer Academic Publishers, 2001. ISBN: 0-792-37178-X. http://www.wtec.org/loyola/nano/societalimpact/nanosi.pdf

"Small Times: Big News in Small Tech": http://www.smalltimes.com/

"The Smalley Group at Rice University": http://smalley.rice.edu/

Stephenson, Neal. *The Diamond Age, or, A Young Lady's Illustrated Primer*. New York: Bantam, 1995. ISBN: 0-553-09609-5.

Stephenson is one of his generation's premier writers of future fiction. This work evokes a little girl growing up in a society shaped by nano and other GRIN technologies. I find most of Stephenson's stuff fascinating (*Snow Crash* and *Cryptonomicon*, e.g.), but this book probably is the one most accessible to a general readership.

The Three Human Evolutions: Biological

Bowler, Peter J. *Evolution: The History of an Idea*. Berkeley: University of California Press, third revised edition, 2003. ISBN: 0-520-23693-9.

Darwin, Charles. *The Works of Charles Darwin: The Origin of Species*. New York: New York University Press, sixth edition, 1988. ISBN: 0-814-71805-1.

Dyson, Freeman J. *Origins of Life*. New York: Cambridge University Press, 1985. ISBN: 0-521-30949-2.

Fortey, Richard. *Life: A Natural History of the First Four Billion Years of Life on Earth*. New York: Knopf, 1998. ISBN: 0-375-40119-9.

Jerison, Harry J. *Evolution of the Brain and Intelligence*. New York: Academic Press, 1973. ISBN: 0-123-85250-1.

Kauffman, Stuart A. *The Origins of Order: Self-Organization and Selection in Evolution*. New York: Oxford University Press, 1993. ISBN: 0-195-05811-9.

————. *At Home in the Universe: The Search for Laws of Self-Organization and Complexity*. New York, Oxford University Press, 1995. ISBN: 0-195-09599-5.

>   The spontaneous emergence of that most spectacular self-organization, life itself, through the lens of complexity and chaos theory, by one of its pioneers.

Leakey, Richard. *The Origin of Humankind*. New York: Basic Books, 1994. ISBN: 0-465-03135-8.

Smith, John Maynard, and Eors Szathmary. *The Origins of Life: From the Birth of Life to the Origin of Language*. New York: Oxford University Press, new edition, 2000. ISBN: 0-192-86209-X.

Wenke, Robert J. *Patterns in Prehistory: Mankind's First Three Million Years*. New York: Oxford University Press, 1980. ISBN: 0-195-02556-3.

Wills, Christopher. *Children of Prometheus: The Accelerating Pace of Human Evolution*. New York: Perseus Books, 1998. ISBN: 0-738-20003-4.

The Three Human Evolutions: Cultural

Copernicus, Nicolaus. *On the Revolutions of Heavenly Spheres*. First privately circulated in outline form in 1514 and published in 1543.

Amherst, NY: Prometheus Books (Great Minds Series), 1995. ISBN: 1-573-92035-5.

Descartes, René. *Six Metaphysical Meditations;Wherein it is Proved that there is a God. And That Mans Mind is really distinct from his Body.* London: Benjamin Tooke, 1680. New York: Cambridge University Press, revised edition, 1996. ISBN: 0-521-55818-2.

"I think, therefore I am." The rise of Cartesian logic is the point at which many historians mark the hard turn Western civilization made toward the technological and rationalist world in which we live today.

Diamond, Jared. *The Third Chimpanzee: The Evolution and Future of the Human Animal.* New York: HarperCollins, 1992. ISBN: 0-060-18307-1.
———. *Guns, Germs, and Steel: The Fates of Human Societies.* New York: Norton, 1997. ISBN: 0-393-03891-2.

The Pulitzer Prize–winning, best-selling modern author on cultural evolution.

Eliade, Mircea, Willard J. Trask, trans. *The Myth of the Eternal Return: Or, Cosmos and History.* New York: Pantheon Books, 1954. Princeton: Princeton University Press, reprint edition, 1971. ISBN: 0-691-01777-8.

Galilei, Galileo. *Dialogue Concerning the Two Chief World Systems.* New York: Modern Library, 2001. ISBN: 0-375-75766-X.

Published in Florence in 1632, this work—demonstrating the truth of the Copernican system in which the earth revolves around the sun—was the most proximate cause of Galileo's trial before the Inquisition. Its influence is incalculable. With a foreword by Albert Einstein.

Giedion, Siegfried. *Mechanization Takes Command: A Contribution to Anonymous History.* New York: Oxford University Press, 1948. New York: Norton, 1969. ASIN: 0-393-00489-9.

How industrialization split our modes of thinking from our modes of feeling, with ideas about how to bridge that gap.

Hobbes, Thomas. *Leviathan; or, The Matter, Forme and Power of a Commonwealth Ecclesiasticall and Civill.* London: Andrew Crooke, 1651. Cambridge, England: Cambridge University Press, student edition, 1996. ISBN: 0-521-56797-1.

At a "time wherein men live without other security than what their own strength and their own invention shall furnish them withal," there is "no arts; no letters; no society; and which is worst of all, continual fear, and danger of violent death; and the life of man, solitary, poor, nasty, brutish, and short."

Kuhn, Thomas S. *The Copernican Revolution: Planetary Astronomy in the Development of Western Thought.* Cambridge, MA: Harvard University Press, 1957. ISBN: 0-674-17103-9.

————. *The Structure of Scientific Revolutions.* Chicago: Chicago University Press, 1970, third edition, 1996. ISBN: 0-226-45808-3.

Kuhn first used the word "paradigm" to suggest that any researchers' then-current worldviews, institutions, and beliefs will shape any body of research. Hence, these paradigms, or worldview structures, are subject to sharp, discontinuous transformations as new structures are found that describe reality better. These breaks, or "paradigm shifts," are commonly referred to as scientific or cultural revolutions. The influence of Kuhn's work has been immense. The acceptance of plate tectonics in the 1960s, for instance, was sped by geologists' reluctance to be on the wrong side of a paradigm shift.

Nisbet, Robert A. *Social Change and History: Aspects of the Western Theory of Development.* New York: Oxford University Press, 1969; reprint edition, 1992. ASIN: 0-195-00042-0.

Petroski, Henry. *The Evolution of Useful Things: How Everyday Artifacts— From Forks and Pins to Paper Clips and Zippers—Came to Be as They Are.* New York: Knopf, 1992. ISBN: 0-679-41226-3.

The process of invention as an artifact of cultural evolution.

Plato. *The Republic.* Mineola, NY: Dover Publications, 2000. ISBN: 0-486-41121-4.

What is justice?

Poundstone, William. *Prisoner's Dilemma: John Von Neumann, Game Theory and the Puzzle of the Bomb.* New York: Doubleday, 1992. ISBN: 0-385-41567-2.

The development of game theory, which arguably allowed us to survive the Cold War, is one of the most important pieces of cultural co-evolution of the twentieth century.

Wilson, Edward O. *Consilience: The Unity of Knowledge.* New York: Knopf, 1998. ISBN: 0-679-45077-7.

The Three Human Evolutions: Engineered

Bostrom, Nick. "How Long Before Superintelligence?" First published

1997, revised 1998 and 2000. Department of Philosophy, Logic and Scientific Method, London School of Economics. http://www.nickbostrom.com/superintelligence.html

Dyson, Freeman J. *The Sun, the Genome, and the Internet: Tools of Scientific Revolution*. New York: Oxford University Press, 1999. ISBN: 0-195-12942-3.

Dyson, George B. *Darwin Among the Machines: The Evolution of Global Intelligence*. Reading, MA: Addison-Wesley, 1997. ISBN: 0-201-40649-7.

Engelbart, Douglas C. *Augmenting Human Intellect: A Conceptual Framework*. Summary Report AFOSR-3223 under Contract AF 49(638)-1024, SRI Project 3578 for Air Force Office of Scientific Research, Stanford Research Institute, Menlo Park, CA, 1962. http://sloan.stanford.edu/mousesite/EngelbartPapers/B5_F18_ConceptFrameworkInd.html
The manifesto of a new discipline, by its founder.

Gelernter, David. *Mirror Worlds: Or the Day Software Puts the Universe in a Shoebox—How It Will Happen and What It Will Mean*. New York: Oxford University Press, 1991. ISBN: 0-195-06812-2.

———. *The Muse in the Machine: Computerizing the Poetry of Human Thought*. New York: Free Press, 1994. ISBN: 0-029-11602-3.

Hayles, N. Katherine. *How We Became Posthuman: Virtual Bodies in Cybernetics, Literature, and Informatics*. Chicago: University of Chicago Press, 1999. ISBN: 0226321460.

Hillis, W. Daniel. "Intelligence as an Emergent Behavior; or, The Songs of Eden." In *Daedalus*, Winter 1988. Proceedings of the American Academy of Arts and Sciences 117, no.1. Available at http://www.kurzweilai.net/articles/art0463.html?printable=1
The pioneer of massively parallel computers, thinking about the songs of apes, speculates on how we might learn to build an intelligence.

Kelly, Kevin. *Out of Control: The Rise of Neo-Biological Civilization*. Reading, MA: Addison-Wesley, 1994. ISBN: 0-201-57793-3.

McLuhan, Marshall. *The Gutenberg Galaxy: The Making of the Typographic Man*. Toronto: University of Toronto Press, 1962.

———. *Understanding Media: The Extensions of Man*. New York: McGraw-Hill, 1964.

———, and Quentin Fiore. *The Medium Is the Massage*. New York: Random House, 1967.

Mitchell, William J. *Me++: The Cyborg Self and the Networked City*. Cambridge, MA: MIT Press, 2003. ISBN: 0-262-13434-9.

Murphy, Michael. *The Future of the Body: Explorations into the Future Evolu-

tion of Human Nature. Los Angeles: J.P. Tarcher, 1992. ISBN: 0874777305, paperback edition.

Michael Murphy is the cofounder of California's Esalen Institute, a New Age guru, and a charming and fascinating dinner companion. This is his massive tome on "the transformative capacities of human nature," focusing on saints, psychics, mystics, geniuses, artists, and the like. I did not find his views about what constitutes evidence useful to my purposes, but your mileage may vary.

The New Atlantis: A Journal of Technology and Society. Washington: Ethics and Public Policy Center. http://www.thenewatlantis.com

My favorite journal at the intersection of ethics, politics and technology.

Rheingold, Howard. *Virtual Reality*. New York: Summit Books, 1991. ISBN: 0-671-69363-8.

————. *The Virtual Community: Homesteading on the Electronic Frontier*. Reading, MA: Addison-Wesley, 1993. ISBN: 0-201-60870-7.

————. *Smart Mobs: The Next Social Revolution*. New York: Perseus Books Group, 2002. ISBN: 0-738-20608-3.

From a pioneering analyst of the impact of information technology on society.

Roberts, Chalmers M. "The Decision of a Lifetime: In His Twilight, Facing the End on His Terms." In *The Washington Post*, August 28, 2004, page A1. http://www.washingtonpost.com/wp-dyn/articles/A40467-2004Aug27.html

A stunningly lucid and personal discussion of a decision many of us may soon be facing at an even more advanced level—whether or not to pass up Enhancement and embrace death.

Roco, Mihail C., and William Sims Bainbridge. *Converging Technologies for Improving Human Performance: Nanotechnology, Biotechnology, Information Technology and Cognitive Science*. Arlington, VA: A National Science Foundation/Department of Commerce sponsored report, June 2002. Kluwer Academic Publishers, 2003. ISBN: 1-402-01254-3.

An amazing work from government agencies. http://www.wtec.org/ConvergingTechnologies/Report/NBIC_frontmatter.pdf

Rose, Michael R. *Evolutionary Biology of Aging*. New York: Oxford University Press, 1991. ISBN: 0-195-06133-0.

Sterling, Bruce. *Holy Fire*. New York: Bantam, 1996. ISBN: 0-553-09958-2.

Sterling is another of his generation's prime writers of future fiction.

This book looks at a society in which an aging woman can choose to be forever young. I find much of Sterling's work rewarding—check out the rollicking *Zeitgeist* or *Heavy Weather,* for example. But this is arguably the book most congenial to a general audience.

Stock, Gregory. *Redesigning Humans: Our Inevitable Genetic Future.* Boston: Houghton Mifflin, 2002. ISBN: 0-618-06026-X.

Turkle, Sherry. *The Second Self: Computers and the Human Spirit.* New York: Simon & Schuster, 1984. ISBN: 0-671-46848-0.

————. *Life on the Screen: Identity in the Age of the Internet.* New York: Simon & Schuster, 1995. ISBN: 0-684-80353-4.

Not about computers, but about people and how information technology is causing us to reevaluate our identities, engaging in new ways of thinking about evolution, relationships, politics, sex and the self.

Von Neumann, John. *The Computer and the Brain.* New Haven, CT: Yale University Press, 1958, 2000. ISBN: 0-300-08473-0.

One of the towering geniuses of the computer age looks at how the brain is, and is not, digital.

Weeks, Linton. "Putting God on Notice: Ready or Not, We're Taking Control of Our Evolution. Gulp." In *The Washington Post,* page F1, February 9, 2003. http://www.washingtonpost.com/ac2/wp-dyn/A39931-2003Feb7?language=printer

Wiener, Norbert. *The Human Use of Human Beings: Cybernetics and Society.* Boston: Houghton Mifflin, 1950. Cambridge: Da Capo Press, 1988. ISBN: 0-306-80320-8.

————. *God and Golem, Inc.: A Comment on Certain Points where Cybernetics Impinges on Religion.* Cambridge: The MIT Press, 1966. ISBN: 0-262-73011-1.

The inventor of cybernetics thinks deeply on the implications of what he hath wrought.

Wolfe, Tom. "Sorry, but Your Soul Just Died." In *Hooking Up.* New York: Farrar, Straus and Giroux, 2000. ISBN: 0374103828. Also: http:// www.brainmachines.com/body_wolf.html

The Heaven Scenario

Bacon, Sir Francis. *New Atlantis.* First published in 1627. http://oregonstate.edu/instruct/phl302/texts/bacon/atlantis.html. Also collected in:

Ideal Commonwealths: Comprising, More's Utopia, Bacon's New Atlantis, Campanella's City of the Sun and Harrinton's Oceana. Sawtry, England: Dedalus, Ltd., 1989. ISBN: 0-946-62626-X.

Bloom, Howard. *The Global Brain: The Evolution of Mass Mind from the Big Bang to the 21st Century.* New York: John Wiley & Sons, Inc., 2000. ISBN: 0-471-29584-1.

Gilder, George. *Microcosm: The Quantum Revolution in Economics and Technology.* New York: Simon & Schuster, 1989. ISBN: 0-671-50969-1.

Haldane, J. B. S. *Daedalus; or, Science and the Future.* A paper given to the Heretics Society in Cambridge in 1923. First published in London in 1924 by Kegan Paul, Trench, Trubner and Co. Out of print, but transcribed text available. http://home.att.net/~p.caimi/Daedalus.PDF

A remarkably prescient and optimistic view.

Kurzweil, Ray. *The Age of Intelligent Machines.* Cambridge, MA: MIT Press, 1990. ISBN: 0-262-11121-7.

————. *The Age of Spiritual Machines: When Computers Exceed Human Intelligence.* New York: Viking, 1999. ISBN: 0-670-88217-8.

————. "Exponential Growth an Illusion?: Response to Ilkka Tuomi." KurzweilAI.net, September 23, 2003. http://www.kurzweilai.net/meme/frame.html?main=/articles/art0593.html

————. The Web site of Kurzweil Technologies, his company: http://www.kurzweiltech.com

More, Sir Thomas. *Utopia.* First published in Latin in 1516. Mineola, NY: Dover Publications, 1997. ISBN: 0-486-29583-4.

Nisbet, Robert A. *History of the Idea of Progress.* New York: Basic Books, 1980. ASIN: 0-465-03025-4.

A work for the ages. Interesting companion reading to "History of the Idea of Decline," cited below.

Postrel, Virginia. *The Future and Its Enemies: The Growing Conflict Over Creativity, Enterprise, and Progress.* New York: Free Press, 1999. ISBN: 0-684-86269-7.

Postrel argues that the great political divide is not between liberals and conservatives, but between "statists," who seek to command and control change, and "dynamists," who do not fear spontaneous evolution.

Rauch, Jonathan. "Will Frankenfood Save the Planet?" In *The Atlantic Monthly,* October 2003. http://www.theatlantic.com/issues/2003/10/rauch.htm

Richards, Jay W., ed. *Are We Spiritual Machines? Ray Kurzweil vs. The Critics of Strong AI.* Seattle: Discovery Institute, 2002.

A debate between Kurzweil and some of his stronger opponents.

The Hell Scenario

Berry, Wendell. *Life Is a Miracle: An Essay Against Modern Superstition.* New York: Counterpoint Press, 2001. ISBN: 1-582-43141-8.

> A book that Bill Joy says has influenced him, by a poet, novelist and farmer who stubbornly insists there is more to reality than science can explain. "At the very end of Wendell Berry's book 'Life Is a Miracle,'" Joy says, "there is this wonderful passage where he talks about science and all these things we're doing. These are not the world. I'm paraphrasing when he says something like—these are tools to make our habitation on earth more comfortable, but it's not about them and we mistake them for the world. We mistake the goal of progress for life in the world."

Bostrom, Nick. "Existential Risks: Analyzing Human Extinction Scenarios and Related Hazards." In the *Journal of Evolution and Technology*, Vol. 9, March 9, 2002; first version: 2001. http://www.nickbostrom.com/existential/risks.pdf

Butler, Samuel. *Erewhon; or, Over the Range.* London: Trübner & Co., 1872; new and rev. ed., London: A.C. Fifield, 1913; Classic Publishers, 1923. ISBN: 1-582-01002-1.

> Originally published privately by its New Zealand author, this is one of the first dystopian novels.

Crichton, Michael. *Prey: A Novel.* New York: HarperCollins, 2002. ISBN: 0-066-21412-2.

> A rip-snorting romp through a day-after-tomorrow future by the author of *Jurassic Park.* Guaranteed to scare the molecules out of you. Criticized for its inaccuracies, especially about nanotechnology, this fiction is nonetheless unusual in that the author in his introduction lays out the nonfiction technological underpinnings of his plot, only heightening the verisimilitude of his drama.

Fukuyama, Francis. *Our Posthuman Future: Consequences of the Biotechnology Revolution.* New York: Farrar, Straus and Giroux, 2002. ISBN: 0-374-23643-7.

> The work that brought the idea of "posthumanism" into the mainstream. Fukuyama hates and fears its prospect. An exemplary and thought-provoking piece.

Goethe, Johann Wolfgang von. *Faust I & II.* (*Goethe: The Collected Works*, Vol. 2) Princeton: Princeton University Press, reprint edition, 1994. ISBN: 0-691-03656-X.

Greenfield, Susan. *Tomorrow's People: How 21st Century Technology Is Changing the Way We Think and Feel.* New York: Penguin Books, 2003. ISBN: 0-713-99631-5.

Herman, Arthur. *The Idea of Decline in Western History.* New York: Free Press, 1997. ISBN: 0-684-82791-3.

A useful, if somewhat ideological, look at why Hell scenarios are so enduringly popular.

Huxley, Aldous. *Brave New World.* Garden City, NY: Doubleday, Doran & Company, Inc., 1932.

Joy, Bill. "Why the Future Doesn't Need Us." In *Wired,* April 2000, 8.04. www.wired.com/wired/archive/8.04/joy.html

Kaczynski, Theodore. *The Unabomber Manifesto: Industrial Society and Its Future.* Jolly Roger Press, 1995. ASIN: 0-963-42052-6.

He's a homicidal maniac, but as a writer and analyst, he didn't totally squander his Harvard education. Worth looking at.

Kass, Leon. *Toward a More Natural Science: Biology and Human Affairs.* New York: Free Press, 1985. ISBN: 0-029-18340-5.

———. "The Moral Meaning of Genetic Technology." In *Commentary,* September 1999. http://www.commentarymagazine.com/Summaries/V108I2P34-1.htm

———. "Preventing a Brave New World: Why We Should Ban Cloning Now." In *The New Republic,* May 21, 2001. http://www.tnr.com/052101/kass052101_print.html

———. *Beyond Therapy: Biotechnology and the Pursuit of Happiness.* The President's Council on Bioethics. New York: ReganBooks, 2003. ISBN: 0-060-73490-6. http://www.bioethics.gov/reports/beyondtherapy

Not too many presidential commissions post on their Web sites a "bookshelf" of recommended readings on fate, suffering and dignity, with literate introductions to selected writings by Ovid, Plutarch, Shakespeare, Swift and Tolstoy. But then, not too many presidential commissions are run by Leon Kass, the conservative moral philosopher. It's a fascinating experiment Kass is running in his President's Council on Bioethics. He is trying to formulate White House policy on things like stem cells, basing doctrine not only on ideas of morality but, more specifically, on eighteenth-century concepts like "natural law," illustrating it with literary antecedents. This is very Jeffersonian. You have to give the program points for style, even if you agree with opponents who believe this endeavor is going to destroy America's economy by giving away its high-tech edge to Asian researchers. Highly readable and thought-provoking.

Kurzweil, Ray. "Promise and Peril of the 21st Century." In *CIO* magazine, Fall/Winter 2003. http://www.cio.com/archive/092203/kurzweil.html
A thoughtful and useful response to McKibben and Joy's Hell scenarios.

Lewis, C.S. *The Abolition of Man*. New York: Simon & Schuster, Touchstone edition, 1996. ISBN: 0-805-42047-9.

Mander, Jerry. *In the Absence of the Sacred: The Failure of Technology and the Survival of the Indian Nations*. San Francisco: Sierra Club Books, 1991. ISBN: 0-871-56739-3.

McKibben, Bill. *Enough: Staying Human in an Engineered Age*. New York: Times Books, Henry Holt and Company, 2003. ISBN: 0-805-07096-6.

Orwell, George. *Nineteen Eighty-Four: A Novel*. New York: Harcourt, Brace, 1949. Plume Books, Centennial edition, 2003, with an introduction by Thomas Pynchon. ISBN: 0-452-28423-6.

Posner, Richard A. *Catastrophe: Risk and Response*. New York: Oxford University Press, 2004. ISBN: 0195178130.

Rees, Martin. *Our Final Hour: A Scientist's Warning: How Terror, Error, and Environmental Disaster Threaten Humankind's Future in this Century on Earth and Beyond*. New York: Basic Books, 2003. ISBN: 0-465-06862-6. Published in Britain as *Our Final Century: The 50/50 Threat to Humanity's Survival*. London: Heinemann, 2003.

Rifkin, Jeremy. *The Biotech Century: Harnessing the Gene and Remaking the World*. New York: Jeremy P. Tarcher/Putnam, 1998. ISBN: 0-874-77909-X.

Russell, Bertrand. *Icarus, or, The Future of Science*. London: Kegan Paul, Trench, Trubner and Co, 1924. The text can be found at http://cscs.umich.edu/~crshalizi/Icarus.html
A response to Haldane's Daedalus lecture, cited above.

Seal, Cheryl. "Frankensteins in the Pentagon: DARPA's Creepy Bioengineering Program: DARPA Bioengineering Program Seeks to Turn Soldiers Into Cyborgs." In *The News Insider*, August 25, 2003. http://www.newsinsider.org/

Shelley, Mary Wollstonecraft. *Frankenstein; or, the Modern Prometheus*. London: Lockington, Hughes, Harding, Mavor & Jones, 1818. St. Martin's Press, 1995. ISBN: 0-312-12461-9.

The Prevail Scenario

Brand, Stewart. How Buildings Learn: What Happens After They're Built. New York: Viking, 1994. ISBN: 0-670-83515-3.

————. The Clock of the Long Now, Time and Responsibility: The Ideas Behind the World's Slowest Computer. New York: Basic Books, 1999. ISBN: 0-465-04512-X.

Brand has written two books about managing change in complex systems, including human management of The Curve. How Buildings Learn proposes that any physical structure is composed of layers which change at different speeds, thus creating shear forces that, if they are bonded too tightly, ultimately tear the structure apart. If they're designed to slip past each other, however, the fast parts adaptively handle innovation and shocks, while the slow parts integrate everything into a long-term working whole. The analogy is to the way geological plates, slipping and sliding past each other, produce earthquakes. The layers, from fast to slow, include the furnishings, the layout of the interior, the mechanical systems, the exterior, the structure that holds the load, and the site itself. See page 13 of How Buildings Learn for the quick version of this idea, and the rest of the book for its elaboration. Brand then expanded the idea to civilization as a whole in The Clock of the Long Now. There the health of civilization depends on encouraging respect and slippage between its layers, which include fashion, commerce, infrastructure, governance, culture and nature. Understanding and managing the shear forces between them thus become crucial and central issues in managing The Curve.

Brin, David. The Transparent Society: Will Technology Force Us to Choose Between Privacy and Freedom? New York: Perseus Books, 1998. ISBN: 0-738-20144-8.

Brockman, John, ed. The New Humanists. New York: Barnes & Noble Books, 2003. ISBN: 0-760-74529-3.

Brown, John Seely, and Paul Duguid. The Social Life of Information. Boston: Harvard Business School Press, 2000. ISBN: 0-875-84762-5.

A remarkable work examining the proposition that information does not exist abstractly as zeroes and ones. It is embedded in, and dependent on, culture and values.

Butterflies and Wheels. "Butterfliesandwheels.com: Fighting Fashionable Nonsense." http://www.butterfliesandwheels.com/about.htm

From its statement of purpose: "There are two motivations for setting up the Web site. The first is the common one having to do with the thought that truth is important, and that to tell the truth about the world it is necessary to put aside whatever preconceptions (ideological, political, moral, etc.) one brings to the endeavour. The second has to do with the tendency of the political Left (which both editors of this site consider themselves to be part of) to subjugate the rational assessment of truth-claims to the demands of a variety of pre-existing political and moral frameworks. We believe this tendency to be a mistake on practical as well as epistemological and ethical grounds."

Capra, Fritjof. *The Hidden Connections: Integrating the Biological, Cognitive, and Social Dimensions of Life into a Science of Sustainability.* New York: Doubleday, 2002. ISBN: 0-385-49471-8.

———. *The Tao of Physics: An Exploration of the Parallels Between Modern Physics and Eastern Mysticism.* Berkeley: Shambhala, 1975. ISBN: 0-877-73077-6.

———. *The Web of Life: A New Scientific Understanding of Living Systems.* New York: Anchor, 1996. ISBN: 0-385-47675-2.

An Indian physicist with enormous guts attempts to link the mystic philosophical traditions of the East to the rationalist scientific traditions of the West.

Carse, James P. *Finite and Infinite Games.* New York: Free Press, 1986. ISBN: 0-029-05980-1.

Hampden-Turner, Charles, and Fons Trompenaars. *Mastering the Infinite Game: How East Asian Values Are Transforming Business Practices.* Oxford: Capstone, 1997. ISBN: 1-900-96108-3.

Lanier, Jaron. "One Half a Manifesto." In *Edge,* September 2000. http://www.edge.org/3rd_culture/lanier/lanier_index.html

A highly readable and intriguing document by the advocate of The Prevail Scenario presented in this book, outlining some of the underpinnings of his thinking.

———. "A Future That Loves Us: An Optimistic One Thousand Year Scenario." Presented at Global Business Network, June 15, 2004. http://www.advanced.org/jaron/lovely/default.htm

An elaboration of his thinking regarding The Prevail Scenario.

Lessig, Lawrence. *The Future of Ideas: The Fate of the Commons in a Connected World*. New York: Random House, 2001. ISBN: 0-375-50578-4.

A Stanford law professor makes a strong case that we must resist the efforts of media and software industries to shut off access to publicly held material, which Lessig sees as a kind of intellectual commons. He persuasively decries any lopsided control of ideas and suggests practical solutions that consider the rights of both creators and consumers, while acknowledging the serious impact of new technologies on old ways of doing business.

Ogilvy, Jay. *Creating Better Futures: Scenario Planning as a Tool for a Better Tomorrow*. New York: Oxford University Press, 2002. ISBN: 0-195-14611-5.

This work by one of the pioneers of scenario planning is unusual in that it discusses how you might achieve the future you desire.

Pacotti, Sheldon. "Are we doomed yet?: the computer-networked, digital world poses enormous threats to humanity that no government, no matter how totalitarian, can stop. A fully open society is our best chance for survival." In Salon.com, March 31, 2003. http://salon .com/tech/feature/2003/03/31/knowledge/print.html

A thoughtful and nuanced argument for spreading knowledge as widely as possible, even in the face of unimaginable new threats.

Singer, Peter. *The Expanding Circle: Ethics and Sociobiology*. New York: Farrar, Straus and Giroux, 1981. ISBN: 0-374-23496-5.

———. *Animal Liberation*. New York: New York Review Books, 1990. ISBN: 0-940-32200-5.

———, and Paola Cavalieri. *The Great Ape Project: Equality Beyond Humanity*. New York: St. Martin's Press, 1994. ISBN: 0-312-10473-1.

———, and Helga Kuhse, eds. *Bioethics: An Anthology*. Malden, MA: Blackwell, 1999. ISBN: 0-631-20310-9.

A massively controversial philosopher, seen in some quarters as testing the limits of freedom of speech, examines the implications of expanding our circle of empathy.

Smith, Merritt Roe, and Leo Marx, eds. *Does Technology Drive History?* Cambridge MA: MIT Press, 1994. ISBN: 0-262-19347-7.

Leo Marx is one of my heroes, but I wish this book answered the title's question better.

Tenner, Edward. *Why Things Bite Back: Technology and the Revenge of Unintended Consequences*. New York: Knopf, 1996. ISBN: 0-679-42563-2.

———. *Our Own Devices: How Technology Remakes Humanity*. New York: Knopf, 2003. ISBN: 0-375-40722-7. Vintage, 2004. ISBN: 0-375-70707-7.

We shape our machines, and then they shape us.

Twain, Mark. *Huckleberry Finn*. New York: Penguin Books, 2002. ISBN: 0-142-43717-4.

> The transcendent cussedness that may see us through.

On Human Nature

Barrow, John D., and Frank J. Tipler. *The Anthropic Cosmological Principle*. New York: Oxford University Press, 1986. ISBN: 0-198-51949-4.

> Five centuries after Copernicus, this collection of ideas puts human nature back front and center in creation, holding that the existence of intelligent observers determines the fundamental structure of the universe.

Bidney, David. "Human Nature and the Cultural Process." In *American Anthropologist*, July-September 1947, 49(3):375–399.

Brown, Donald E. *Human Universals*. New York: McGraw-Hill, 1991. ISBN: 0-070-08209-X.

> The modern evidence that such a thing as human nature exists.

Calvin, William H. *A Brain for All Seasons: Human Evolution and Abrupt Climate Change*. Chicago: University of Chicago Press, 2002.

———. *A Brief History of the Mind: From Apes to Intellect and Beyond*. New York: Oxford University Press, 2004.

———. Web site: www.WilliamCalvin.com

> From the theoretical neurophysiologist who is my favorite because of his relentless quest for meaning.

Campbell, Joseph. *The Hero with a Thousand Faces*. New York: Pantheon, 1949.

———, with Bill Moyers, Betty Sue Flowers, ed. *The Power of Myth*. New York: Doubleday, 1988. ISBN: 0-385-24773-7.

Crick, Francis. *The Astonishing Hypothesis: The Scientific Search for the Soul*. New York, Scribner, 1994. ISBN: 0-684-19431-7.

> We don't need no stinking soul. Crick is the Nobel Prize–winning co-discoverer, with James Watson, of the double-helix structure of DNA.

Fernández-Armesto, Felipe. *So You Think You're Human?: A Brief History of Humankind*. Oxford: Oxford University Press, 2004. ISBN: 0-192-80417-0.

Freeman, Derek. *Margaret Mead and Samoa: The Making and Unmaking of an Anthropological Myth*. Cambridge, MA: Harvard University Press, 1983. ISBN: 0-674-54830-2.

Fromm, Erich, ed. *Marx's Concept of Man: With a Translation from Marx's Economic and Philosophical Manuscripts by T.B. Bottomore.* New York: F. Ungar, 1961. ISBN: 0804461619.

Graves, Robert. *The Greek Myths, Complete Edition.* London: Penguin Books, 1992; reprint edition, 1993. ISBN: 0-140-17199-1.

My most-consulted work on myth.

Hume, David. *A Treatise of Human Nature: Being an Attempt to Introduce the Experimental Method of Reasoning into Moral Subjects.* Oxford: Oxford University Press, Oxford Philosophical Texts, new edition, 2000. ISBN: 0-198-75172-9.

The 1740 work basing philosophy on an observationally grounded study of human nature.

Lewis, Thomas, Fari Amini, and Richard Lannon. *A General Theory of Love.* New York: Random House, 2000. ISBN: 0-375-50389-7.

Love as a fundamental, hard-wired human need.

Mead, Margaret. *Coming of Age in Samoa: A Psychological Study of Primitive Youth for Western Civilisation.* New York: William Morrow, 1928. Perennial Classics edition, 2001. ISBN: 0-688-05033-6.

The high point in the popular mind of the hypothesis that bad human outcomes are produced far more by the nasty aspects of industrial civilization than by biological nature.

Pinker, Steven. *The Language Instinct: The New Science of Language and Mind.* New York: William Morrow, 1994. ISBN: 0-688-12141-1.

———. *How the Mind Works.* New York: Norton, 1997. ISBN: 0-393-04535-8.

———. *The Blank Slate: The Modern Denial of Human Nature.* New York: Viking, 2002. ISBN: 0-670-03151-8.

The current era's best-selling author on human nature.

Stevenson, Leslie, and David L. Haberman. *Ten Theories of Human Nature: Confucianism, Hinduism, The Bible, Plato, Kant, Marx, Freud, Sartre, Skinner, Lorenz.* New York: Oxford University Press, 1998. ISBN: 0-19-512040-X, hardcover.

A fascinating and meticulous examination that is somewhat misnamed. Stevenson and Haberman are most interested in the ideologies that have grown up around varying definitions of human nature. Christianity and Marxism, to take two of their examples, come up with background theories about the world, a diagnosis of what is wrong with us and prescriptions for putting it right. They make subordinate to that investigation the actual theories of human nature on which these ideologies are based—the Christian view that mankind is made in the image of God who has a definite purpose for our life, for example, or the Marxist notion that the real nature of man is the totality of social relations.

————, ed. *The Study of Human Nature: A Reader.* New York: Oxford University Press, second edition, 2000. ISBN: 0-19-512715-3, paperback.

Tiger, Lionel, and Robin Fox. *The Imperial Animal.* New York: Holt, Rinehart & Winston, 1971. ISBN: 0-030-86582-4.

Tooby, John, and Leda Cosmides. "On the Universality of Human Nature and the Uniqueness of the Individual: The Role of Genetics and Adaptation." In *Journal of Personality,* 1990, 58:1:17-67.

Wilson, Edward O. *Sociobiology:The New Synthesis.* Cambridge, MA: Belknap Press of Harvard University Press, 1975. ISBN: 0-674-81621-8.

————. *On Human Nature.* Cambridge MA: Harvard University Press, 1978. ISBN: 0-674-63441-1.

————. "Reply to Fukuyama." In *The National Interest,* no. 56, Spring 1999. From the godfather of sociobiology, which is the scientific study of the biological basis of all forms of social behavior in all kinds of organisms, including man.

Wright, Robert. *The Moral Animal: Evolutionary Psychology and Everyday Life.* New York: Pantheon Books, 1994. ISBN: 0-679-40773-1.

Transcendence

Armstrong, Karen. *A History of God:The 4000-Year Quest of Judaism, Christianity and Islam.* New York: Ballantine Books, 1994. ISBN: 0-345-38456-3.

Bateson, Gregory. *Steps to an Ecology of Mind.* San Francisco: Chandler Pub. Co., 1972. ISBN: 0-810-20447-9.

Bellah, Robert N., et al. *Habits of the Heart: Individualism and Commitment in American Life.* Berkeley: University of California Press, 1985. ISBN: 0-520-05388-5.

Boyer, Pascal. *Religion Explained:The Evolutionary Origins of Religious Thought.* New York: Basic Books, reprint edition, 2002. ISBN: 0-465-00696-5.

Durkheim, Émile. "Elementary Forms of Religious Life," 1912. In Robert N. Bellah, ed., *Émile Durkheim: On Morality and Society.* Chicago: University of Chicago Press, 1973.

Dyson, Freeman J. *Infinite in All Directions.* New York: Harper & Row, 1988. ISBN: 0-060-39081-6.

Giovannoli, Joseph, et al. *The Biology of Belief: How Our Biology Biases Our Beliefs and Perceptions.* Rosetta Press, Inc., 2001. ISBN: 0-970-81371-6.

Havel, Václav. "The Need for Transcendence in the Postmodern World," delivered at Independence Hall, Philadelphia, July 4, 1994. http://www.worldtrans.org/whole/havelspeech.html

James, William. *The Varieties of Religious Experience: A Study in Human Nature.* New York: Longmans, Green, & Co., 1902. New York: Routledge, Centennial edition, 2002. ISBN: 0415278090.

Morowitz, Harold J. *The Emergence of Everything: How the World Became Complex.* New York: Oxford University Press, 2002. ISBN: 0-19-513513-X.

With an amazing last chapter on "Science and Religion," from a leader in the science of complexity and cochair of the science board of the Santa Fe Institute.

Sartre, Jean-Paul, Hazel E. Barnes, trans. *Being and Nothingness: An Essay on Phenomenological Ontology.* First published 1943. New York: Washington Square Press, reprint edition, 1993. ISBN: 0-671-86780-6.

Shermer, Michael. *Why People Believe Weird Things: Pseudoscience, Superstition, and Other Confusions of Our Time.* New York: Owl Books, second revised edition, 2002. ISBN: 0-805-07089-3.

———. *How We Believe: Science, Skepticism, and the Search for God.* New York: Owl Books, second edition, 2003. ISBN: 0-805-07479-1.

———. The Skeptics Society and *Skeptic* magazine, "Dedicated to the promotion of science and critical thinking, and to the investigation of extraordinary claims and revolutionary ideas." http://www.skeptic.com/

One of the great difficulties humans encounter with rapid change is separating the marvelous from the magical, and myths from madness. Therefore, one of the great challenges for our generation is going to be reconciling belief systems. If we cannot come up with an agreed-upon frame for what constitutes something as basic as reality, it's hard to see how we will manage to respond to change together. Predictions to the contrary, rationalism has not banished religiosity. Religion is bigger than ever, possibly because it addresses issues that science can't, possibly because there is something hardwired in the human brain that inclines us to believe. Thus, I have great admiration for those who are attempting to deal with this schism not by rejecting out of hand the validity of various ways of knowing, but by trying to construct this larger frame. Shermer is one of those who, like Boyer and Giovannoli, attempts from the rationalist perspective to respectfully understand what is going on with faith-based systems. Perfection, alas, has not been achieved. These are worthwhile starts, however. We've got a long way to go and a short time to get there.

Stock, Gregory. *Metaman: The Merging of Humans and Machines into a Global Superorganism.* New York: Simon & Schuster, 1993. ISBN: 0-671-70723-X.

Stock is best known for his work exploring genetically engineered evolution. *Metaman,* however, is really his take on transcendence.

Teilhard de Chardin, Pierre; Bernard Wall, trans. *The Phenomenon of Man.* New York: Harper, 1959. Ursula King, trans., Maryknoll, NY: Orbis Books, 1999. ISBN: 1570752486.

————, Norman Denny, trans. *The Future of Man.* New York: Harper & Row, 1964. New York: Perennial, 1969. ISBN: 0061303860.

Remarkably prescient writing that was faith-based speculation when he wrote it. Hardheaded evidence that he might be right, however, continues to emerge. If the test of future writing is whether it looks better the older it gets, that explains why De Chardin is getting ever-increasing attention.

Wright, Robert. *NonZero: The Logic of Human Destiny.* New York: Pantheon Books, 2000. ISBN: 0-679-44252-9.

For those firmly embedded in the rationalist tradition, as am I, who nonetheless are persuaded by the evidence that we might be headed toward a transcendence of human nature, Wright has done a cheerful and impressive job of trying to reconcile enormous contradictions.

On Scenario Planning

Ringland, Gill. *Scenario Planning: Managing for the Future.* Chichester, England: John Wiley & Sons, 1998. ISBN: 0-471-97790-X.

————. *Scenarios in Business.* Chichester, England: John Wiley & Sons, 2002. ISBN: 0-470-84382-9.

————. *Scenarios in Public Policy.* Chichester, England: John Wiley & Sons, 2002. ISBN: 0-470-84383-7.

Schwartz, Peter. *The Art of the Long View.* New York: Doubleday, 1991. ISBN: 0-385-26731-2.

This is the most readable and compact work detailing the methods and benefits of scenario planning for hardheaded, bet-the-company decision making, written by the head of the pioneering scenario-planning firm Global Business Network.

———. *Inevitable Surprises: Thinking Ahead in a Time of Turbulence*. New York: Gotham Books, 2003. ISBN: 1-59240-027-2.

Thinking about what will be regarded as bolts-out-of-the-blue by those who haven't prepared for them.

Van der Heijden, Kees. *Scenarios: The Art of Strategic Conversation*. Chichester, England: John Wiley & Sons, 1996. ISBN: 0-471-96639-8.

Wack, Pierre. "Scenarios: Shooting the Rapids." In *Harvard Business Review,* November/December 1985, 139–50.

———. "Scenarios: Uncharted Waters Ahead." In *Harvard Business Review,* September/October 1985, 72–79.

From the godfather of scenario planning.

DARPA

DARPA's home page: http://www.darpa.mil/
The Defense Sciences Office home page: http://www.darpa.mil/dso/
Both are worth mining in depth. In the Defense Sciences Office home page, for example, check out "Technology Thrusts" and "Future Areas of Interest."

Technology Transition, Defense Advanced Research Projects Agency, January 1997. http://www.darpa.mil/body/pdf/transition.pdf
DARPA's brag list of what it has actually shipped to the troops from its founding in the wake of the Sputnik shock, through the NATO intervention in Bosnia.

Strategic Plan, Defense Advanced Research Projects Agency, February 2003. http://www.darpa.mil/body/strategic.html
A broad overview of DARPA's direction.

Bridging the Gap, Defense Advanced Research Projects Agency, March 2004. http://www.darpa.mil/DARPATech2004/proceedings.html
The proceedings of the DARPATech symposium, March 9–11, 2004, Anaheim, California, displaying the organization's thinking.

Transhumanism

Bostrom, Nick. "Transhumanist Values." Department of Philosophy, Yale University, version of April 18, 2001. http://www.nickbostrom.com/tra/values.html

———. "The Transhumanist FAQ: A General Introduction." World Transhumanist Association, version 2.1, 2003. http://www.transhumanism.org/resources/faq.html

> Interesting and serious documents, worth reading.

Esfandiary, F.M. *Up-Wingers.* New York: John Day, 1973. ISBN: 0-381-98243-2.

> Fair warning: The work of many transhumanists, especially the early ones, is so far out-there it has taken the movement decades to begin to be taken seriously. This, a founding document, is a case in point. Fereidoun M. Esfandiary wound up preferring to be known as FM-2030.

Extropy Institute: http://www.extropy.org

"Extropy Institute's 'Extropian Principles.'" http://www.extropy.org/ideas/principles.html

> Home of the more, but by no means the most, out-there transhumanists.

Hughes, James J. "The Politics of Transhumanism." Originally prepared for the 2001 Annual Meeting of the Society for Social Studies of Science, Cambridge, MA, November 1–4, 2001. Version 2.0, March 2002. http://www.changesurfer.com/Acad/TranshumPolitics.htm

———. "Democratic Transhumanism." Originally published in *Transhumanity,* April 28, 2002. Version 2.0 at: http://www.changesurfer.com/Acad/DemocraticTranshumanism.htm

———. *Citizen Cyborg.* Boulder, CO: Westview Press, 2004. ISBN: 0-813-34198-1.

> Serious and courageous political thinking.

Nietzsche, Friedrich. *The Portable Nietzsche,* Walter Kaufmann, ed. New York: Viking, 1954. Penguin Books, new edition, 1977. ISBN: 0-140-15062-5.

———. *Thus Spake Zarathustra.* Originally published 1883–1885. Mineola, NY: Dover Publications, 1999. ISBN: 0-486-40663-6.

Regis, Ed. *Great Mambo Chicken and the Transhuman Condition: Science Slightly over the Edge.* Reading, MA: Addison-Wesley, 1990. ISBN: 0-201-09258-1.

————. "Meet the Extropians." In *Wired*, 2:10, October 1994. http://www.wired.com/wired/archive/2.10/extropians_pr.html

Amusing looks at those who, especially at the time these were written, were viewed as "over the edge" of transhumanism.

Vita-More, Natasha: home page: http://www.extropic-art.com

————. *Create: Recreate: The Third Millennial Culture*. Los Angeles: More-Art, second ed., 1999.

"World Transhumanist Association: For the Ethical Use of Technology to Extend Human Capabilities," home page: http://www.transhumanism.org/index.php/WTA/index/

————. "The Transhumanist Declaration." World Transhumanist Association, 2002. http://www.transhumanism.org/declaration.htm

The WTA, generally, is that portion of the movement least easily dismissed by the open- but serious-minded.

Other Future Visions

Brockman, John, ed. *The Next Fifty Years: Science in the First Half of the Twenty-first Century*. New York: Vintage Books, 2002. ISBN: 0-375-71342-5.

Broderick, Damien. *The Spike: How Our Lives Are Being Transformed by Rapidly Advancing Technologies*. New York: Forge, 2001. ISBN: 0-312-87781-1.

Broderick, a senior research fellow at the University of Melbourne, is considered the dean of Australian science fiction. People who dislike science fiction and/or who would not dream of describing themselves as "extropian" will probably not find this volume to their taste. However, the first appendix, "Summary: Paths and Time-Lines to the Spike"—Broderick's term for The Singularity—offers an exhaustive range of scenarios on the subject.

Denning, Peter J., ed. *The Invisible Future: The Seamless Integration of Technology Into Everyday Life*. New York: McGraw-Hill, 2002. ISBN: 0-071-38224-0.

Didsbury, Howard F. Jr. *Twenty-first Century Opportunities and Challenges: An Age of Destruction or an Age of Transformation*. Bethesda, MD: World Future Society, 2003. ISBN: 0-930-24258-0.

"Global Consciousness Project." Roger D. Nelson, director. http://noosphere.princeton.edu

An odd project that would have amazing consequences were it to succeed. It attempts no less than to define and measure, in scientific terms, the mind's extended reach—global consciousness.

Kelly, Eamonn, Peter Leyden, and Members of the Global Business Network. *What's Next: Exploring the New Terrain for Business.* Cambridge, MA: Perseus, 2002. ISBN: 0-738-20760-8.

Sterling, Bruce. *Tomorrow Now: Envisioning the Next Fifty Years.* New York: Random House, 2002. ISBN: 0-679-46322-4.

Wells, H.G. *Anticipations of the Reaction of Mechanical and Scientific Progress upon Human Life and Thought.* New York: Harper and Brothers, 1902. Mineola, NY: Dover Publications, 1999. ISBN: 0-486-40673-3.

An astonishing tour de force. In the graduate classes I've taught on long-term thinking, I have used this collection of nonfiction essays by Wells—as well as the less-prescient fictions of Jules Verne (*Paris in the Twentieth Century,* written in 1863) and Edward Bellamy (*Looking Backward: 2000–1887*)—as existence proof that imagining the distant future is possible and useful. What Wells missed is fascinating and instructive (women's liberation, e.g.). Yet the innumerable things he figured out correctly are impressive, given that he was thinking one hundred years ahead.

Notes

Chapter One PROLOGUE: THE FUTURE OF HUMAN NATURE

4 *allow warriors to run at Olympic sprint speeds for 15 minutes on one breath of air:* Interview, Michael Goldblatt, Defense Advanced Research Projects Agency (DARPA), December 17, 2002.

4 *Call them the GRIN technologies:* A variety of acronyms is being used to refer to the idea that several master technologies are intertwining exponentially to drive change. They all boil down to the same idea, although their emphases may vary. Take, for example, Mihail C. Roco and William Sims Bainbridge, eds., *Converging Technologies for Improved Human Performance: Nanotechnology, Biotechnology, Information Technology and Cognitive Science* (Arlington, VA: National Science Foundation/Department of Commerce, 2002) and (New York: Kluwer Academic Publishers, 2003). ISBN: 1-402-01254-3. http://wtec.org/ConvergingTechnologies They refer to the "NBIC" technologies—nanotechnology, biological engineering, information technology, and cognitive engineering. Douglas Mulhall in *Our Molecular Future*, (Amherst, NY: Prometheus Books, 2002), ISBN: 1573929921, refers to them as the GRAIN technologies—genetics, robotics, artificial intelligence and nanotechnology. Bill Joy in "Why the Future Doesn't Need Us," *Wired*, April 2000—http://www.wired.com/wired/archive/8.04/joy.html—refers to them as the GNR technologies: genetics, nanotechnology, and robotics. So does Ray Kurzweil in his work. There are doubtless others.

5 *"The current doping agony is a kind of very confused referendum on the future of human enhancement":* Michael Sokolove, "In Pursuit of Doped Excellence; The Lab Animal," *New York Times Magazine*, January 18, 2004. http://www.nytimes.com/2004/01/18/magazine/18SPORTS.html

5 *Competitive bodybuilding is already divided:* See, for example, Krista Scott-Dixon, "The Biggest Lies in the Gym," Bodybuilding.com. http://bodybuilding.about.com/cs/women/a/aa041003a.htm

5 *"If someone said, 'Here's $10 million' . . . you could get pretty imaginative":* For a detailed report on his work, see H. Lee Sweeney, "Gene Doping: Gene therapy for restoring muscle lost to age or disease is poised to enter the clinic, and elite athletes are eyeing it to enhance performance. Can it be long before gene doping changes the nature of sport?," *Scientific American*, July 2004, page 63. http://www.sciamdigital.com/

5 *a functioning prototype exoskeleton:* Interview, Homayoon "Hami" Kazerooni, professor of mechanical engineering, University of California at Berkeley, March 20, 2003, and other dates. See also Duncan Graham-Rowe, "Artificial Exoskeleton Takes the Strain," NewScientist.com news service, March 5, 2004. http://www.newscientist.com/news/news.jsp?id=ns99994750

5 *allow soldiers to leap tall buildings with a single bound:* Interview, Jean-Louis "Dutch" DeGay, Equipment Specialist, Objective Force Warrior Technology Program Office, U.S. Army SBCCOM, Natick Soldier Center, June 25, 2003.

5 *"Just five years from now the boundary will be breached":* Rodney A. Brooks, *Flesh and Machines: How Robots Will Change Us* (New York: Pantheon Books, 2002). ISBN: 0-375-42079-7, page 5.

5 *"The next frontier is our own selves":* At a conference on "The Adaptable Human Body: Transhumanism and Bioethics in the 21st Century," Yale University, June 27, 2003.

6 *stop making any more fraudulent claims:* Jack Uldrich, " 'Exponential' Thinking for the Future," Tech Central Station, January 21, 2004. http://www2.techcentralstation. com/1051/printer.jsp?CID=1051-012104D

7 *the challenging stuff of One L fame:* Scott Turow, *One L: The Turbulent True Story of a First Year at Harvard Law School* (New York: Warner Books, 1997). ISBN: 0446673781

7 *They have amazing thinking abilities:* This is part of what DARPA originally referred to as the Brain-Machine Interface program, later called the Human Assisted Neural Devices program. See Brett Giroir, "Beyond the Bio-Revolution, Maintaining Soldier Performance," presentation prepared for DARPATech 2004, March 9–11, 2004, Anaheim, CA. http://www.darpa.mil/DARPAtech2004/pdf/scripts/GiroirScript. pdf See also "Biological Sciences" under the heading "Technology Thrusts" in the DSO section of the DARPA Web site. http://www.darpa.mil/dso/thrust/biosci/ biosci.htm For further information on the repositioning of various DARPA programs, see Chapter Eight: Epilogue. For more on neurological brain augmentation, see Laura Beil, "Brain-Boosting 'Cosmetic Neurology' On The Horizon," *Dallas Morning News*, November 6, 2004. http://www.parkinsons-information-exchange-network-online.com/parkmail1/2004d/msg00452.html

7 *They have photographic memories:* Ibid. See also: Lakshmi Sandhana, "Chips Coming to a Brain Near You," *Wired News*, October 22, 2004. http://www.wired.com/ news/print/0,1294,65422,00.html

7 *They're beautiful, physically:* See, for example, Jon Entine, "The Coming of the Über-Athlete," Salon.com, March 21, 2002. http://www.salon.com/news/sports/ 2002/03/21/genes/index_np.html See also Christen Brownlee, "Gene Doping: Will Athletes Go for the Ultimate High?," *Science News*, October 30, 2004, page 280. http://www.sciencenews.org/articles/20041030/bob9.asp See also "GM 'Marathon' Mice Break Distance Records," NewScientist.com News Service, August 23, 2004. http://www.newscientist.com/article.ns?id=dn6310

7 *They talk casually about living a very long time, perhaps being immortal:* See, for example, Ray Kurzweil, *The Age of Spiritual Machines* (New York: Penguin, 1999). ISBN: 0-670-88217-8, page 280 and elsewhere.

7 *Within minutes it simply stopped bleeding:* The program originally referred to by DARPA as part of Persistence in Combat, later subsumed as part of Peak Soldier Performance.

7 *vaccinated against pain:* The program originally referred to by DARPA as part of Persistence in Combat, later subsumed as part of Soldier Self-Care.

8 *They call it "silent messaging":* See, for example, Vernor Vinge, "Synthetic Serendipity," IEEE Spectrum Online, July 8, 2004. http://www.spectrum.ieee.org/ WEBONLY/publicfeature/jul04/0704/far.html

8 *It almost seems like telepathy:* Originally referred to by DARPA as part of the Brain-Machine Interface Program. Later work that was described as a step toward detecting and transmitting signals from the language areas of the brain was reported in *Science*, volume 305, page 258, and referred to in Duncan Graham-Rowe, "Brain Implants 'Read' Monkey Minds," NewScientist.com news service, July 8, 2004. http://www.newscientist.com/news/print.jsp?id=ns99996127

8 *They have this odd habit:* Ibid.

8 *For a week or more at a time, they don't sleep:* The program originally referred to by DARPA as part of the Continuous Assisted Performance program, later largely subsumed under the Preventing Sleep Deprivation program.

9 *"Forget fiction, read the newspaper":* Damien Cave, "Killjoy," Salon.com, April 10, 2000. http://dir.salon.com/tech/view/2000/04/10/joy/index.html?pn=2

9 *more computers than she has lightbulbs:* As of 2001, there were 24 processors in a Ford Taurus and 60 in a high-end Mercedes.

9 *"We shape our buildings, and afterwards our buildings shape us":* Regarding how the

House of Commons chamber should be rebuilt after it was bombed in 1941. http://www.bbc.co.uk/history/lj/churchlj/stephen_09.shtml

10   *"The Organization Man":* William Hollingsworth Whyte, *The Organization Man* (Philadelphia: University of Pennsylvania Press, reprint edition, 2002). ISBN: 0812218191.

10   *"The Man in the Gray Flannel Suit":* Sloane Wilson, *Man in the Gray Flannel Suit* (New York: Four Walls Eight Windows, 4th edition, 2002). ISBN: 1568582463.

11   *ways to increase their child's SAT scores:* See, for example, "Supercharging the brain: Biotechnology: New drugs promise to improve memory and sharpen mental response. Who should be allowed to take them?," *The Economist,* September 16, 2004. http://www.economist.com/science/tq/PrinterFriendly.cfm?Story_ID=3171454 See also Mary Carmichael, "Medicine's Next Level: With insight into the mechanisms that help keep your brain sharp, neurological researchers move closer to improving your recall with a 'memory pill,' " *Newsweek,* December 6, 2004, page 46. http://www.msnbc.msn.com/id/6595798/site/newsweek/

11   *memory enhancers: The Economist,* "Supercharging the Brain" and *Newsweek,* "Medicine's Next Level." Keep an especially close eye on the clinical trial results of companies such as Memory Pharmaceuticals, Sention, Helicon Therapeutics, Saegis Pharmaceuticals, and Cortex Pharmaceuticals.

12   *medical robots traveling the human bloodstream:* Rick Weiss, "For Science, Nanotech Poses Big Unknowns," *Washington Post,* February 1, 2004, page A1. http://www.washingtonpost.com/ac2/wp-dyn/A1487-2004Jan31?language=printer

12   *two gay males . . . to make a baby:* Rick Weiss, "In Laboratory, Ordinary Cells Are Turned into Eggs," *Washington Post,* May 2, 2003, page A1.

12   *the last fifty years . . . a guide to the next fifty years:* Interview, Ray Kurzweil, April 14, 2003.

12   *90-percent-male alpha-geek population:* I am indebted to Chris Anderson, editor of *Wired* magazine, for this insight into the demographics of the digerati.

Chapter Two   BE ALL YOU CAN BE

15   *"The future is already here":* William Gibson, the author of *Neuromancer*—wherein he coined the word *cyberspace*—for the life of him cannot remember where or precisely when he first used this line, although he knows it was in the late eighties. Personal communication, June 7, 2003.

17   *This particular January:* Interviewed January 3, 2003.

19   *The first telekinetic monkey:* Miguel A. L. Nicolelis and John K. Chapin, "Controlling Robots with the Mind," *Scientific American,* October 2002, pages 47–53. For how this works in humans see, for example, Robert Lee Hotz, "Device for the Paralyzed Turns Thinking to Doing," *Los Angeles Times,* December 7, 2004, page A1. http://news .orb6.com/stories/latimests/20041207/devicefortheparalyzedturnsthinkingtodoing .php See also Rick Weiss, "Mind Over Matter: Brain Waves Guide a Cursor's Path: Biomedical Engineers Create Devices That Turn Thoughts Into Action and Could Help the Paralyzed Move Their Limbs," *Washington Post,* December 13, 2004, page A8. http://www.washingtonpost.com/ac2/wp-dyn/A59791-2004Dec12?language= printer

20   *thinner than the finest sewing thread:* 25–50 microns. Alan Rudolph, DARPA.

20   *In the 1930s and 1940s . . . transcend the mortal bounds of everyday humanity:* See, for example, Joel Garreau, "The Next Generation: Biotechnology May Make Superhero Fantasy a Reality," *Washington Post,* April 26, 2002, page C1. http://www.washingtonpost. com/ac2/wp-dyn?pagename=article&contentId=A50958-2002Apr25&notFound=true

20   *He was a Depression-era orphan:* Jeff Rovin, *The Encyclopedia of Superheroes* (New York: Facts on File Publications, 1985). ISBN: 0-8160-1168-0, pages 57–58.

21 *think of it as a wearable robot suit:* See, for example, Gregory T. Huang, "Wearable Robots: Robotics Inventor Stephen Jacobsen Demonstrates an Exoskeleton That Provides Superhuman Strength," *Technology Review,* July/August 2004. http://www. technologyreview.com/articles/print_version/demo0704.asp

21 *now in development as part of a $50 million program:* Interview, Jean-Louis "Dutch" De Gay, Equipment Specialist, Objective Force Warrior Technology Program Office, U.S. Army SBCCOM, Natick Soldier Center, June 25, 2003.

21 *The weakling was reborn:* Rovin, *The Encyclopedia of Superheroes,* pages 46–47.

21 *what evil lurks in the hearts of men:* See, for example, Mark Peplow, "Brain imaging could spot liars: Tests reveals patches in the brain that light up during a lie," News@ Nature.com, November 29, 2004. http://www.nature.com/news/2004/041129/full/ 041129-1.html Also, "Lie Detection: Making Windows in Men's Souls: The Science of Lie Detection Has a Chequered Past. But It Is Becoming More Reliable," *The Economist,* July 8, 2004. http://www.economist.com/printedition/PrinterFriendly. cfm?Story-ID=2897134

21 *X-ray vision:* An Israeli company, Camero, says it has devised an ultra-wide-band radar system that can produce three-dimensional pictures of what lies behind a wall from a distance of up to 20 meters. The pictures resemble those produced by ultra-sound. "Camero develops radar system to see thru walls," Business Line, Internet edition, July 2, 2004. http://www.thehindubusinessline.com/businessline/blnus/ 14021106.htm

22 *There were no computer science departments:* Technology Transition, DARPA, 2003, page 40. http://darpa.mil/body/pdf/transition.pdf

22 *"the main and essential medium of informational interaction for governments, institutions, corporations, and individuals":* J. C. R. Licklider, memo to "the members and affiliates of the Intergalactic Computer Network," April 25, 1963. Cited in: M. Mitchell Wal-drop, *The Dream Machine: J.C.R. Licklider and the Revolution That Made Computing Per-sonal* (New York: Viking Penguin, 2001). ISBN: 0 14 20.0135 X (paperback), pages 6–7.

22 *"Soldiers having no physical, physiological, or cognitive limitations":* Final draft of Gold-blatt's DARPATech 2002 presentation as of June 25, 2002, pages 6–7.

23 *"science action, not science fiction":* Ibid., pages 1–2.

23 *"bio-revolution" program represents only a fraction of DARPA's overall agenda:* Strategic Plan, DARPA, 2003, page 17. http://www.darp.mil/body/strategic.html

23 *"only charter is radical innovation":* Ibid., page 3.

23 *DSO's deputy director:* This was Wax's title as of 2003. He later became director of DSO.

23 *"We try not to violate any of the laws of physics":* Goldblatt and Wax interview, December 6, 2002.

23 *DARPA has a track record:* Technology Transition, various pages. See note in the "Sug-gested Readings" chapter.

23 *It was a key player:* Goldblatt, interview, April 1, 2003.

24 *the first robot known to incinerate a human being:* There is debate about what level of autonomy characterizes a true robot. Should you count guided and cruise missiles as robots? For that matter, since a human remotely guided the unmanned Predator and told it to fire the missile, can it be said that we actually have murderous robots yet? Reasonable people differ. The fellows in the SUV, of course, might find these dis-tinctions overly fine.

24 *"accelerate the future into being":* Strategic Plan, pages 3–4.

24 *DARPA invests 90 percent:* Ibid., pages 7–8.

24 *Academic centers . . . coalesced because of DARPA:* Technology Transition, page 42.

24 *If it feels companies need to exist:* Ibid., pages 42–43.

24 *If standards need to exist, DARPA sometimes steps in:* Ibid., page 43.

24 *President Eisenhower created DARPA:* Goldblatt interview, April 1, 2003.

24 *"the ultimate high ground":* DARPA, *Strategic Plan,* page 11.

24 *But most of all it wanted never again to be surprised:* Technology Transition, page 12.

25 *enhancing human performance, the program managers of DSO see a "golden age":* Strategic Plan, page 17.

25 *the old Army slogan " 'Be All You Can Be' takes on a new dimension":* Final draft of Goldblatt's DARPATech presentation as of June 25, 2002, pages 3–4.

26 *Tales have been told about them now for 3,300 years:* The tale dates to the 13th century B.C. Robert Graves, *The Greek Myths, Complete Edition* (London: Penguin Books, 1992), page 585.

26 *They included:* Ibid., pages 579–80.

26 *Jason yoked two fire-breathing bulls:* Ibid., pages 599–601.

26 *"We do not fear the unknown":* Final draft of Goldblatt's DARPATech 2002 presentation as of June 25, 2002, page 3.

27 *The hope is that it will also mend wounds to skin:* Persistence in Combat original program description, later subsumed into Soldier Self Care Program. www.darpa.mil/dso

27 *spinal cord injuries, Parkinson's disease and brain tumors:* Jonathan Sidener, "LED Therapy Shows Promise in Studies," *San Diego Union,* June 23, 2003, page A1.

27 *"physiostimulator":* "Federation Facts." http://www.nexus1.net/FederationFactsPage. html

27 *stop bleeding can be triggered by signals from the brain:* Bielitzki's DARPATech 2002 presentation, slide 9.

28 *Then they switch brains:* Ibid., slide 6.

29 *Think what this will do for college students and medical residents:* Ibid., slide 6.

29 *"how not to sleep":* Carney interview, December 17, 2002.

29 *CAP's major [sleep] research efforts:* DARPATech presentation, 2002, slide 5.

30 *the essential part of life:* Goldblatt in Goldblatt and Wax interview, December 6, 2002.

30 *Nothing in the cell gets through:* Ibid.

30 *Will these approaches throw out some side effects:* I am indebted to Lawrence Osborne for raising this interesting question in a different context in *New York Times Magazine,* December 15, 2002, "The Year in Ideas: Genetically Modified Saliva." http://query. nytimes.com/gst/abstract.html?res=F60812FD3C5E0C768DDDAB0994DA404482

31 *140 decision makers united by a common travel department:* Robert Mullan Cook-Deegan, "Does NIH Need a DARPA?" Issues in Science and Technology Online, winter 1996. http://www.nap.edu/issues/13.2/cookde.htm

32 *Hunger, exhaustion and despondency:* Bielitzki interview, December 19, 2002.

32 *"Be all that you can be and a lot more":* Bielitzki presentation, DARPATech 2002, slide 12.

34 *a PhD . . . and an MBA:* DARPA bio. http://www.darpa.mil

34 *He has 15 patents:* Ibid.

34 *It makes it an interesting place:* Rudolph interview, December 14, 2002.

35 *computer eyes:* See, for example, Erika Jonietz, "Demo: Artificial Retina: An Electronic Device Implanted in the Eye Could Restore the Sight of Millions," *Technology Review,* September 2004. http://www.technologyreview.com/articles/04/09/demo0904.asp?p=1

36 *Cyberkinetics:* http://www.cyberkineticsinc.com

39 *"Can you imagine traversing":* DeGay interview, June 25, 2003.

40 *It would allow them to survive even without oxygen:* Bielitzki presentation, DARPATech 2002, slide 10.

40 *The Bioinspired Dynamic Robotics program:* Per Alan Rudolph interview, July 12, 2003, predecessor program, Controlled Biological and Biomimetic Systems.

40 *These include electronics:* Goldblatt DARPATech 2002 presentation, slide 13, Meso-. scopic Integrated Conformal Electronics (MICE) program description. http://www. darpa.mil

40 *[living creatures] as "remote sentinels":* Goldblatt interview, July 11, 2003; program description. http://www.darpa.mil

40 *The Brain-Machine Interface program:* Alan Rudolph interview, July 11, 2003.

40 **"that truly know what they're doing":** Abstract of a talk presented by Ron Brachman, "Developing Cognitive Systems," delivered May 3, 2004, at Sarnoff Corporation, Princeton, NJ, before the IEEE Signal Processing Society, Dr. Hui Cheng, organizer.

41 *Time Reversal Methods:* DARPA BAA03-02. http://www.darpa.mil

42 *The bulk of DARPA's projects:* Goldblatt in Goldblatt and Wax interview, December 6, 2002.

42 *"That's Veterans Affairs":* Goldblatt interview, December 17, 2002. The Pentagon has been showing greatly increased interest in tissue regeneration since the flow of wounded from the Iraq war commenced, DARPA insiders report.

42 *"we will be the first species to control our own evolution":* Amy Harmon, "Technology Elite Are Focusing Next on Human Body," *New York Times,* June 16, 2003.

42 *most common word you hear is fun:* Goldblatt and Wax interview, December 6, 2002, and elsewhere.

43 *"There's potential for contradictions in all of science":* Bielitzki interview, December 19, 2002.

Chapter Three THE CURVE

48 *When its orchards erupted into bloom:* History San Jose. http://www.historysanjose. org/Valley.html

48 *Since the 1920s, this area:* M. Mitchell Waldrop, *The Dream Machine: J. C. R. Licklider and the Revolution That Made Computing Personal* (New York: Viking Penguin, 2001). ISBN: 0-670-89976-3, page 338.

48 *Their first big customer was Walt Disney:* Ibid.

48 *referred to . . . as Silicon Valley:* In January, 1971, a three-part series in *Electronic News* was entitled "Silicon Valley USA," Ibid.

48 *He was thoughtful and cautious:* T. R. Reid, *The Chip: How Two Americans Invented the Microchip and Launched a Revolution* (New York: Random House, 1985). ISBN: 0-375-75828-3, page 8.

49 *"He just looked so normal":* Interview, Howard I. High, July 3, 2003.

49 *Sure enough, in 2002 . . . 27th doubling:* George Gilder, "Moore's Quantum Leap." *Wired,* January 2002, page 104. http://www.wired.com/wired/archive/10.01/gilder _pr.html

50 *from 23 miles to 2,808:* Wil McCarthy, personal communication based on the research notes for his article "Runaway Train," *Wired,* January 2002, page 103. http:// www.wired.com/wired/archive/10.01/tracks_pr.html

50 *It took another 36 years:* The Railroad Museum of Pennsylvania, Railroad History Timeline, compiled by Kurt R. Bell, librarian/archivist. http://www.rrmuseumpa. org/education/historytimeline1.htm

50 *Not for nothing do historians still celebrate the driving of the Golden Spike:* See, for example, the Museum of the City of San Francisco. http://www.sfmuseum.org/ hist1/rail.html

51 *A voyage to a new life cost 25 cents:* I am indebted to Wil McCarthy for some of these observations.

51 *The last transcontinental railroad:* History Link: "Chicago, Milwaukee and St. Paul Railroad lays the last rail." http://www.historylink.org/_output.cfm?file_ID=930

51 *one millionth Model T:* Joel Garreau, *Edge City: Life on the New Frontier* (New York: Doubleday 1991). ISBN: 0-385-26249-3, page 104.

52 *you can get it for free:* Even more impressive, in 2004, J.C. Penney offered a free DVD player—which is a computer and a laser—to anyone spending $100 at its stores on Father's Day.

52 *Passports come equipped:* John Leyden, "U.S. Names the Day for Biometric Passports," *The Register,* August 22, 2003. http://www.theregister.co.uk/content/55/31885.html

52 *All for under $5:* Evan I. Schwartz, "How You'll Pay: Smart Cards, Radio Tags, And Microchip Buttons Are Going to Revolutionize the Way You Buy Things," *Technology Review,* December 2002–January 2003, page 50. http://www.technologyreview.com/articles/02/12/schwartz1202.asp?p=0

52 *The cost of shipping a ton of grain:* Nathan Myhrvold, former technology chief of Microsoft, in an interview with MIT's *Technology Review,* June 2002. http://www.technologyreview.com/articles/print_version/qa0602.asp

52 *"the housewife will sit at home":* Jeffrey Zygmont, *Microchip: An Idea, Its Genesis, and the Revolution It Created* (New York: Perseus Publishing, 2003). ISBN: 0-7382-0561-3, page 82.

53 *"If this computer unlocks":* "IBM Announces $100 Million Research Initiative to Build World's Fastest Supercomputer: 'Blue Gene' to Tackle Protein Folding Grand Challenge," IBM Research, Yorktown Heights, NY, December 6, 1999. http://domino.research.ibm.com/comm/pr.nsf/pages/news.19991206_bluegene.html

53 *Even the slightest change:* Ibid.

53 *He expects it soon to cost:* "Myhrvold's Exponential Economy," *Technology Review,* June 2002. http://www.technology review.com/articles/02/06/qa0602.asp?p=1

53 *It also offers the possibility:* I am indebted to Nick Kristof for his thoughts on this. See Nicholas D. Kristof, "Where Is Thy Sting," *New York Times,* August 12, 2003, page A17. http://query.nytimes.com/gst/abstract.html?res=FB0E1FFD3A550C718DDDA10894DB404482

55 *Seventy percent of its merchandise:* Jerry Useem, "One Nation Under Wal-Mart," *Fortune,* February 18, 2002, page 65. http://www.fortune.com/fortune/mostadmired/articles/0,15114,423053,00.html

55 *"Hi, I'm here":* Claudia H. Deutsch, and Barnaby J. Feder, "A Radio Chip in Every Consumer Product," *New York Times,* February 25, 2003, page C1. http://query.nytimes.com/gst/abstract.html?res=F20D13F634590C768EDDAB0894DB404482

55 *By 2006 Wal-Mart expects:* Interview, Tom Williams, Wal-Mart spokesman, August 29, 2003.

55 *Did you know one of the most boosted:* For those of you diligent enough to read the endnotes, a little reward: The reason for the high theft rate of Preparation H, according to a Wal-Mart spokesman, is that its tissue-shrinking properties are allegedly useful to junkies trying to make their injection tracks less obvious. You're welcome.

55 *"The choice facing Dell's rivals":* Timothy J. Mullaney, et al., "Special Report—the E-Business Surprise," *Business Week,* May 12, 2003, page 62. http://www.businessweek.com/@48FPjocQDaRXSQIA/magazine/content/03_19/b3832601.htm

55 *Travel agency locations closed in one year:* Ibid.

56 *the sick, the otherwise healthy with a critical need and the rest of us:* See, for example, Gregory M. Lamb, "Strange Food for Thought: The Brain-Gain Revolution Is Already Under Way. But Will These 'Neural Enhancement' Drugs Turn Us Into Einsteins or Frankensteins?", *Christian Science Monitor,* June 17, 2004. http://www.csmonitor.com/2004/0617/p14s01-stct.htm

56 *"nobody sleeps in New York or Washington":* Joel Garreau, "The Great Awakening: With a Pill Called Modafinil, You Can Go 40 Hours Without Sleep—and See into

the Future," *Washington Post*, June 17, 2002, page C1. http://www.washingtonpost. com/ac2/wp-dyn/A61282-2002Jun16?language=printer

56 *He is the first human confirmed:* Rob Stein, "Muscle-Bound Boy Offers Hope for Humans: Scientists Work to Isolate Secrets of a Genetic Mutation That Could Alleviate Weakness Accompanying Disease and Aging," *Washington Post*, June 28, 2004, page A7. http://www.washingtonpost.com/wp-dyn/articles/A10196-2004Jun27.html

57 *"Athletes find a way of using just about anything. This, unfortunately, is no exception":* E. M. McNally, "Powerful Genes—Myostatin Regulation of Human Muscle Mass," *New England Journal of Medicine* 350, 26 (June 24, 2004), pages 2642–4. http://content.nejm.org/cgi/content/extract/350/26/2642?hits=20&andorexactfulltext=and &where=fulltext&searchterm=mcnally&sortspec=Score%2Bdesc%2BPUB-DATE_SORTDATE%2Bdesc&sendit=GO&excludeflag=TWEEK_element&searc hid=1089587844579_8502&FIRSTINDEX=0&journalcode=nejm

57 *The traditional music industry is being gutted. . . . Sales were down 20 percent:* Business Week, "The E-Business Surprise."

57 *It took more than 18 centuries:* Angus Maddison, *Dynamic Forces in Capitalist Development: A Long-Run Comparative View* (Oxford: Oxford University Press, 1991). ASIN: 0198283989, pages 1–29.

58 *The world's gross domestic product doubled almost three times:* grew by a factor of 7, according to the World Trade Organization, International Trade Statistics, 2003. http://www.wto.org/english/res_e/statis_e/its2003_e/it203_70c_e.htm

58 *The world's exports doubled six and a half times:* grew by a factor of 91. Ibid.

58 *4 billion years:* See, for example, the evolutionary and geological timelines in Ray Kurzweil, *The Age of Spiritual Machines: When Computers Exceed Human Intelligence* (New York: Viking, 1999). ISBN: 0-670-88217-8, page 261, or at the Talk Origins archive. http://www.talkorigins.org/origins/geo_timeline.html

59 *brain-scanning devices—doubling time, 12 months:* Source: Ray Kurzweil.

59 *robots—doubling time, 9 months:* The United Nations Economic Commission predicted a nearly 10-fold growth of the personal and service robotics sector, from $600 million in 2002 to $5.2 billion in 2005. These were digitally intelligent devices with sensing and/or mobility qualities controlled by the equivalents of p.c.s.

59 *The number of scientific journals:* Ilkka Tuomi, "Kurzweil, Moore, and Accelerating Change," working paper, Joint Research Centre, Institute for Prospective Technological Studies, August 27, 2003. http://www.jrc.es/~tuomiil/articles/Kurzweil.pdf

59 *It expects to ship its second billionth chip:* "Intel Ships Billionth Chip," *Silicon Valley/ San Jose Business Journal*, June 9, 2003. http://sanjose.bizjournals.com/sanjose/stories/2003/06/09/daily13html

59 *"gut cam":* Rob Stein, "Patients Find 'Gut Cam' Technology Easy to Swallow," *Washington Post*, December 30, 2002, page A1.

60 *Ryan and Gross were not the most obvious academics:* Everett M. Rogers, *Diffusion of Innovations,* fourth edition (New York: The Free Press, 1995). ISBN: 0-02-874074-2.

61 *"What people mean by the word technology":* Stewart Brand, *The Clock of the Long Now: Time and Responsibility* (New York: Basic Books, 1999). ISBN: 0-465-04512-X, page 16.

61 *He found himself staring:* Interview with Fee, April 28, 2000.

62 *Suffered the classic anxiety attack:* Joel Garreau, "PC Be With You: The New Technology Doesn't Just Want to Be Your Friend. It Wants to Be Your Brain," *Washington Post*, August 29, 2000, page C1. http://www.washingtonpost.com/wp=dyn/articles/a4573=2000aug29.html

62 *They have become part of us and we part of them:* See, for example, Sherry Turkle, *The Second Self: Computers and the Human Spirit* (New York: Simon & Schuster, 1984), AISN: 0671468480. Also Sherry Turkle, *Life on the Screen: Identity in the Age of the Internet* (New York: Simon & Schuster, 1997). ISBN: 0684833484.

62 *"lost her soul":* Janet Rae-DuPree, "Help for Hard Drives," *San Jose Mercury-News,* April 28, 1996. http://www.drivesavers.com/media/sjmercury.htm

62 *"You can hear the white knuckles":* Stange interview, August 17, 2000.

63 *Thomas Lewis, a psychiatrist:* Lewis interview, April 28, 2000.

63 *He is the co-author of:* Fari Amini, Richard Lannon, Thomas Lewis, *A General Theory of Love* (New York: Vintage, 2001). ISBN: 0375709223.

63 *two NASA scientists:* Manfred E. Clynes and Nathan S. Kline, "Cyborgs and Space," *Astronautics,* September 1960, pp. 26–27 and 75–75; reprinted in Chris Habels Gray et al, eds., *The Cyborg Handbook,* (New York: Routledge, 1995). ISBN: 0415908493, pages 29–34.

64 *When it vibrated in the middle of the night:* Deanna Kosma interview, August 21, 2000.

64 *He talks about the two weeks:* Montgomery Kosma interview, August 21, 2000.

65 *a guide to the next decade and a half:* Calculations by Ray Kurzweil.

65 *Below is a list of 15 events:* I am indebted to Erik Smith of Global Business Network for creating most of this list.

66 *exploring and explaining ourselves by telling stories of our future:* I am indebted to Henry Jenkins for his thoughts on this subject. See Henry Jenkins, "Science Fiction and Smart Mobs," *Technology Review,* January 31, 2003. http://www.technology-review.com/articles/wo_jenkins013103.asp?p=1

70 *If you had been in Steve Jobs' garage:* The Apple Museum. http://www.theapple-museum.com/index.php?id=tam&page=history&subpage=1970

71 *In a seminal academic paper:* Vernor Vinge, "The Coming Technological Singularity: How to Survive in the Post-Human Era." The original version was presented at the VISION-21 Symposium sponsored by NASA Lewis Research Center and the Ohio Aerospace Institute, March 30–31, 1993. A slightly changed version appeared in the winter 1993 issue of *Whole Earth Review.* http://www-rohan.sdsu.edu/faculty/vinge/misc/singularity.html

72 *"At this singularity the laws of science and our ability to predict the future would break down":* Stephen Hawking, *The Illustrated A Brief History of Time* (New York: Bantam, 1996). ISBN: 0553937715, page 114. Cited in Brand, *The Clock of the Long Now..*

72 *It is also the point beyond which you cannot see:* See, for example, the discussion at the University of Illinois' Science Center. http://archive.ncsa.uiuc.edu/Cyberia/Num-Rel/BlackHoleAnat.html

72 *The sheer magnitude of each doubling becomes unfathomable:* The Singularity is also re-ferred to as "the Spike" in Damien Broderick, *The Spike: How Our Lives Are Being Transformed by Rapidly Advancing Technologies* (New York: Forge, 2001). ISBN: 0312877811.

WHAT ARE SCENARIOS?

78 *What are scenarios?:* This discussion of scenarios is adapted from the work of Global Business Network of Emeryville, California, the pioneering scenario planning firm.

Chapter Four HEAVEN

87 *where modern America began:* Founding of the Slater Mill, Pawtucket, RI. http://www.slatermill.org

87 *No matter how crazy:* Joel Garreau, *The Nine Nations of North America* (Boston: Houghton-Mifflin, 1981). ISBN: 0-395-29124-0, page 39.

87 *In his book Walden, or Life in the Woods:* Henry David Thoreau, *Walden* (New York: New American Library, 1960).

87 *"men have become the tools of their tools":* Ibid.

88 *As Edison famously said:* The Jerome and Dorothy Lemelson Center for the Study of

Invention and Innovation at the National Museum of American History of the Smithsonian Institution, citing Robert S. Halgrim *Thomas Edison/Henry Ford Winter Estates, Ft. Meyers, Florida* (Kansas City: Terrell Publishing Co., 1993). ISBN: 0-935031-67-7. http://www.si.edu/harcourt/nmah/lemel/edison/html/his_thoughts.html

89 *Then he whispered:* "A biography of Ray Kurzweil," Kurzweil Technologies. http://www.kurzweiltech.com/raybio.htm

89 *He invented the first practical flatbed scanner:* Ibid.

90 *He's received eleven honorary degrees:* "A Brief Career Summary of Ray Kurzweil," *Kurzweil Technologies.* http://www.kurzweiltech.com/aboutray.html

92 *"Religion in my household was ideas and knowledge":* Kurzweil interview, April 14, 2003.

96 *"Well, that was all very well and good":* The outcome of Kurzweil's predictions in *The Age of Intelligent Machines* is discussed in Kurzweil, *The Age of Spiritual Machines,* pages 170–78.

97 *In* The Age of Spiritual Machines: Kurzweil, *The Age of Spiritual Machines* (New York: Penguin, 1999). ISBN: 0-670-88217-8.

97 *He even collaborated on a book:* Jay W. Richards, ed., *Are We Spiritual Machines?: Ray Kurzweil vs. the Critics of Strong AI* (Seattle, Discovery Institute Press, 2002). ISBN: 0-963854-3-9.

97 *In his narrative describing:* Kurzweil, *The Age of Spiritual Machines,* pages 189–201.

98 *Intelligent assistants with animated personalities:* A well-reviewed book and CD package enabling the creation of virtual humans with a convincing illusion of personality was already on the market by 2003. See Peter Plantec, *Virtual Humans: A Build-It-Yourself Kit, Complete With Software and Step-by-Step Instructions* (New York: AMACOM, 2003). ISBN: 0814472214.

98 *Human musicians routinely jam with computer-generated musicians:* On October 14, 2004, the celebration of the 25th anniversary of Carnegie Mellon's renowned Robotics Institute culminated with a concert by the multimedia artist Laurie Anderson, who incorporates robotics and other leading-edge technologies into her art and concerts. "Carnegie Mellon Prepares To Celebrate 25th Anniversary of its Robotics Institute," Carnegie Mellon Media Relations, August 23, 2004. http://www.cmu.edu/PR/releases04/040823_robotics.html

99 *People generally communicate with [computers] . . . as in . . .* Minority Report: Joel Garreau, "Washington as Seen in Hollywood's Crystal Ball," *Washington Post,* June 21, 2002, page C1. http://www.washingtonpost.com/ac2/wp-dyn?pagename=article&node=&contentId=A20469-2002Jun20&notFound=true

100 *This turned on its head Andy Warhol's line:* Public Broadcasting Service, American Masters: "Andy Warhol." http://www.pbs.org/wnet/americanmasters/database/warhol_a.html

100 *"an undercurrent of concern":* Kurzweil, *The Age of Spiritual Machines,* page 206.

101 *"standing in a room up to our knees in a flammable fluid":* Ibid., page 217.

102 *100,000 annual deaths:* Ibid., page 230. See also Jason Lazarou, Bruce H. Pomeranz and Paul N. Corey, "Incidence of Adverse Drug Reactions in Hospitalized Patients: A Meta-analysis Prospective Studies," *Journal of the American Medical Association* 279 (1998) 1200–5; Dorothy Bonn, "Adverse Reactions Remain a Major Cause of Death," *The Lancet* 351 (1998) 118: "Adverse drug reactions (ADR) are the fourth commonest cause of death in the United States, with more than 100,000 deaths per year, after heart disease, cancer and stroke. In a meta-analysis of 39 prospective studies, the incidence of serious and fatal ADRs was 6.7% among patients admitted to hospital because of an ADR. This impressive figure is higher than expected, and suggests that ADRs are considerably under-reported."

103 *smile in every possible way:* Kurzweil, *The Age of Spiritual Machines,* page 227.

104 *"unable to meaningfully participate":* Ibid., page 234.

104 *"accused of preferring younger women":* Ibid., page 235.

104 *"This is quite a technology":* Ibid., page 236.

105 *Kurzweil introduces the idea of MOSHs:* Ibid., page 228.

105 *To get to a point of peace:* Ibid., page 251.

105 *She lives for moments of spiritual experience:* Ibid., page 252.

106 *"I wasn't trying to reverse-engineer a religious vision":* Kurzweil interview, April 14, 2003.

106 *Sorcerers would create:* Michael Denton, "Organism and Machine: The Flawed Analogy," in *Are We Spiritual Machines?,* page 78.

106 *In our earliest epic:* Isaac Mendelsohn, ed., *Religions of the Ancient Near East* (New York: Liberal Arts Press, 1955), pages 47–115; translation by E. A. Speiser, in *Ancient Near East Texts* (Princeton: Princeton University Press, 1950), pages 72–99, notes by Mendelsohn. See also http://alexm.here.ru/mirrors/www.enteract.com/jwalz/Eliade/159.html

106 *"No! You will not die!":* Genesis 3:5 from Alexander Jones, ed., *The Jerusalem Bible* (Garden City, NY: Doubleday, 1966). Library of Congress Catalog Card Number 66-24278.

106 *Prometheus not only created humans:* Robert Graves, *The Greek Myths* (London: Moyer Bell Ltd., 1955); see also Sara Baase, "The Prometheus Myth," http://www-rohan.sdsu.edu/faculty/giftfire/prometheus.html

106 *Daedalus confounded:* Graves, *The Greek Myths.*

106 *We could shape ourselves:* I am indebted to Nick Bostrom of the Faculty of Philosophy of Oxford University for his presentation "Introduction to Transhumanism" at Yale University, June 26, 2003, that inspired many of the historical connections offered here. See also Bonnie Kaplan and Nick Bostrom, "A Somewhat Whiggish and Spotty Historical Background," The Ethics, Technology and Utopian Visions Working Group, Yale University, 2002. It was prepared for a course they taught, "Ethics and Policy of New Technologies," at Yale. "Whig history," by the way, is a pejorative term for history writing that depicts the past as a march of progress to the current "correct" state. "We use it with self-irony," Bostrom says. See http://www.transhumanism.org/resources/Syllabi/YaleHistory.htm

106 *Pico della Mirandola's:* Paul Brians, et al., eds., *Reading About the World,* vol. 1 (Pullman, VA: Harcourt Brace Custom Books, n.d.). http://www.wsu.edu:8080/~wldciv/world_civ_reader/world_civ_reader_1/pico.html

107 *Indeed, in 1580, the kabbalistic Jews:* See especially Rabbi Loew, 1525–1609.

107 *"Philosophy," he wrote:* Galileo Galilei, *Opere Il Saggiatore,* (The Assayer) (Rome: Accademia dei Lincei, 1623), page 171. Cited by J. J. O'Connor and E. F. Robertson, "Galileo Galilei," School of Mathematics and Statistics, University of St. Andrews, Scotland, November 2002. http://www-gap.dcs.st-and.ac.uk/~history/Mathematicians/Galileo.html

107 *elevated scientists over priests:* Arthur Herman, *The Idea of Decline in Western History* (New York: The Free Press, 1997). ISBN: 0-684-82491-3, page 20.

107 *"clearing away of idols":* Francis Bacon, *Novum Organum,* XL. http://www.ac.wwu.edu/~jimi/450/bacon.pdf

107 *It was able convincingly to waddle:* Sigvard Strandh, *A History of the Machine* (New York: Dorset Press, 1979). ASIN: 0894790250. http://music.calarts.edu/~sroberts/articles/DeVaucanson.duck.html

107 *"No doubt man will not become immortal":* Paul Halsall, "Condorcet: The Future Progress of the Human Mind," Modern History Sourcebook, 1997. http://www.fordham.edu/halsall/mod/condorcet-progress.html

107 *In 1780, Benjamin Franklin wrote:* I am indebted to Bruce N. Ames, who researches

the mechanisms of aging at the University of California at Berkeley, for alerting me to this correspondence from Benjamin Franklin.

108 *"The Frankenstein Principle":* I am indebted to the network theorist J. C. Herz for memorably articulating the Frankenstein Principle at a Global Business Network WorldView Meeting on "The Causes and Consequences of Cultural Change," Santa Monica, California, December 9, 2002.

108 *The still-controversial* Origin of Species: Richard Dawkins, "An Early Flowering of Genetics," *The Guardian*, February 8, 2003. http://books.guardian.co.uk/print/ 0,3858,4610613-101750,00.html

108 *With his series about extraordinary voyages:* Petri Liukkonen, "Jules Verne (1828– 1905)," Pegasos, Finland. http://www.kirjasto.sci.fi/verne.htm

109 *In 1958 that vessel:* "History of USS Nautilus (SSN 571)," Historic Ship Nautilus and and Submarine Force Museum. http://www.ussnautilus.org/history.htm

109 **Around the World in Eighty Days:** The film was remade in 2004, to considerably less acclaim.

109 *A few decades after Verne:* Petri Liukkonen, "H(erbert) G(eorge) Wells (1866–1946)," Pegasos, Finland. http://www.kirjasto.sci.fi/hgwells.htm

109 *"more and more a race":* H. G. Wells, *The Outline of History* (New York: Macmillan, 1921).

110 *Three Laws of Robotics:* From the fictional *Handbook of Robotics, 56th Edition, 2058 A.D.,* as quoted in Isaac Asimov, *I, Robot* (New York: Spectrum, 1991). ISBN: 0553294385. http://www.asimovonline.com/asimov_FAQ.html#series13

110 *Dick's sensibility is all:* Frank Rose, "The Second Coming of Philip K. Dick," *Wired*, December 2003, page 198. http://www.wired.com/wired/archive/11.12/philip.html

111 *The ones on the right:* Jonathan K. Cooper, "The Complete Tom Swift Jr. Home Page." http://www.geocities.com/Area51/Vault/3712

111 *"There is no problem that you":* Kurzweil interview, April 14, 2003.

111 *"To expect the unexpected shows":* Oscar Wilde, *An Ideal Husband*, Act 3. http:// www.online-literature.com/booksearch.php

111 *"Why make people inquisitive":* Neil Gaiman and Terry Pratchett, *Good Omens* (New York: Workman Publishing Company, 1990). ISBN: 0441003257, pages 345–346.

111 *"His high pitched voice already":* Andrew Hodges, *Alan Turing: The Enigma of Intelligence* (New York: HarperCollins, 1985). ASIN: 0045100608, page 251.

112 *a Washington policy document:* Mihail C. Roco and William Sims Bainbridge, eds., *Converging Technologies for Improved Human Performance: Nanotechnology, Biotechnology, Information Technology and Cognitive Science* (Arlington, VA: National Science Foundation/Department of Commerce, 2002) and (New York: Kluwer Academic Publishers, 2003). ISBN: 1-402-01254-3. http://wtec.org/Converging Technologies

112 *"It is time to rekindle":* Ibid., page 3.

112 *"a golden age that will be":* Ibid., page 5.

112 *The bullet points in part say that in the next 10 to 20 years:* Ibid., pages 4–5.

114 *"like a single, distributed and interconnected 'brain'":* Ibid., page 6.

115 *"the spirit of the Renaissance":* Ibid., page 3.

115 *"biology's bid to keep pace":* Gregory Stock, *Redesigning Humans: Our Inevitable Genetic Future* (Boston: Houghton Mifflin, 2002). ISBN: 0-618-06026-X, page 33.

115 *"No one really has the guts":* Ibid., page 12.

115 *An earlier one is called:* Gregory Stock, *Metaman: The Merging of Humans and Machines into a Global Superorganism* (New York: Simon & Schuster, 1993). ISBN: 0-671-70723-X.

115 *"humans . . . as . . . divergent as 'poodles and Great Danes'":* *Redesigning Humans,* page 34.

115 *"Homo sapiens would spawn its own successors":* Ibid., page 4.

115 *It's a whole lot closer:* Ibid.

115 *"We are not about to turn away":* Ibid., page 13.

116 *It's intended to fix genes:* Ibid., page 37.

116 *especially when that person:* The September 17, 1999, death of Jesse Gelsinger was the first treatment-related fatality involving adenoviral vectors, which are viruses specially modified to carry genes into certain cells. Ibid., page 36.

116 *Even the Amish use it:* Ibid., page 39.

116 *"Our blindness about the consequences":* Ibid., page 11.

116 *If, however, we add a new chromosome:* Ibid., page 66.

117 *It just holds plug-in points:* Ibid.

117 *"Parents will want the most up-to-date genetic modifications":* Ibid., pages 69–70.

117 *ethical argument, offered by the Council of Responsible Genetics:* Ibid., page 70. See also Council for Responsible Genetics, "Position Paper on Human Germline Manipulation," 1992, updated Fall 2000. http://www.gene-watch.org/programs/cloning/germline-position.html

117 *"People say it would be terrible if we made all girls pretty":* Shaoni Bhattacharya, "Stupidity Should Be Cured, Says DNA Discoverer," NewScientist.com news service, February 28, 2003. http://www.newscientist.com/news/news.jsp?id=ns99993451

117 *But he also sees his own Heaven Scenario as "tame":* Redesigning Humans, page 18.

118 *Rudimentary artificial chromosomes already exist:* Ibid., page 67.

118 *"we should have a fair idea of the size of the task":* Ibid., page 63.

118 *"traditional reproduction may begin to seem antiquated":* Ibid., page 56.

118 *He sees his projections as not at all out of touch:* Ibid., page 67.

118 *"huge leap of faith":* Ibid., page 21.

118 *It is the length of five carbon:* Ray Kurzweil, personal communication, July 7, 2004.

118 *the distance your fingernail grows:* Chris Phoenix, director of research, Center for Responsible Nanotechnology, personal communication, July 7, 2004.

118 *If a nanometer were the size of your nose:* Adam Keiper, "The Nanotechnology Revolution," *The New Atlantis,* summer 2003, pages 17–34. http://www.thenewatlantis.com/archive/2/keiper.htm

118 *By 2003, hundreds of tons of nanomaterials:* Rick Weiss, "For Science, Nanotech Poses Big Unknowns," *Washington Post,* February 1, 2004, page A1. http://www.washingtonpost.com/ac2/wp-dyn/A1487-2004Jan31?language=printer

118 *Such nanotechnology is expected to be a $1 trillion business:* California House Representative Mike Honda, quoted in R. Colin Johnson, "Nanotech R&D Act Becomes Law," *EE Times,* December 3, 2003. http://www.eetimes.com/printableArticle?doc_id=OEG20031203S0025

118 *the gross national product of Canada:* 2002 estimate, U.S. Central Intelligence Agency, *The World Fact Book.* http://www.odci.gov/cia/publications/factbook/rankorder/2001rank.html

119 *On December 29, 1959, . . . Feynman gave:* Richard P. Feynman, "There's Plenty of Room at the Bottom: An Invitation to Enter a New Field of Physics," *Engineering and Science,* California Institute of Technology, February 1960. http://www.zyvex.com/nanotech/feynman.html

119 *At a time when the audience:* For this and other aspects of the history of nanotechnology I am indebted to Adam Keiper, managing editor of the marvelously level-headed yet provocative *The New Atlantis: A Journal of Technology and Society.* Much of the history cited here comes from Keiper, "The Nanotechnology Revolution," *The New Atlantis,* summer 2003, pages 17–34. http://www.thenewatlantis.com/archive/2/keiperprint.htm

119 *In his lecture, Feynman described a world:* Chris Phoenix, "Of Chemistry, Nanobots, and Policy," Center for Responsible Nanotechnology, December 2003. http://crnano.org/Debate.htm

119 *Buckyballs and their cousins:* Richard E. Smalley, "The Smalley Group," http://smalley.rice.edu. See also "Nanotechnology: Molecular Manufacturing: Buckyballs

in Support of Nanotechnology." http://www.riverdeep.net/current/2000/03/front.030300.nano.jhtml

119 *General Electric, Motorola, DuPont:* "The Next Small Thing: A Bright Little Idea Comes to Market," *The Economist,* January 17, 2004, page 52. http://www.economist.com/displaystory.cfm?story_id=S%27%298%20%29P1%5F%25%20%20%20D%0A

120 *Nano may be the basis of half of all pharmaceuticals:* Mihail C. Roco, cited in Jack Uldrich, "'Exponential' Thinking for the Future," Tech Central Station, edited by James F. Glassman, January 21, 2004. http://www2.techcentralstation.com/1051/printer.jsp?CID=1051-012104D

120 *This could be the breakthrough to the stars:* "Audacious and Outrageous: Space Elevators," Science at NASA, September 7, 2000. http://science.nasa.gov/headlines/y2000/ast07sep_1.htm

120 *It is conscious.:* John Searle, professor of philosophy at the University of California at Berkeley, is credited with coining the phrase "strong AI" in his 1980 article "Minds, Brains, and Programs," in *The Behavioral and Brain Sciences,* vol. 3. (Cambridge University Press, 1980). There he wrote: "According to strong AI, the computer is not merely a tool in the study of the mind; rather, the appropriately programmed computer really *is* a mind, in the sense that computers given the right programs can be literally said to *understand* and have other cognitive states. In strong AI, because the programmed computer has cognitive states, the programs are not mere tools that enable us to test psychological explanations; rather, the programs are themselves the explanations." See http://members.aol.com/NeoNoetics/MindsBrainsPrograms.html

122 *Weak machine intelligence can:* Don Phillips, "For Some Airline Pilots, Flying Gets Boring," *Washington Post,* December 17, 2003, page A1. http://www.washingtonpost.com/wp-dyn/articles/A6368-2003Dec16.html

122 *Marvin Minsky is the grand old man:* John Brockman, "Consciousness Is a Big Suitcase: A Talk with Marvin Minsky," Edge: The Third Culture. http://www.edge.org/3rd_culture/minsky/index.html

122 *In fact, he was present when:* Hans Moravec, *Mind Children: The Future of Robot and Human Intelligence* (Cambridge: Harvard University Press, 1988). ISBN: 0-674-57618-7 (paper), page 8.

123 *"bit by bit we'll rebuild ourselves":* Brockman, "Consciousness Is a Big Suitcase: A Talk with Marvin Minsky."

123 *comparable to "only two earlier inventions":* Marvin Minsky, "Foreword," in K. Eric Drexler, *Engines of Creation* (Garden City, NY: Anchor, 1986). ISBN: 0-385-19973-2, page vii.

123 *"of beings whose making we'll supervise":* Marvin Minsky, "What Comes After Minds?" in John Brockman, ed., *The New Humanists: Science at the Edge* (New York: Barnes and Noble Books, 2003). ISBN: 0760745293, pages 198–199.

124 *Japanese Aibo owners have a more emotional relationship:* Tony McNicol, "The Rise of the Machines: Japan's Leading the Robotics Charge, But to Where," *Japan Times,* November 25, 2003. http://www.japantimes.co.jp/cgi-bin/getarticle.pl5?fl20031125zg.htm

124 *Whack it on the nose:* G. Jeffrey MacDonald, "If You Kick a Robotic Dog, Is It Wrong?" *Christian Science Monitor,* February 5, 2004. I love this question. http://www.csmonitor.com/2004/0205/p18s01-stct.html

124 *More than half the owners of these robots give them names:* Erik Baard, "Cyborg Liberation Front: Inside the Movement for Posthuman Rights," *Village Voice,* July 30, 2003. http://www.villagevoice.com/issues/0331/baard.php

124 *In 1988 this pioneer published:* Hans Moravec, *Mind Children: The Future of Robot and Human Intelligence* (Cambridge: Harvard University Press, 1988). ISBN: 0-674-57618-7 (paper).

125 *He meant the intelligent children:* Hans Moravec, *Robot: Mere Machine to Transcendent Mind* (New York: Oxford University Press, 1999). ISBN: 0-19-513630-6 (paper). page viii.

125 *the first time in history somebody tried to surrender to a robot:* Richard A. Muller, "Weapons of Precise Destruction," *Technology Review,* May 10, 2002. http://www. technologyreview.com/articles/print_version/wo_muller051002.asp

125 *the size of sparrows:* 15 cm. long.

125 *such as the Black Widow:* "UAV Forum: Black Widow." http://www.uavforum.com/ vehicles/developmental/blackwidow.htm

125 *The dog would be there to bite the pilot:* "Future of Flight: High Times," *The Economist,* December 13, 2003, page 79. http://www.economist.com/displaystory.cfm?story_ id=2282185

125 *"a robot out of Jello":* John Brockman, "Beyond Computation: A Talk With Rodney Brooks," Edge: The Third Culture. http://www.edge.org/3rd_culture/brooks_ beyond/beyond_index.html

126 *"just because Bill Joy was afraid of them":* Ibid.

126 *These are emergency and military robots:* iRobot home page: http://www.irobot.com

126 *DARPA loves the experimental versions:* See especially the work of Robert S. Full of the Poly-PEDAL Lab at Berkeley. http://polypedal.berkeley.edu

126 *As a result "we will become a merger between flesh and machines":* Rodney A. Brooks, *Flesh and Machines: How Robots Will Change Us* (New York: Pantheon Books, 2002). ISBN: 0-375-42079-7, page x.

Chapter Five HELL

133 *"Technological progress is like an axe":* This is the pithiest and most widely quoted version of Einstein's aphorism. It is also rendered as "Our entire much-praised technological progress, and civilization generally, could be compared to an axe in the hand of a pathological criminal" in Albert Folsing, *Albert Einstein: A Biography* (New York: Viking Press, 1997). ISBN: 0670855456.

133 *"Human nature will be the last":* C. S. Lewis, *The Abolition of Man* (New York: Simon & Schuster, 1996). ISBN: 0-8054-2047-9, pages 69–70.

135 *To snag the end of the runway:* I am indebted to Jim Elwood, director of aviation, Aspen–Pitkin County Airport, for this calculation.

136 *From 1877 to 1893:* Malcolm Rohrbough, *Aspen: The History of a Silver Mining Town 1879–1893* (New York: Oxford University Press, 1986). ISBN: 0-19-564064-3. Also Frank L. Wentworth, *Aspen on the Roaring Fork* (Denver: Sundance Publications, 1976). ISBN: 1112829911 (earlier editions exist). See also, "Aspen, Colorado," Wikipedia. http://en2.wikipedia.org/wiki/Aspen,_Colorado

136 *Joy is in Aspen because:* Joy interview, May 5–6, 2003.

136 *whistle up for him a Citation X:* Joy interview, February 6, 2004.

136 *the world's fastest private jet:* "Citation X." http://citationx.cessna.com/home.chtml

136 *He settled on Aspen, population 5,914:* U.S. Census, year 2000.

136 *He was the commencement speaker:* Joy interview, May 5–6, 2003.

137 *Some houses run to 40,000 square feet:* Joy interview, February 6, 2004.

137 *To get to the Internet, he fires up a satellite dish:* Joy interview, May 5–6, 2003.

137 *no man is an island:* John Donne. "Devotions Upon Emergent Occasions: Meditation XVII, No Man Is an Island."

137 *"We're both from Michigan":* See, for example, Gerald Beals: "Biography of Thomas Alva Edison," Thomasedison.com. http://www.thomasedison.com/biog.htm

137 *"It was fun":* Joy interview, May 5–6, 2003.

137 *In 1982, Joy married that system:* Joy interview, February 6, 2004.

138 *In it, intelligence would be embedded:* Joy launched this ecosystem with the concept known as "peer-to-peer," an advanced form of distributed computing. In it, the instructions that allowed the sharing were kept small and simple. Any device could

quickly and efficiently turn to any other. To accomplish this, Joy invented computer languages and networking technologies and protocols that he called Java, Jini, and JXTA (as in juxtaposition) and offered them to the world for free.

138 *In 1988 this crystallized:* "The Other Bill," *The Economist*, Technology Quarterly, September 19, 2002, page 25. http://www.economist.com/PrinterFriendly.cfm?Story_ID=1324644

138 *He intended his warning to be reminiscent of Albert Einstein's:* Joel Garreau, "From Internet Scientist, a Preview of Extinction," *Washington Post*, March 12, 2000, page A15. http://www.sfgate.com/cgi-bin/article.cgi?file=/chronicle/archive/2000/03/13/MN108057.DTL&type=printable

138 *In a vast, 24-page spread:* Bill Joy, "Why the Future Doesn't Need Us," *Wired*, April 2000. http://www.wired.com/wired/archive/8.04/joy_pr.html

139 *"gray goo" end-of-the-world scenario:* K. Eric Drexler, *Engines of Creation: The Coming Era of Nanotechnology* (New York: Anchor, 1986). ISBN: 0-385-19973-2, pages 172–175. Drexler has backed off considerably on the gray goo scenario since first offering it, arguing that it would be much more simple and efficient to create assemblers that can't autonomously self-replicate than to create ones that can. He also believes there are much bigger things to worry about, such as using nano to create a new arms race, as was actually advocated by the president of India. See, for example: Chris Phoenix and Eric Drexler, "Safe Exponential Manufacturing," *Nanotechnology* 15:8, August 2004, pages 869–72. http://www.iop.org/EJ/abstract/0957-4484/15/8/001/

139 *"Gray goo would surely be":* In his *Wired* piece, Joy inserts the following footnote here: "In his 1963 novel *Cat's Cradle*, Kurt Vonnegut imagined a gray-goo-like accident where a form of ice called ice-nine, which becomes solid at a much higher temperature, freezes the oceans."

140 *His dad, William:* Joy interview, May 5–6, 2003.

140 *By the time he was 3:* Joy. "Why the Future Doesn't Need Us."

141 *He believes they are unprepared:* Joy interview, May 5–6, 2003.

141 *As the right of each sentient species:* "Prime Directive," http://www.70disco.com/startrek/primedir.htm

141 *At the age of 16, in 1971:* Joy interview, May 5–6, 2003.

142 *"breaking the marble spell":* In discussing this in "Why the Future Doesn't Need Us," Joy inserts the following footnote:
Michelangelo wrote a sonnet that begins:

> Non ha l' ottimo artista alcun concetto
> Ch' un marmo solo in sè non circonscriva
> Col suo soverchio; e solo a quello arriva
> La man che ubbidisce all' intelleto.

Stone translates this as:

> The best of artists hath no thought to show
> which the rough stone in its superfluous shell
> doth not include; to break the marble spell
> is all the hand that serves the brain can do.

Stone describes the process: "He was not working from his drawings or clay models; they had all been put away. He was carving from the images in his mind. His eyes and hands knew where every line, curve, mass must emerge, and at what depth in the heart of the stone to create the low relief." *The Agony and the Ecstasy* (New York: Doubleday, 1961), page 144.

142 *The turning point for Joy came on September 17, 1998:* Date from Kurzweil's records.

142 *Financiers with what seemed:* Gary Rivlin, "The Madness of King George: George Gilder listened to the technology, and became guru of the telecosm. The markets lis-

tened to his newsletter, and followed him into the Global Crossing abyss. Yet he's never stopped believing," *Wired*, July 2002. http://www.wired.com/wired/archive/10.07/gilder_pr.html

144 *He discovered he was reading:* Theodore Kaczynski, *The Unabomber Manifesto: Industrial Society & Its Future* (Jolly Roger Press, 1995). ASIN: 0963420526. The history of the original publication of this work is summarized by Court TV Online at http://www.courttv.com/trials/unabomber/manifesto:

> The document that would come to be known as the Unabomber's manifesto was first mentioned in a letter to *The New York Times* editor Warren Hoge. This April 24, 1995 letter proposed a "bargain": if the *Times* would publish a lengthy article, penned by the letter's author—then known only as a representative of "FC," a presumed acronym for a terrorist group—"FC" would end a terrorist campaign which, the letter claimed, included several of the attacks attributed to the Unabomber. That letter was followed a little over a month later by a copy of the 65-page manuscript described in the April letter. The attached letter to Hoge laid out additional terms for publication.
>
> Michael Getler of *The Washington Post* received a similar letter on June 24, 1995, along with a copy of the manuscript. The same day, Bob Guccione of *Penthouse* magazine received a letter, responding to an earlier offer to publish the work in his magazine. The author—"FC"—indicated he would rather publish the work in a more "respectable" publication.
>
> Almost three months later on September 19, the *Times* and the *Post* split costs on a special section of the *Post* that reprinted the manifesto in full. It was that special publication that led David Kaczynski to draw a comparison between the Unabomber and his estranged brother Ted.

145 *He was a speaker:* First Foresight Conference on Nanotechnology in October 1989, a talk titled "The Future of Computation." Published in B. C. Crandall and James Lewis, eds., *Nanotechnology: Research and Perspectives* (Cambridge, MA: MIT Press, 1992), page 269. See also www.foresight.org/Conferences/MNT01/Nano1.html

145 *"would some of them go off":* Joy interview, May 5–6, 2003.

145 *"extraordinary claims require extraordinary proof":* The quote is by the late astronomer Carl Sagan, winner of the Pulitzer Prize and believer in the possibility of extraterrestrial life and intelligence.

146 *Such an artificial life form:* Barton Gellman, "Iraq's Arsenal Was Only on Paper: Since Gulf War, Nonconventional Weapons Never Got Past the Planning Stage," *Washington Post*, January 7, 2004, page A1. http://www.washingtonpost.com/ac2/wp-dyn/A60340-2004Jan6?language=printer

146 *flesh-eating . . . nanobots:* Sheldon Pacotti, "Are We Doomed Yet?: The computer-networked, digital world poses enormous threats to humanity that no government, no matter how totalitarian, can stop. A fully open society is our best chance for survival," Salon.com, March 31, 2003. http://salon.com/tech/feature/2003/03/31/knowledge/print.html

147 *Cowpox virus also has been genetically altered:* Debora MacKenzie, "US Develops Lethal New Viruses," *New Scientist*, October 20, 2003. http://www.newscientist.com/news/news.jsp?id=ns99994318

147 *"At first, they'd bear about":* Lucretius (Titus Lucretius Carus), *Of the Nature of Things*, Book VI: The Plague of Athens, trans. William Ellery Leonard, University of Adelaide Library. http://etext.library.adelaide.edu.au/l/l940/chap32.html

148 *mass starvation by 1975:* Paul Ehrlich, *Population Bomb* (New York: Random House, 2000). ASIN: 0871560194.

149 *theories of collapse:* Arthur Herman, *The Idea of Decline in Western History* (New York: The Free Press, 1997). ISBN: 0-684-82791-3, page 13.

149 *"Accursed be the soil because of you":* Jerusalem Bible, Genesis 3:17–19.

149 *In Greek myth, Pandora:* Roy Willis, gen. ed., *World Mythology* (New York: Henry Holt and Company, 1993). ISBN: 0-8050-4913-4 (paper).

149 *The Hindus, too, have a saga:* Herman, *The Idea of Decline,* page 14.

149 *"To whom can I speak today?":* H. Frankfort, *Ancient Egyptian Religion* (New York: Columbia University Press, 1948), page 143. Cited in Herman, *The Idea of Decline,* page 14.

149 *When she spurned his advances:* "Cassandra in Greek Mythology," *Mythography.* http://www.loggia.com/myth/cassandra.html

149 *To "make a name for ourselves":* Jerusalem Bible (Garden City, NY: Doubleday), Genesis 11:4.

150 *"This is but the start of their undertakings":* Jerusalem Bible, Genesis 11:6.

150 *hard to argue against rationality:* Herman, *The Idea of Decline,* page 22.

150 *The idea of progress was inseparable:* François Guizot, *Historical Essays and Lectures,* Stanley Mellon, ed. (Chicago: University of Chicago Press, 1972), cited in Herman, *The Idea of Decline,* page 25.

150 *"Dependency, especially on political":* Herman, *The Idea of Decline,* page 24.

150 *The perfection of human reason seemed to equal transcendence:* Ibid., page 26.

150 *[Bacon] describes a society:* See, for example, the explanation of why the editors of *The New Atlantis: A Journal of Technology and Society* named it after his work. http://www.thenewatlantis.com/about/

150 *They have no illusions:* Eric Cohen, "Biotech Loses Its Innocence: War and Peace in the Brave New World," *The Weekly Standard,* June 24, 2002. http://www.weeklystandard.com/Content/Public/Articles/000/000/001/369zqxvp.asp

150 *"Virtue and truth produced strength":* Herman, *The Idea of Decline,* page 27.

151 *Malthus, by the way:* Ibid., page 28.

151 *In 1808, Johann Wolfgang von Goethe:* Alfred Bates, ed., *The Drama: Its History, Literature and Influence on Civilization* (London: Historical Publishing Company, 1906), vol. 11, pages 5–7. Reprinted in "The Faust Legend," TheatreHistory.com. http://www.theatrehistory.com/german/goethe003.html

151 *a sensational leap of imagination:* Mary Wollstonecraft Shelley, *Frankenstein* (London: Henry Colburn and Richard Bentley, 1851), page vii.

151 *"It was on a dreary night":* Ibid., page 43.

151 *"When I placed my head":* From the introduction to the 1831 edition, pages x–xi. See Electronic Text Center, University of Virginia Library. http://etext.us.virginia.edu/

152 *"principles of human nature":* Ibid., Preface, page 1.

152 *In 1932, Aldous Huxley addressed:* Aldous Huxley, *Brave New World and Brave New World Revisited* (New York: Perennial, 1942). ISBN: 0060901012.

152 *contentment is more important than freedom or truth:* Jack Coulehan, "Literary Annotations: Huxley, Aldous, Brave New World," New York University Literature, Arts, and Medicine Database, January 1998. http://endeavor.med.nyu.edu/lit-med/lit-med-db/webdocs/webdescrips/huxley1256-des-.html

153 Orwellian *has become the gold standard adjective:* Glenn Frankel, "A Seer's Blind Spots: On George Orwell's 100th, a Look at a Flawed and Fascinating Writer," *Washington Post,* June 25, 2003, page C1.

153 *Human behavior was just the sum total of the effect of external forces:* "People and Discoveries: B. F. Skinner," Public Broadcasting System. http://www.pbs.org/wgbh/aso/databank/entries/bhskin.html

153 *Harvard Law of Animal Behavior:* Steven Pinker, *The Blank Slate: The Modern Denial of Human Nature* (New York: Viking, 2002). ISBN: 0-670-03151-8, page 177.

154 *Informing all of Carson's work:* Brian Payton, "On the Shoulders of Giants: Rachel Carson (1907–1964)," Earth Observatory Library, National Aeronautics and Space Administration. http://earthobservatory.nasa.gov/Library/Giants/Carson/Carson2.html

154 *change is specifically* not *progress:* Joel Garreau, George Cook, et. al., "Progress: Can Americans Ever Agree on What It Means?" School of Public Policy, George Mason University, course syllabus, 1994–1998.

155 *It could cause the extinction of a quarter of the planet's species:* J. Alan Pounds and Robert Puschendorf, "Ecology: Clouded Futures: Global warming is altering the distribution and abundance of plant and animal species. Application of a basic law of ecology predicts that many will vanish if temperatures continue to rise," *Nature,* January 8, 2004.

155 *Those who writhe in icy agony:* Dante Alighieri, *The Inferno,* trans. Robert Pinsky, (New York: Noonday Press, 1996), Circle 9, cantos 31–34. ISBN: 0374524521. See also "Danteworlds," University of Texas, Austin. http://danteworlds.laits.utexas.edu/circle9.html

155 *"Human nature exists":* Francis Fukuyama, *Our Posthuman Future: Consequences of the Biotechnology Revolution* (New York: Farrar, Straus and Giroux, 2002). ISBN: 0-374-23643-7, page 7.

155 *"happy slaves":* Leon Kass, *Toward a More Natural Science: Biology and Human Affairs* (New York: Free Press, 1985). ISBN: 0029170710, page 35.

156 *no end to history . . . if there were no end to science:* Francis Fukuyama, "Second Thoughts: The Last Man in a Bottle," *The National Interest,* summer 1999, pages 16–33. http://www.findarticles.com/p/articles/mi_m2751/is_56/ai_55015107

156 *technology could destroy democracy:* Fukuyama, *Our Posthuman Future,* page 7.

156 *"breed some people with saddles on their backs":* Ibid., page 10.

156 *The Enhanced, The Naturals, and The Rest:* Focusing more on genetics than the other GRIN technologies, Lee M. Silver refers to people with many of the same capabilities as The Enhanced as "the GenRich" in *Remaking Eden: How Genetic Engineering and Cloning Will Transform the American Family* (New York: Avon, 1997). ISBN: 0-380-79243-5. In the uncommonly thoughtful 1997 film *Gattaca,* featuring Uma Thurman, those who qualify for positions at prestigious corporations such as Gattaca by virtue of their genetic engineering are called "Valids"; the naturally born are called "In-Valids."

157 *like fundamentalists eschewing modern pleasures:* If you really want to scare yourself, don't obsess about fundamentalists who want to remain Naturals. Worry about the ones who want to become Enhanced so they can seem to create miracles.

157 *If her family travels to Africa:* Personal observation, Nigeria and Ghana, May 2004.

158 *"self-awareness may be overrated":* Vernor Vinge interview, January 24, 2003.

158 *dial this creature back to 1603:* Terry A. Gray, "A Shakespeare Timeline Summary Chart." http://shakespeare.palomar.edu/timeline/summarychart.htm

158 *he created both Othello and Caliban:* Alden T. and Virginia Mason Vaughan, *Shakespeare's Caliban: A Cultural History* (New York: Cambridge University Press, 1993). ISBN: 052145817X. http://books.cambridge.org/052145817X.htm

159 *he would have no trouble recognizing the astronauts as adventurers:* I am indebted to M. Mitchell Waldrop for this observation.

159 *"The thing I'm worried the most about":* Fukuyama interview, November 20, 2002.

159 *He loves the work of the anthropologist:* Donald E. Brown, *Human Universals* (New York: McGraw-Hill, 1991). ISBN: 007008209X.

159 *If there is prompt recognition:* Fukuyama, *Our Posthuman Future,* page 128.

159 *He defines human nature as:* Ibid., page 130.

160 *That essence, whatever it is:* Ibid., pages 149–151.

160 *"the idea that one could exclude any group of people":* Ibid., page 155.

160 *human nature has not been static:* Ibid., page 156.

160 *"I mean, the Romans were unbelievably cruel":* Fukuyama interview, November 20, 2002.

160 *He recalls a passage:* Alexis De Tocqueville, *Democracy in America* (New York: Alfred A. Knopf, 1945), Volume II, Third Book, Chapter I, "Influence of Democracy on Manners Properly So Called: How Customs Are Softened," pages 174–77.

161 *effort to fix its shortcomings:* Fukuyama, *Our Posthuman Future,* page 13.

161 *The cotton gin was bad:* Ibid., page 15.

161 *The ultimate result was the bloodiest conflict:* Merritt Roe Smith and Leo Marx, eds., *Does Technology Drive History?: The Dilemma of Technological Determinism* (Cambridge, MA: MIT Press, 1994). ISBN: 0-262-19347-7, page x.

162 *"full-scale class war":* Fukuyama, *Our Posthuman Future,* page 16.

162 *"human nature is what gives us a moral sense":* Ibid., pages 101–2.

162 *He's terrified that The Enhanced:* Ibid., page 157.

162 *"actually picking up guns and bombs":* Ibid., page 158.

162 *"I think the answer is no":* Fukuyama interview, November 20, 2002.

162 *"You can't have real compassion":* Ibid.

163 *"literally, people dying off":* Ibid.

163 *unless robots learn to dream:* The title of the 1968 Philip K. Dick book on which the film *Blade Runner* was based, is *Do Androids Dream of Electric Sheep?* (New York: Del Rey, 1996). ISBN: 0345404475.

163 *Fukuyama, who was born in 1952:* "Biography: Francis Fukuyama." http://www.sais-jhu.edu/fukuyama/biograph.htm

163 *"[Raising children] plays a role in the socialization of the parents":* Fukuyama interview, November 20, 2002.

163 *"sex becomes a fairly minor part of life":* Ibid.

163 *"can people conceive of dying for a cause":* Ibid.

164 *"I'm not sure that I'd be happy" [about living for a very long time.]:* Ibid.

164 *"what the world needs is more regulation":* Fukuyama, *Our Posthuman Future,* page 183.

164 *He thinks little of scientific self-regulation:* Ibid., page 184.

164 *"Science cannot by itself establish the ends":* Ibid., page 185.

164 *as much revulsion as we did after Hiroshima:* Ibid., page 216.

165 *"Scientists do not believe":* Personal communication, August 25, 2004.

165 *If you view an organism as so dangerous:* "Secondary Barriers in Representative P4 Facility," Donald S. Fredrickson Papers, National Library of Medicine. http://profiles.nlm.nih.gov/FF/B/B/M/D

165 *"It's just in two different forms":* Joy interview, May 5–6, 2003.

166 *"an ultimate 'Doomsday' catastrophe":* Martin Rees, *Our Final Hour. A Scientist's Warning* (New York: Basic Books, 2003). ISBN: 0-465-06862-6, pages 1–2.

166 *"society more vulnerable to disruption":* Ibid., page 21.

166 *He quotes the odds:* Ibid., page 8.

166 *"rather be red than dead":* Ibid., page 28.

166 *"even if the alternative was a certainty of a Soviet takeover":* Ibid.

167 *As the renowned British satirist:* Terry Pratchett, *Thief of Time* (New York: Harper Collins, 2001). ISBN: 0-06-103132-1, page 82.

167 *"this rock on that head":* Ibid, page 175.

167 *"should support be withdrawn from a line of 'pure' research":* Rees, *Our Final Hour,* page 80.

167 *a "brain gain":* Ibid., page 81.

167 *"Singapore and China aim to leapfrog the competition":* Ibid., page 81.

167 *"The difficulty with a dirigiste":* Ibid.

168 *"To say that a few grams could in principle kill millions":* Ibid., page 50.

168 *a novel way to colonize Mars:* Ibid., page 174.

168 *Actually, in the American Film Institute:* "100 Heroes and Villains," American Film Institute. http://www.afi.com/tvevents/100years/handv.aspx

168 *Susan Greenfield:* Susan Greenfield, *Tomorrow's People: How 21st-Century Technology Is Changing the Way We Think and Feel* (London: Allen Lane: The Penguin Group, 2003). ISBN: 0-713-99631-5.

168 *"permanently 'blow' our minds":* Ibid., page 264.

168 *couldn't handle the problems of plot:* Ibid., page x.

168 *So she tried to make her book nonfiction:* Ibid., page ix.

168 *Since "large-scale death":* Ibid., page 264.

169 *"Men's collective passions are mainly for evil":* Ibid., page 216. Also cited in Melvin Konner, "Fuzzy Vision: A Dark View of a Future Where the Lines of Self Are Blurred," *Nature*, November 27, 2003. http://www.nature.com/cgi-taf/DynaPage.taf?file=/nature/journal/v426/n6965/full/426385a_fs.html

169 *"the celebration of individuality":* Greenfield, *Tomorrow's People*, page 271; Konner, "Fuzzy Vision."

169 *"The central problem of an intelligent species is the problem of sanity":* Greenfield, *Tomorrow's People*, page 270. Konner, "Fuzzy Vision."

169 *"On a visit to North-East Asia, I saw this future":* Greenfield, *Tomorrow's People*, page 266.

170 *In developing countries the proportion of people with access to a phone:* "Mobiles Narrow Information Gap: Mobile Phones Are Helping to Bridge the Communications Divide Between the World's Rich and Poor, a Report Says," BBC News, December 23, 2003. http://newsvote.bbc.co.uk/mpapps/pagetools/print/news.bbc.co.uk/1/hi/technology/3344437.stm

170 *an organization devoted to "an environmentally sustainable":* Worldwatch Institute. http://www.worldwatch.org

170 *Judging from the billboards in the megacity of Lagos:* Personal observation, May 2004.

170 *intentionally slow the revolution:* "Poor Connections: Trouble on the Internet Frontiers," *RAND Review*, December 2002. http://www.rand.org/publications/randreview/issues/rr.12.02/connections.html

170 *Some Middle Eastern societies recoil:* Ian Buruma, "The Origins of Occidentalism," The Chronicle of Higher Education, February 6, 2004. http://chronicle.com/free/V50/i22/22b01001.htm

170 *Singapore researchers examining:* Hao Xiaoming and Chow Seet Kay, "Factors Affecting Internet Development: An Asian Survey," *First Monday* (February 2004). http://firstmonday.org/issues/issue9_2/hao/

170 *The International Telecommunication Union:* "ITU Digital Access Index: World's First Global ICT Ranking: Education and Affordability Key to Boosting New Technology Adoption," International Telecommunication Union, November 20, 2003. http://www.itu.int/newsarchive/press_releases/2003/30.html

171 *In a postliterate world:* Xiaoming and Kay, "Factors Affecting Internet Development."

171 *The digital divide seems to be narrowing:* Wenhong Chen and Barry Wellman, "Charting and Bridging Digital Divides: Comparing Socio-economic, Gender, Life Stage, and Rural-Urban Internet Access and Use in Eight Countries," Netlab, Center for Urban and Community Studies, University of Toronto, October 31, 2003. http://www.chass.utoronto.ca/~wellman/publications/index.html

171 *This technology is getting to the masses a lot faster:* I am indebted to Kevin Kelly for this observation. See John Brockman, *Digerati: Encounters With the Cyber Elite* (San Francisco: HardWired, 1996). ISBN: 1-888869-04-6, page 158.

171 *It has been translated into 20 languages:* Carl T. Hall, "The Unnatural Man: A Search for Meaning in a Genetically Engineered Future," *San Francisco Chronicle*, May 12, 2003. http://sfgate.com/cgi-bin/article.cgi?f=/c/a/2003/05/12/MN20770.DTL

171 *inspired by the Bill Joy alert:* McKibben interview, February 4, 2004.

171 *"one of the great Paul Revere moments"*: Bill McKibben, *Enough: Staying Human in an Engineered Age* (New York: Henry Holt and Company, 2003). ISBN: 0-8050-7096-6, page 92.

171 *he argues in favor of humans embracing their limits:* Ibid., page 216.

171 The End of Human Nature: McKibben interview, February 4, 2004.

172 *"When it was done I had a clearer sense of myself":* McKibben, *Enough,* page 2.

172 *"It's not the personal* challenge *that will disappear":* Ibid., page 7.

172 *The Chinese turned against global maritime exploration:* Ibid., page 169.

172 *and the Japanese gave up guns:* Ibid., page 172.

172 *But, as a result, . . . the West . . . was to dominate:* Ibid., page 171, citing Nicholas D. Kristof, "1492: The Prequel," *New York Times,* January 1, 2000.

172 *The Middle East, too, is still shaped:* See, for example, Bernard Lewis, *What Went Wrong: Western Impact and Middle Eastern Response* (New York: Oxford University Press, 2002). ISBN: 0-19-514420-1.

173 *The first licensed gene therapy:* "China Aproves First Gene Therapy," *Nature Biotechnology,* January 2004, page 3.

173 *The first cloned human embryos came out of South Korea:* Helen R. Pilcher, "Cloned Human Embryos Yield Stem Cells: Study Brings Therapeutic Cloning One Step Closer," *Nature Science Update,* February 12, 2004. http://www.nature.com/nsu/040209/040209-12.html

173 *more interesting case is that of the Amish:* McKibben, *Enough,* page 166.

173 *In 1991, for example, a team of Amish carpenters:* Ibid., page 167, citing Stock, *Redesigning Humans,* page 39.

173 *"actually rein in these technologies":* McKibben, *Enough,* page 162.

173 *"translate into effective political resistance":* Ibid., page 163.

174 *Among those deeply skeptical of human enhancement:* James J. Hughes, "The Politics of Transhumanism," paper originally prepared for the annual meeting of the Society for Social Studies of Science, Cambridge, MA, 2001, page 8. http://www.changesurfer.com/Acad/TranshumPolitics.htm

174 *But those finding some things about which to agree:* Ronald Bailey, "Rage Against the Machines: Witnessing the Birth of the Neo-Luddite Movement," *Reason,* July 2001. http://reason.com/0107/fe.rb.rage.shtml

174 *people who don't believe we are descended from monkeys:* McKibben, *Enough,* page 195.

174 *off-shoots of the Christian anti-abortion movement:* Hughes, "The Politics of Transhumanism," page 26.

174 *McKibben describes as more pro-choice:* Nigel Cameron and Lori Andrews, "Cloning and the Debate on Abortion," *Chicago Tribune,* August 8, 2001, cited by McKibben, *Enough,* page 196.

174 *and Prince Charles:* Scott Rhodie, "Charles Fears Science Could Kill Life on Earth," *Scotland on Sunday.* April 27, 2003. http://scotlandonsunday.scotsman.com/uk.cfm?id=481682003

174 *Yet Friends of the Earth:* McKibben, *Enough,* page 196.

174 *William Kristol, editor of the conservative:* Ibid., page 197.

175 *grudgingly forced to acknowledge . . . technology funding is federal:* Hughes, "The Politics of Transhumanism," page 6.

175 *In the Heaven enthusiasts' Web sites:* World Transhumanist FAQ, cited in Hughes, "The Politics of Transhumanism," page 16.

175 *Greens are imaginatively represented:* "The Viridian Design Movement," http://www.viridiandesign.org

175 *Disabled people, who are among the most technology-dependent:* Hughes, "The Politics of Transhumanism," page 22.

175 *You can find feminists:* Ibid., page 25.

175 *it's the least all those robots . . . can do:* Damien Broderick, "The Spike," page 254, cited in Hughes, "The Politics of Transhumanism," page 21.

175 *You can find them among cultural liberals and conservatives:* James Hughes, "Democratic Transhumanism," *Transhumanity,* April 28, 2002. http://www.changesurfer.com/Acad/DemocraticTranshumanism.htm

175 *"keeps us more or less human":* McKibben, *Enough,* page 197.

176 *"the rush of technological innovation . . . can finally slow":* Ibid, page 198.

176 *what used to be known as "vitalism":* Michael Denton, "Organism and Machine: The Flawed Analogy" in Joy W. Edwards, ed., *Are We Spiritual Machines?* (Seattle: Discovery Institute Press, 2002), page 79.

177 *His larger goal is to create:* Emma Young, "Venter Gets Go-ahead to Build Lifeform," NewScientist.com news service, November 21, 2002. http://www.newscientist.com/news/news.jsp?id=ns99993094

177 *consciousness created by the effects of quantum physics:* See, for example, Roger Penrose, *Shadows of the Mind: A Search for the Missing Science of Consciousness* (New York: Oxford University Press, 1994). ISBN: 0-198-53978-9.

177 *John Searle is such a critic:* John Searle, "I Married a Computer," in Edwards, ed., *Are We Spiritual Machines?*, pages 56–76.

177 *Indeed, they will be "spiritual":* Ibid., page 71.

177 *"They have to do a lot more":* Personal communication, September 17, 2003.

178 *Kurzweil believes that:* Ellen Ullman, "Engineering as a Quantum Act: Mistaking the Pointing Finger for the Moon," presented at "Pop!Tech 2000: Being Human in the Digital Age," October 27, 2000.

178 *writes Ellen Ullman, author of:* Ellen Ullman, *Close to the Machine: Technophilia and Its Discontents* (San Francisco: City Lights Books, November 1997). ISBN: 0872863328; *The Bug* (New York: Nan A. Talese, 2003). ISBN: 0385508603.

178 *"Poetry is what gets lost":* http://www.quoteland.com/author.asp?AUTHOR_ID=127

178 *Then there is Steven Pinker of Harvard:* At the time of this presentation, Pinker was still with his former employer, the Massachusetts Institute of Technology.

179 *genetic enhancement . . . is not particularly likely:* Steven Pinker, "Why Genetic Enhancement Is Too Unlikely to Worry About," *Boston Globe,* June 1, 2003, page D1. This article is very similar to the paper he presented at Boston University. http://pinker.wjh.harvard.edu/globe_better_babies.html

180 *A more sophisticated analysis:* In *Flesh and Machines: How Robots Will Change Us* (New York: Pantheon, 2002). ISBN: 0-375-42079-7, page 206, Rodney Brooks delivers a report on this immortality prediction phenomenon, a slightly condensed version of which is: "In 1993 I attended a conference where Pattie Maes gave a talk entitled 'Why Immortality Is a Dead Idea.' She took as many people as she could find who had publicly predicted downloading of consciousness into silicon, and plotted the dates of their predictions, along with when they themselves would turn seventy years old. Not too surprisingly, the years matched up for each of them. Three score and ten years from their individual births, technology would be ripe for them to download their consciousnesses into a computer. Just in the nick of time! They were each, in their own minds, going to be remarkably lucky, to be in just the right place at the right time." I am indebted to Geoff Cohen, Center for Business Innovation, Cap Gemini Ernst & Young, for drawing this to my attention.

180 *"A traditional utopia is a good society":* Joy interview, May 5–6, 2003.

180 *Dyson has written many:* "Freeman Dyson," The Third Culture. http://www.edge.org/3rd_culture/bios/dysonf.html Also, "Freeman Dyson," The George Garnow Memorial Lecture Series, University of Colorado at Boulder. http://spot.colorado.edu/~gamow/george/1991bio1.html

180 *He replied to Joy's provocative article:* Freeman J. Dyson, "The Future Needs Us!" *New York Review of Books,* February 13, 2003. http://www.nybooks.com/articles/16053

181 *On 9/11, the fourth airplane never made it:* See, for example, Sheldon Pacotti, "Are We Doomed Yet?" Salon.com, March 31, 2003. http://www.salon.com/tech/feature/2003/03/31/knowledge/print.html

181 *An even more hopeful argument:* John Seely Brown and Paul Duguid, "A Response to Bill Joy and the Doom-and-Gloom Technofuturists," *The Industry Standard,* April 13, 2000, reprinted in The American Association for the Advancement of Science, *Science and Technology Yearbook, 2001.* http://www.aaas.org/spp/rd/ch4.pdf

181 *They wrote a very wise book:* John Seely Brown and Paul Duguid, *The Social Life of Information* (Boston: Harvard Business School Press, 2000). ISBN: 0-87584-762-5.

181 *he can't see any other forces:* Brown and Duguid, "A Response to Bill Joy," page 78.

181 *"Technological and social systems shape each other":* Ibid., page 79.

182 *"social systems . . . shape, moderate and redirect the raw power of technologies":* Ibid.

182 *nanotechnology . . . almost wholly on the drawing board:* Ibid.

182 *an immune system could co-evolve:* Ray Kurzweil, "Promise and Peril of the 21st Century," *CIO Magazine,* fall/winter 2003. http://www.cio.com/archive/092203/kurzweil.html

182 *"The thing that handicaps robots most is their lack of a social existence":* Brown and Duguid, "A Response to Bill Joy," page 81.

182 *"We must shore up the foundations of civilization":* William H. Calvin, " 'Phi Beta Kappa Book Prize for Science' acceptance speech," Washington, DC, December 6, 2002. http://williamcalvin.com/2002/PBK.htm See also *A Brain for All Seasons: Human Evolution and Abrupt Climate Change* (Chicago: University of Chicago Press, 2002). ISBN: 0226092011

183 *Whenever Joy and I:* Joy interview, May 5–6, 2003.

Chapter Six PREVAIL

187 *"I decline to accept":* William Faulkner officially earned the Nobel Prize in literature for the year 1949, but he did not receive it until the following year, because the Nobel Prize committee could not reach a consensus in 1949. Hence, two Nobel Prizes were awarded in 1950, for each year. The speech Faulkner delivered was not intelligible to his listeners, both because of his southern accent and because his mouth was too far from the microphone. When it was printed in newspapers the following day, however, it was immediately hailed as one of the more significant addresses ever delivered at a Nobel ceremony. The text here is reprinted from William Faulkner, *Essays, Speeches, and Public Letters,* ed. James B. Meriwether, 2nd edition, (New York: Modern Library, 2004). ISBN: 081297137X. I am indebted to the author Bruce Sterling for reminding me of Faulkner's words, and for appending the following observation about Faulkner's breathtaking line, "when the last ding-dong of doom has clanged and faded from the last worthless rock hanging tideless in the last red and dying evening, that even then there will still be one more sound: that of his puny inexhaustible voice, still talking." Sterling notes: "You know, the most interesting part about that speech is that part right there, where William Faulkner, of all people, is alluding to H. G. Wells and the last journey of the Traveler from *The Time Machine.* It's kind of a completely heartfelt, probably drunk mishmash of cornball crypto-religious literary humanism and the stark, bonkers, apocalyptic notions of atomic Armageddon, human extinction, and deep Darwinian geological time. Man, that was the 20th century all over." Personal communication, March 5, 2004.

189 *In New Mexico, El Camino Real:* Christine Preston, et al., *The Royal Road: El Camino Real from Mexico City to Santa Fe* (Albuquerque: University of New Mexico Press, 1998). ISBN: 082631936X.

189 *But the boundaries move a lot:* "History of Mesilla," ¡Viva Mesilla! http://www.oldmesilla.org/html/history_of_mesilla.html See also: Ellen Dornan, "Discovering

the New World," American Frontiers: A Public Lands Journey. http://www.ameri-canfrontiers.net/history "Mesilla," El Camino Real de Tierra Adentro National Historic Trail. http://elcaminoreal.org Bill Kelly, "La Mesilla, New Mexico—The Last 100 Years," SouthernNewMexico.com, 2003. http://www.southernnewmex-ico.com/Articles/Southwest/Dona_Ana/LaMesilla/LaMesillaNewMexico-thelas.html

190 *Billy the Kid:* Marcelle Brothers, "About Billy the Kid," BillytheKid.com, 2004. http://www.aboutbillythekid.com/ See also " 'Outlaw' William Henry McCarty," The Rockin Cherokee Ranch. http://www.rockincherokee.com/Billy.htm Also, "Billy the Kid Legend," NMIA.com. http://www.nmia.com/~btkog/billy_the_kid.htm There you can experience the impressive audio of a Winchester lever-action rifle—like the one Billy the Kid is holding in the celebrated tintype for which he posed—being cocked every time you change the page.

190 *Mesilla has seen little change:* Kelly, "La Mesilla."

190 *Learned journals have published:* Lanier interview, May 9–11, 2003.

190 *"virtual reality" as a shared experience:* Virtual reality head mounts for a single user were first developed by Ivan Sutherland.

191 *made the cover of Scientific American:* Lawrence G. Tesler, "Programming Languages: They offer a great diversity of ways to specify a computation. A language transforms the computer into a 'virtual machine' whose features and capabilities are determined by the software," *Scientific American,* September 1984, pages 70–78. Cited in Oliver Burkeman, "The Virtual Visionary," *The Guardian,* December 29, 2001. http://education.guardian.co.uk/academicexperts/story/0,1392,625423,00.html

191 *It aimed to create alternative worlds:* "Brief Biography of Jaron Lanier," The Well. http://www.well.com/user/jaron/general.html See also "National Tele-Immersion Initiative." http://www.advanced.org/tele-immersion See also Jaron Lanier, "Phenotropics, or Prospects for Protocol-adverse Computing," The International Computer Science Institute, 2003. http://www.icsi.berkeley.edu/talks/Lanier.html

192 *Virtual reality "represents a kind of new contract between humans and computers":* Howard Rheingold, "Virtual Reality: The Faustian Bargain?" *Noetic Sciences Review,* autumn 1991, page 17. http://www.noetic.org/Ions/publications/review_archives/19/issue19_17.html Quoted in Burkeman, "The Virtual Visionary."

192 *"It was a constant struggle":* Lanier interview, May 9–11, 2003.

192 *"There were diseases that":* Burkeman, "The Virtual Visionary."

192 *his classmates had murdered the child:* Lanier interview, May 9–11, 2003.

193 *He recalls little of man's first landing on the moon:* Lanier, personal communication, March 20, 2004.

193 *"my father made some disastrous financial decisions":* Lanier interview, May 9–11, 2003.

194 *The White Sands Missile Range:* Keith Gaudet, "The Cultural History of Los Alamos and Nuclear Matters," Atomic America: Technology, Representative and Culture in the 20th Century, University of New Mexico. http://www.unm.edu/~abqteach/atomicamerica/00-01-01.htm

195 *"an inconspicuous movement":* Jon Katz, *Geeks: How Two Lost Boys Rode the Internet Out of Idaho* (New York: Broadway, 2001). ISBN: 0767906993. Cited by Burkeman, "The Virtual Visionary." Excerpt: "Post-Nerds, Part I: The Geek Ascension," July 1, 1997, HotWired.com. http://hotwired.wired.com/synapse/katz/97/26/katz1a_text.html

195 *"an avant-garde seismic engineer":* Lanier interview, May 9–11, 2003.

195 *a postage stamp honoring him:* Burkeman, "The Virtual Visionary."

195 *"the connection I lost":* Ibid.

195 *"It's such a gizmo outcome":* Lanier interview, May 9–11, 2003.

196 *"The universe doesn't provide":* Ibid.

197 *in 1986 published a book:* James P. Carse, *Finite and Infinite Games* (New York: Ballantine Books, 1987). ISBN: 0345341848.

197 *They are serious and determined about getting that outcome:* Flemming Funch, "Finite and Infinite Games," World Transformation. http://www.worldtrans.org/pos/infinitegames.html

197 *"Life, liberty, and the pursuit":* Lanier interview, May 9–11, 2003.

198 *If you treat a computer like a person:* Peter Leyden, "Around the World of Ideas: A GBN Interview with Jaron Lanier," Global Business Network, September 2001, page 26. http://www.gbn.com/public/gbnstory/downloads/gbn_world_ideas.pdf

198 *We model ourselves after our technologies:* "What Keeps Jaron Lanier Awake at Night: Artificial Intelligence, Cybernetic Totalism, and the Loss of Common Sense," interview with Alex Steffen, *Whole Earth* magazine, spring 2003. http://www.wholeearthmag.com/ArticleBin/111-5.pdf

198 *"It shuts down the game":* Lanier interview, May 9–11, 2003.

199 *Will a bot ever get to know you:* I am indebted to Adrienne Cook Garreau for providing this example of her experience with Royal Thai Tailors and Cleaners of New Baltimore, VA.

199 *"slow suicide through nerdification":* Lanier interview, May 9–11, 2003.

199 *"the religion of the elite technologists":* Jaron Lanier, "One Half of a Manifesto," Edge.org, no. 74, September 25, 2000. http://www.edge.org/3rd_culture/lanier/lanier_index.html Also published in *Wired,* December 2000. http://www.wired.com/wired/archive/8.12/lanier_pr.html

199 *"computers are becoming . . . a successor species":* Lanier interview, May 9–11, 2003.

204 *You're making this up:* Lanier is not making this up. See, for example, Roger T. Hanlon and John B. Messenger, *Cephalopod Behaviour* (Cambridge: Cambridge University Press, 1996). ISBN: 0521645832. See also Eric Scigliano, "Through the Eye of an Octopus: An Exploration of the Brainpower of a Lowly Mollusk," photography by Jennifer Tzar, *Discover,* October 10, 2003. http://www.discover.com/issues/oct-03/features/feateye Roland Anderson and Jennifer Mather, "Octopuses Are Smart Suckers." http://is.dal.ca/~ceph/TCP/smarts.html James Wood's Cephalopod site has scientific articles, a wealth of information about different species, and excellent FAQ pages. http://www.dal.ca/~ceph/TCP For more hard-core cephaloscience, go to CephBase. http://www.cephbase.utmb.edu

205 *"nation of shopkeepers":* "England is a nation of shopkeepers," Napoleon, Napoleonic Guide. http://www.napoleonguide.com/maxim_britain.htm "Let Pitt then boast of his victory to his nation of shopkeepers," Bertrand Barere before the National Convention, WorldofQuotes.com. http://www.worldofquotes.com/topic/England/1 "To found a great empire for the sole purpose of raising up a nation of shopkeepers, may at first sight appear a project fit only for a nation of shopkeepers. It is, however, a project altogether unfit for a nation of shopkeepers, but extremely fit for a nation whose government is influenced by shopkeepers," Adam Smith, *Wealth of Nations* (1776), Giga-USA.com. http://www.giga-usa.com/gigaweb1/quotes2/quautsmith1adamx001.htm

206 *"hero stories . . . have survival value":* Personal communication, March 1, 2004.

206 *"I might exterminate you":* Jerusalem Bible, Exodus 33:3.

206 *Six hundred thousand families:* Exodus 12:37.

206 *kvetching and wailing:* See, for example, Exodus 16.

206 *spend forty years:* Exodus 16:35.

207 *"That's just the way: a person does a low-down thing":* Mark Twain, *The Adventures of Huckleberry Finn* (New York: Harper & Brothers, Publishers, Authors National Edition, 1884), pages 277–79.

209 *the ragged human convoy:* I am indebted to Mary Luti for this observation. See Mary J.

Luti, "Muddling Through (II Kings)," *The Christian Century*, September 23–30, 1998, page 859. http://www.religion-online.org/cgi-bin/relsearchd.dll/showarticle?item_id=629

210   *"Americans can always be counted on to do the right thing":* Quotations Database. http://www.quotedb.com/quotes/2313

210   *Genghis Kahn's Mongols killed nearly as many people as did all of World War II:* See, for example, Colin Mcevedy and Richard Jones, *Atlas of World Population History* (Harmondsworth: Penguin Books Ltd, 1978). ASIN: 0713910313.

210   *streets . . . "greasy with the fat of the slain":* The words of an eyewitness with a talent for detail cited in Felipe Fernández-Armesto, "Steppes Towards the Future," *The Independent*, March 12, 2004. http://enjoyment.independent.co.uk/books/reviews/story.jsp?story=500264

210   *a third ramp exists . . . of increased connection between people:* See, for example, "The Evolution of Everyday Life: Cooperation has brought the human race a long way in a staggeringly short time," *The Economist*, August 12, 2004. http://www.economist.com/finance/PrinterFriendly.cfm?Story_ID=3084745

211   *"I'm serious about that":* Personal communication, March 4, 2004.

211   *"buy a telephone for less than $10 and you expect it to work":* Joab Jackson, "DARPA Takes Aim at IT Sacred Cows," *Government Computer News*, March 11, 2004. http://www.gcn.com/cgi-bin/udt/im.display.printable?client.id=gcndaily2&story.id=25240

211   *"planet of the help desks":* Lanier, "One Half of a Manifesto."

211   *In 1950, in the article:* Waldemar Kaempffert, "Miracles You'll See in the Next 50 Years," *Popular Mechanics*, February 1950, page 112.

212   *Inventors fundamentally misunderstood human behavior:* It's also why we have so few watches with digital readouts. There is something esthetically compelling about the recurring cycle of time displayed by the sweep of the hands of an analog watch.

212   *"a really bad sign for people who expect a real-soon-now Singularity":* Eamoun Kelly, Peter Leyden and members of Global Business Network, *What's Next: Exploring the New Terrain for Business* (New York: Perseus Publishing, 2002). ISBN: 0-7382-0760-8, page 178.

212   *software . . . still an exponential increase:* Ray Kurzweil, "One Half of an Argument: A Counterpoint to Jaron Lanier's Dystopian Visions of Runaway Technology Cataclysm in 'One Half of a Manifesto,'" KurzweilAI.net, July 31, 2001. http://www.kurzweilai.net/articles/art0236.html

213   *If they can write the software, they can prove I'm wrong:* Lanier interview, May 9–11, 2003.

213   *This altercation goes around and around:* See, for example, the responses of George Dyson, Freeman Dyson, Cliff Barney, Bruce Sterling, Rodney Brooks, Henry Warwick, Kevin Kelly, Margaret Wertheim, John Baez, Lee Smolin, Stewart Brand, Rodney Brooks, Daniel C. Dennett, and Philip W. Anderson to "One Half of a Manifesto" at Edge.org. http://www.edge.org/discourse/jaron_manifesto.html See also Lanier's responses to them. http://www.edge.org/discourse/jaron_answer.html

213   *"I want to believe that moral progress has been real":* Jaron Lanier, "Postscript Re: Ray Kurzweil," Edge.org. http://www.edge.org/discourse/jaron_answer.html It is also available at KurzweilAI.net. http://www.kurzweilai.net/articles/art0233.html

213   *"exponential rate of expansion of the 'circle of empathy,'":* Ibid.

213   *"want to draw the circle pretty large":* Lanier interview, May 9–11, 2003.

214   *at the University of St. Andrews:* Garreau, Joel. "Cell Biology: Like the Bee, This Evolving Species Buzzes and Swarms," *Washington Post*, July 31, 2002, Page C1. http://www.washingtonpost.com/ac2/wp-dyn/A23395-2002Jul30?language=printer

214 *"A quite sophisticated text messaging network has sprung up":* "Girls Hot on Will's Trail: Girls Chase Prince by Phone," *Scottish Daily Record,* November 5, 2001, page 9.

215 *He helped pioneer virtual communities:* Howard Rheingold, *The Virtual Community: Homesteading on the Electronic Frontier,* revised edition (Cambridge, MA: MIT Press, 2000). ISBN: 0262681218.

216 *Rheingold . . . author of:* Howard Rheingold, *Smart Mobs: The Next Social Revolution* (New York: Basic Books, 2002). ISBN: 0738208612.

216 *"There go the people":* Attributed to Alexandre Auguste Ledru-Rollin, during the revolution of 1848 in France by, among others, Dan Baum, "Annals of War: Battle Lessons: What the Generals Don't Know," *The New Yorker,* January 17, 2005. http://www.newyorker.com/fact/content/?050117fa_fact

216 *"You can get a rally in 30 minutes":* Rajiv Chandrasekaran, "Philippine Activism, at Push of a Button: Technology Used to Spur Political Change," *Washington Post,* December 10, 2000, page A44.

217 *John Arquilla co-authored:* John Arquilla and David Ronfeldt, *Swarming and The Future of Conflict* (Santa Monica: RAND: National Defense Research Institute, 2000). ISBN: 0-8330-2885-5.

220 *"That wouldn't even be funny once":* I am indebted to Evangeline Reed Garreau for noting that she would find it equally useless to encounter eighth-grade boy classmates with such wings; she is certain they would use them only to project dirty words.

222 *A local sternly lectures:* I am indebted to Pamela McCorduck of Santa Fe for this insight. Personal communication, December 20, 2003.

222 *behind the counter is:* I am indebted to Dave Sment, publisher of the *Hatch Citizen,* for helping me track down Felipe Mendoza after I left Hatch.

Chapter Seven TRANSCEND

227 *"Transcendence is the belief":* Personal communication, December 6, 2002. Shirky, who thinks about the behavior of networks, notes that his formulation borrows from restatements of the laws of thermodynamics as in, for example, John D. Barrow, *The World Within the World* (New York: Oxford University Press, 1990). ASIN: 0192861085.

229 *The precautionary principle is just a way to block research:* Ronald Bailey, "Making the Future Safe," *Reason,* July 2, 2003. http://reason.com/rb/rb070203.shtml

230 *try this on twenty generations of primates first:* Ibid.

230 *the precautionary principle . . . to prevent future calamities:* Ibid.

230 *"like arguing in favor of the plow":* Erik Baard, "Cyborg Liberation Front: Inside the Movement for Posthuman Rights," *Village Voice,* July 30, 2003. http://www.villagevoice.com/issues/0331/baard.php

230 *"It is realism":* Ibid.

230 *"these issues are on the table today":* Ibid.

231 *The colors are striking:* Judith Ann Schiff, "Old Yale: That Wonderful Window," *Yale Magazine,* March 1993, page 15. I am indebted to Nancy Lyon of the Yale archives for unearthing this document.

231 *Transhumanism is a loosely defined movement that started in the 1970s:* "The Transhumanist Movement," Transhumanist Arts and Culture World Center, 2003. http://www.extropic-art.com/history.htm

231 *"posthumans," defined:* Nick Bostrom, "Introduction to Transhumanism," talk delivered at Yale University, June 26, 2003. http://www.nickbostrom.com/ppt/introduction.ppt

232 *Transhumanists view human nature as "a work-in-progress":* Nick Bostrom, "Transhumanist Values," Department of Philosophy, Yale University, April 2001. http://www.nickbostrom.com/tra/values.html

232 *Natasha Vita-More, the artist formerly known as Nancie Clark:* "The Transhumanist Movement."

232 *She offers a conceptual model of an optimized human:* Erik Baard, "Cyborg Liberation Front."

232 *Critics of The Singularity:* "The Rapture of the nerds" is a phrase coined by Ken MacLeod, of Scotland, in his 1998 pulp novel, *The Cassini Division* (New York: Tor, 2000). ISBN: 0812568583. MacLeod's characters transcend their mortal limits by uploading their consciousnesses into android bodies or virtual environments, a process that one character describes as "the Rapture of the nerds." Jeremy Smith describes the plot in "Engines of Light: The Gnostic Potboilers of Ken MacLeod," *Strange Horizons,* January 2003. http://www.strangehorizons.com/2003/20030113/macleod.shtml

One such group of post-humans set themselves up as false gods seeking to dominate the solar system, only to be destroyed in a cometary bombardment orchestrated by a soldier named Ellen May Ngewthu. Ellen May's hatred of the post-humans extends to all artificial and digitized intelligence, until she is forced to transfer her consciousness from her body to an electronic environment and back again, displacing her original consciousness with a copy.

"Something happened then," says the character Ellen May. "In that brief, eternal moment when I sparked across the gap between [the computer] and the skull. . . . I saw a galaxy of green and gold, its starlight filtered through endless, countless habitats; the federation of our dreams. And behind it all, in the walls of all our worlds, an immense but finite benevolence, a great engine of protection and survival; a god on our side, a terror to our enemies and a friend to us, worlds without end."

"Rapture of the nerds" is a play on words. The Rapture refers to the belief of some fundamentalist Christians that in the final days, the just will be assumed bodily into heaven—perhaps even as they are driving down the road. John Nelson Darby, a 19th century Irish lawyer turned Anglican preacher, was "the father of the rapture doctrine," according to "History of the Rapture Theory." http://www.rusearching.com/leftbehind/leftrapturehistory.htm He was the first to develop a full-blown theology that incorporated the teaching that Jesus would return secretly (his Second Coming) to Rapture his true followers, leaving the rest behind to be ruled by an evil Antichrist for seven years, and then return again (his Third Coming) in a visible, glorious coming to destroy the Antichrist, save those who were converted during the seven-year tribulation, and establish his own kingdom.

The jibe is that The Heaven Scenario is merely transcendentalism for people who have replaced a faith in God with a faith in technology, and is no more based in reason.

233 *Indeed, the gathering will be written up:* Carl Elliott, "Humanity 2.0," *The Wilson Quarterly,* autumn 2003. http://www.tc.umn.edu/~ellioo23/documents/humanity.pdf Elliott is the author of *Better Than Well: American Medicine Meets the American Dream* (New York: W.W. Norton & Company, 2003). ISBN: 0-393-05201-X (hardcover).

233 **Reason** *magazine:* Bailey, "Making the Future Safe."

233 **Village Voice:** Baard, "Cyborg Liberation Front."

234 *"we are called to be architects of the future":* BrainyQuote. http://www.brainyquote.com/quotes/quotes/r/rbuckmins153435.html

234 *warlike surplus of frustrated young males:* Leon R. Kass, chairman, *Beyond Therapy: Biotechnology and the Pursuit of Happiness,* Report of the President's Council on Bioethics, pre-publication version, page 61. http://www.bioethics.gov/reports/beyondtherapy

234 *If a child is created:* Ibid., page 55.

234 *If a child is created:* My daughter Evangeline, at the age of thirteen, expressed her view that if parents genetically select for an Olympic athlete or a Yale valedictorian

and that's not what they get, they won't hate the kid; they'll sue the genetic engineering company. I think she's right.

234 *Suppose our desire for [high-performance] children . . . is fulfilled by drugs:* "Beyond Therapy," page 73.

234 *Kass is a medical doctor:* Ibid., page xi.

235 *"Life is not just behaving, performing, achieving":* Ibid., page 94.

235 *light out for the Territory:* Twain, *Huckleberry Finn*, page 375.

235 *"What is a man?":* David Zindell, *The Broken God* (Amherst, MA: Acacia Press, 1994). ISBN: 0586211896.

236 *nasty, brutish and short:* Leslie Stevenson, *The Study of Human Nature: A Reader*, second edition (New York: Oxford University Press, 2000). ISBN: 0-19-512715-3, page 90; and Thomas Hobbes, *Leviathan*, John Plamenatz, ed. (London: Fontana, 1962), excerpted pages 91–97 in Stevenson's work.

237 *embryos should be recognized as humans:* Felipe Fernández-Armesto, *So You Think You're Human? A Brief History of Humankind* (New York: Oxford University Press, 2004). ISBN: 0192804170, page 148.

237 *"We have to face challenges to our concept of humankind":* Ibid., page 168.

237 *He likes the dynamic view:* Justin Stagl, "Anthropological Universality: On the Validity of Generalisations About Human Nature," pages 25–46 in Neil Roughley, ed., *Being Humans: Anthropological Universality and Particularity in Transdisciplinary Perspectives* (Berlin: Walter de Gruyter, 2000). ISBN: 3110169746.

237 *"There is still no agreement about what 'human nature' is":* Fernández-Armesto, *So You Think You're Human*, pages 169–170.

238 *Mead, author of:* Margaret Mead, *Coming of Age in Samoa: A Psychological Study of Primitive Youth for Western Civilisation*, originally published in 1928 (New York: Harper Perennial, 2001). ISBN: 0688050336.

239 *Each is free to shape who he or she is going to be:* Leslie Stevenson and David L. Haberman, *Ten Theories of Human Nature: Confucianism, Hinduism, The Bible, Plato, Kant, Marx, Freud, Sartre, Skinner, Lorenz* (New York: Oxford University Press, 1998). ISBN: 0-19-512040-X (hardcover), pages 3–10.

239 *Edward O. Wilson in his book:* Edward O. Wilson, *On Human Nature* (Cambridge, MA: Harvard University Press, 1978). ISBN: 0-674-63442-X (paperback), pages 3–4.

239 *the human species . . . is indeed quite new:* Melvin Konner, "Why We Did It: An Account of the Driving Forces Behind the Unfolding of Human Civilisation," a review of Michael Cook, *A Brief History of the Human Race* (New York: W.W. Norton, 2003). ISBN: 0393052311, in *Nature*, March 11, 2004.

239 *We share 98 percent of our genome with chimpanzees:* See, for example, Jared M. Diamond, *The Third Chimpanzee: The Evolution and Future of the Human Animal* (New York: Perennial, 1992). ISBN: 0060984031 (paperback).

239 *The genetic difference that makes us human is barely significant:* Fernández-Armesto, *So You Think You're Human*, page 147.

239 *Humans respond with shared tendencies:* See, for example, Jared M. Diamond, *Guns, Germs, and Steel: The Fates of Human Societies* (New York: W.W. Norton, 1997). ISBN: 0393317552.

239 *thousands of human universals:* See, for example, George Peter Murdock, "The Common Denominator of Cultures," in Ralph Linton, ed., *The Science of Man in the World Crisis* (New York: Columbia University Press, 1945), pages 123–42. Cited in Donald E. Brown, *Human Universals* (Boston: McGraw-Hill, 1991). ISBN: 007008209X, pages 69–70.

239 *"The future always comes too fast":* Alvin Toffler, *Future Shock* (New York: Bantam, 1984). ISBN: 0553277375, page 4.

240 *He was born in Helsingborg:* Bostrom interview, June 28, 2003.

241   *It was* Thus Spake Zarathustra: Friedrich Nietzsche, *Thus Spake Zarathustra* (Mineola, NY: Dover, 1999). ISBN: 0486406636.

241   *"God is dead":* The Merriam-Webster Encyclopedia of Literature (Springfield, MA: Merriam-Webster, 1995). ISBN: 0877790426.

241   *Bostrom read* Zarathustra: Bostrom interview, June 28, 2003.

242   *In 1948, T. S. Eliot . . . as he was writing the play:* "The Cocktail Party," in T. S. Eliot, *The Complete Poems and Plays: 1909–1950* (New York; Harcourt, 1952). ISBN: 015121185X, cited in "The Transhumanist Movement." See also "The Cocktail Party: Introduction," Enotes.com. http://www.enotes.com/cocktail-party/12787

242   *in which he coined the term* transhuman: Eliot referred to the human journey as a "process by which the human is Transhumanised."

242   *"we might actually change the human nature":* Personal communication, May 7, 2004.

243   *"Mother Nature . . . in jail for child abuse":* Nick Bostrom, "In Defense of Posthuman Dignity," 2003, accepted for publication in *Bioethics*. http://www.nickbostrom.com/ethics/dignity.html

243   *human, but it is not humane:* Bostrom, "Transhumanist Values."

243   *Pence, who presented:* Gregory E. Pence, *Classic Cases in Medical Ethics: Accounts of Cases That Have Shaped Medical Ethics* (New York: McGraw-Hill, 1999). ISBN: 0073039861. See also Gregory E. Pence, *Brave New Bioethics* (Lanham, MD: Rowman & Littlefield, 2003). ISBN: 0742514366.

243   *more talent in the visual and performing arts:* By 2004, classical musicians were already arguing about the Enhanced in their midst. Those who used anti-fear beta blocking drugs to overcome their paralyzing stage fright were compared to baseball players on steroids. Blair Tindall, "Better Playing Through Chemistry," *New York Times*, October 17, 2004. http://www.listproc.ucdavis.edu/archives/mlist/logo410/0003.html

244   *triple the remaining life span:* At the University of Michigan Medical School in 2004 there already was a genetically engineered dwarf mouse named Yoda who was the human equivalent of 136 years old and described as still sexually active and "looking good." "Genetically Modified Mouse Celebrates Fourth Birthday, Human Equivalent of 130 Plus," Associated Press, April 12, 2004. http://www.signonsandiego.com/news/science/20040412-1544-geriatricmouse.html

244   *Some fear that The Enhanced will see those at the bottom:* See, for example, Michael J. Sandel, "The Case Against Perfection: What's Wrong With Designer Children, Bionic Athletes, and Genetic Engineering," *Atlantic Monthly*, April 2004. http://www.theatlantic.com/issues/2004/04/sandel.htm

244   *Life is unfair:* Lee M. Silver, "The Inevitability of Human Genetic Enhancement and Its Impact on Humanity," presented at "The Future of Human Nature: A Symposium on the Promises and Challenges of the Revolutions in Genomics and Computer Science," Frederick S. Pardee Center for the Study of the Longer-Range Future at Boston University, April 10, 2003. http://www.bu.edu/pardee/conferences/sprg03.Silver.htm

244   *He is the author of:* Lee M. Silver, *Remaking Eden: How Genetic Engineering and Cloning Will Transform the American Family* (New York: Avon, 1997). ISBN: 0-380-79243-5.

245   *"A flourishing human life is not a life lived with an ageless body":* Beyond Therapy, page 299.

245   *compete . . . it usually ends badly:* The common wisdom in paleoanthropology is that ecological niche competition, or the lack thereof, is the reason chimpanzees still exist, while Cro-Magnons and Neanderthals do not. The chimpanzees that survived stayed up in the trees in their own niche and didn't compete with the humans. William Calvin, University of Washington, personal communication, October 11, 2004.

245   *"There could be herds of almost posthumans":* Bostrom interview, June 28, 2003.

246   *"The goal is peaceful coexistence":* Panel discussion at "The Future of Human Nature:

A Symposium on the Promises and Challenges of the Revolutions in Genomics and Computer Science."

246   *"societies not trying to . . . prevent each other from reproducing":* Hughes in a presentation at The Adaptable Human Body: Transhumanism and Bioethics in the 21st Century Conference, June 27–29, 2003, Yale University.

246   *"actually get a decrease in inequality":* Bostrom interview, June 28, 2003.

246   *"Man is a rope, tied between beast and Overman":* Zarathustra, page 126, cited in Robert C. Solomon and Kathleen M. Higgins, *What Nietzsche Really Said* (New York: Shocken Books, 2000). ISBN: 0-805-24157-4, page 20.

246   *aim at becoming such an admirable posthuman:* Solomon and Higgins, *What Nietzsche Really Said,* pages 47, 215.

247   *the five clear and haunting:* Listen to the theme music at, for example, http://www.crystalinks.com/2001z.html

247   *flexibility in timing motherhood:* Rick Weiss, "Procedure Could Revitalize Women's Fertility Hopes: Process of Freezing Eggs and Reimplanting Them When Time Is Right May Allow Flexibility in Timing Motherhood," *Washington Post,* March 15, 2004, page A8. http://www.washingtonpost.com/ac2/wp-dyn/A58458-2004Mar14?language=printer

248   *an issue once far on the fringes:* Alan Cooperman and David Von Drehle, "Vatican Instructs Legislators on Gays: Backing Marriages Called 'Immoral,'" *Washington Post,* August 1, 2003, page A1.

248   *Blaise Pascal was the brilliant:* Mary Bellis, "Blaise Pascal (1623–1662)," About.com, 2004. http://inventors.about.com/library/inventors/blpascal.htm

249   *it pays to play the odds:* Alan Hájek, "Pascal's Wager," *Stanford Encyclopedia of Philosophy,* Edward N. Zalta, ed., spring 2004. http://plato.stanford.edu/archives/spr2004/entries/pascal-wager/

249   *"Genetic engineering":* Sandel, "The Case Against Perfection."

249   *Michael J. Sandel, a professor:* Sandel is the author of *Democracy's Discontent: America in Search of a Public Philosophy* (Cambridge, MA: Belknap Press, 1996). ISBN: 0674197445.

249   *"humans would play God as God does":* Ted Peters, *Playing God? Genetic Determinism and Human Freedom,* second edition (New York: Routledge, 2003). Cited in "Future Survey: A Monthly Abstract of Books, Articles, and Reports Concerning Forecasts, Trends, and Ideas about the Future, a World Future Society Publication," Michael Marien, ed. Monthly newsletter from: World Future Society, 7910 Woodmont Ave., Suite 450, Bethesda, MD 20814, USA. www.wfs.org/fsseptember,-2003,item25:9/449A

250   *She had become a predator . . . at a distance:* William H. Calvin, *The Throwing Madonna: Essays on the Brain,* originally published in 1983, 1991 (New York: Backinprint.Com, 2001). ISBN: 0595160492. See especially Chapter 1, "The Throwing Madonna," which is especially remarkable in that it suggests that most humans are right-handed because women carry babies on their left arm. The infants stay quieter—not scaring off the prey—where they can best hear the thumping beat of the left ventricle of the heart, thus making the preferred pitching hand the right. http://williamcalvin.com/bk2/bk2ch1.htm See also Chapter 4, "Did Throwing Stones Lead to Bigger Brains?" http://williamcalvin.com/bk2/bk2ch4.htm

251   *He has been studying innovation:* See, for example, Don E. Kash, *Perpetual Innovation: The New World of Competition* (New York: Basic Books, 1989). ISBN: 0-465-05533-8 (paperback).

251   *"No one person knows precisely how the organization accomplished it":* Kash interview, April 22, 2004.

251   *"we try sumptin' else":* Ibid.

251   *"If you don't innovate, you die'":* Ibid.

251 *"That's. What. They. Experience."*: Kash, faculty seminar, "The Future of Human Nature," The School of Public Policy, George Mason University, July 9, 2003.

252 *"we have a conceptual model"*: Ibid.

252 *"not scarcity, it's surplus"*: Ibid.

252 *"how to engineer and change it. Including human nature"*: Ibid.

252 *"there isn't anything that God can do that we can't do"*: Kash interview, April 22, 2004.

254 *It points toward a gift economy:* Joel Garreau, "Bearing Gifts, They Come from Afar: The Internet Offers One More Way to Connect with the Season's Spirit," *Washington Post,* December 21, 2000, page C1. http://www.washingtonpost.com/ac2/wp-dyn?pagename=article&node=&contentId=A33711-2000Dec20&notFound=true

254 *says Lewis Hyde:* Lewis Hyde, *The Gift: Imagination and the Erotic Life of Property* (New York: Vintage, 1983). ISBN: 0394715195.

254 *"the selfless offering of value"*: Jim Mason, personal communication, December 11, 2000.

254 *"It's work-as-gift rather than work-as-commodity"*: Richard Barbrook, personal communication, December 9, 2000.

254 *"Cast thy bread upon the waters"*: Ecclesiastes 11:1.

254 *five loaves and two fishes:* Matthew 14:13–21.

254 *"E-mail is the oddest thing"*: Hyde interview, December 6, 2000.

256 *"The Internet has accelerated"*: Joel Garreau, "Flocking Together Through the Web: Bird Watchers May Be a Harbinger of a True Global Consciousness," *Washington Post,* May 9, 2001, page C1. http://www.washingtonpost.com/ac2/wp-dyn?pagename=article&contentId=A1132-2001May8&notFound=true

256 *evolution is aimed in a positive direction:* Robert Wright, *Nonzero: The Logic of Human Destiny* (New York: Pantheon Books, 2000). ISBN: 0679442529. See also Robert Wright, *The Moral Animal: Evolutionary Psychology and Everyday Life* (Mangolia, MA: Peter Smith Publishers, 1997). ISBN: 084466927X, which deals more directly with Teilhard de Chardin.

257 *"The ability to learn faster than your competition"*: World of Quotes.com. http://www.worldofquotes.com/author/Arie-de-Geus/1/

257 *Until you can come up with a new grand story:* Kash interview, April 22, 2004.

257 *but the American people overwhelmingly are:* An October 2003 NBC–*Wall Street Journal* poll conducted by the Gallup Organization with a margin of error of 3 percent found 90 percent of all Americans wanting to keep the words "In God we trust" on U.S. money and 78 percent favoring prayers in public schools, no matter what the courts rule.

257 *Among the most eminent:* Karen Armstrong, *Muhammad: A Biography of the Prophet* (San Francisco: Harper SanFrancisco, 1993). ISBN: 0062508865; *Buddha* (New York: Viking Press, 2001). ISBN: 0670891932; *A History of God: The 4,000-Year Quest of Judaism, Christianity and Islam* (New York: Alfred A. Knopf, 1993). ISBN: 0-345-38456-3.

257 *In 1969, at the age of 24:* Dave Welch, "Karen Armstrong, Turn, Turn, Turn," Powells.com, March 20, 2004. http://www.powells.com/authors/armstrong.html

258 *designer Missoni scarves:* I am indebted to Sheila O'Shea of Knopf for this observation.

258 *the Axial Age:* "Karl Jaspers," Mythos and Logos. http://www.mythosandlogos.com/Jaspers.html

258 *"The search for spiritual breakthrough"*: Harvey Blume, "Divine Reticence: A Conversation with Karen Armstrong, Biographer of the Enlightened One," *Atlantic Unbound,* March 21, 2001. http://www.theatlantic.com/unbound/interviews/int2001-03-21.htm

258 *"search for ultimate meaning"*: Welch, "Karen Armstrong, Turn, Turn, Turn."

258 *"If there is an axis in history"*: Karl Jaspers, *Way to Wisdom: An Introduction to Philosophy,* trans. Ralph Manheim (New Haven: Yale University Press, 1954). ASIN: B00005X50L, pp. 99ff.

259 *an essential human need:* Blume, "Divine Reticence."

259 *"Human beings cannot endure emptiness and desolation":* Armstrong, *A History of God*, page 399.

259 *editor-in-chief of* Skeptic: Michael Shermer, *Why People Believe Weird Things: Pseudoscience, Superstition, and Other Confusions of Our Time* (New York: Owl Books, 2002). ISBN: 0805070893.

259 *rationality is hardly a secret:* Shermer interview, July 5, 2001. See also Joel Garreau, "Science's Mything Links: As the Boundaries of Reality Expand, Our Thinking Seems to Be Going over the Edge," *Washington Post*, July 23, 2001, page C1. http://www.washingtonpost.com/ac2/wp-dyn?pagename=article&contentId=A35319-2001Jul22&notFound=true

260 *"That's the next Enlightenment":* Flowers interview, June 29, 2001. See also Garreau, "Science's Mything Links."

260 *an increase in happiness:* I find it interesting and useful that happiness is a metric all sides of this debate seem to agree on. Leon Kass' work, for example, is called "Beyond Therapy: Biotechnology and the Pursuit of Happiness."

260 *Seligman is president:* Martin E. P. Seligman, *Authentic Happiness: Using the New Positive Psychology to Realize Your Potential for Lasting Fulfillment* (New York: Free Press, 2004). ISBN: 0743222989.

261 *"That's the Hollywood view":* John Brockman, "Eudaemonia, the Good Life: A Talk with Martin Seligman," Edge: The Third Culture, 2004. http://www.edge.org/3rd_culture/seligman04/seligman_index.html

261 *"the exercise of vital powers":* As quoted, for example, by Timothy J. Sullivan, president of the College of William and Mary, May 16, 1999. http://web.wm.edu/president/addresses/commencement_1999.php

262 *"Religion isn't about believing things":* Welch, "Karen Armstrong, Turn, Turn, Turn."

262 *"Without more kindliness":* Bertrand Russell, *Icarus; or, the Future of Science* (London: Kegan Paul & Co., 1924). Cited in Bonnie Kaplan and Nick Bostrom, "A Somewhat Whiggish and Spotty Historical Background," The Ethics, Technology and Utopian Visions Working Group, Yale University, 2002. http://www.transhumanism.org/resources/Syllabi/YaleHistory.htm

262 *"evolution moves inexorably toward our conception of God":* Ray Kurzweil, "As Machines Become More Like People, Will People Become More Like God? Thoughts on Where Technology Is Taking Us," *Talk*, April 2001, page 153.

262 *"Someday after mastering winds":* Pierre Teilhard de Chardin, *The Phenomenon of Man* (New York: Harper & Row, 1961). ISBN: 0-06-090495-X.

263 *"Dignity is something people have to create":* Lanier interview, May 9–11, 2003.

264 *"To be in Hell is to drift; to be in Heaven is to steer":* George Bernard Shaw, *Man and Superman: A Comedy and a Philosophy*, Act Three (New York: Penguin USA, 2001). ISBN: 0140437886. Extracts courtesy of the Concept Exchange Society. http://www.sonic.net/chesters/marvin/CES/september97/abstract.html

264 *our first primitive enhancements:* From 2000 to 2003, for example, in the United States, the number of tummy tuck procedures increased by 61 percent, buttock lifts by 78 percent, and Botox injections by 267 percent. The number of all cosmetic and reconstructive procedures increased by 17 percent from 2002 to 2003. Source: American Society of Plastic Surgeons, National Plastic Surgery Statistics, Cosmetic and Reconstructive Procedure Trends 2000/2001/2002/2003. http://www.plasticsurgery.org/public_education/loader.cfm?url=/commonspot/security/getfile.cfm&PageID=12552

264 *there long will have been several means:* I am indebted to Roger Brent of the Molecular Sciences Institute, Berkeley, CA, for the specifics of this scenario.

265 *"transcendence as the only real alternative to extinction":* Václav Havel, "The Need for Transcendence in the Post Modern World," delivered at Independence Hall, Philadelphia, July 4, 1994. http://ww.worldtrans.org/whole/havelspeech.html

Chapter Eight EPILOGUE

267   *"Que sera, sera. Whatever will be, will be":* "Que Sera Sera," written by Jay Livingston and
Ray Evans for Alfred Hitchcock's 1956 remake of his 1934 film called, remarkably
enough, *The Man Who Knew Too Much*, starring Doris Day and James Stewart. http://ntl.
matrix.com.br/pfilho/html/lyrics/w/whatever_will_be_will_be_que_sera_sera.txt

267   *"We can only see a short distance":* This is the last line of his legendary paper propos-
ing the Turing test to determine whether a machine has demonstrated true intelligence. Alan M. Turing, "Computing Machinery and Intelligence," *Mind* 49: (1950),
pages 433–60. http://cogprints.ecs.soton.ac.uk/archive/00000499/00/turing.html

270   *The peas of quite a few programs:* See, for example, Brett Giroir, "Beyond the
Bio-Revolution, Maintaining Soldier Performance," presentation prepared for DARPA-
Tech 2004, March 9–11, 2004, Anaheim, CA. http://www.darpa.mil/DARPAtech2004/
pdf/scripts/GiroirScript.pdf See also "Biological Sciences" under the heading "Technology Thrusts" in the DSO section of the DARPA Web site. http://www.darpa.mil/dso/
thrust/biosci/biosci.htm If the peas and the shells have moved once again by the time you
read this, go to www.darpa.mil and try putting "biology" into the search engine.

270   *"as bold or bolder":* Although all my DARPA interviews were on the record and tape-
recorded, in this rare instance I choose to avoid creating unnecessary trouble for a
trusted source by not printing the individual's name.

270   *is fueled by the plight of his daughter:* In 2004, the Food and Drug Administration
gave a Foxboro, MA, company called Cyberkinetics approval to test a version of
the telekinetic monkey technology in humans. The hope was that the five paralyzed
humans in the test group would be able, with their thoughts, to control machines
that would allow them to move their limbs. The system is called BrainGate, and
Cyberkinetics hoped that it would be commercially available in 2007 or 2008.
Andrew Pollack, "With Tiny Brain Implants, Just Thinking May Make It So,"
New York Times, April 13, 2004. http://query.nytimes.com/gst/abstract.html?res=
F00D12FD355C0C708DDDAD0894DC404482

270   *Referring to the two men:* DARPA biographies. http://www.darpa.mil

271   *" 'He'll be the watchdog' ":* Goldblatt interview, August 11, 2003.

272   *Cohen has received the National Medal of Science:* "Scientific Advisory Board: Stanley N.
Cohen, MD," Functional Genetics. http://www.functional-genetics.com/about/sci_
adv_board.htm

272   *Functional Genetics has developed:* "Science and Technology: Our Genetic Advantage," Functional Genetics. http://www.functional-genetics.com/sci_tech/genetic_
advantage.htm

273   *"That's right, it's very similar":* Goldblatt interview, August 11, 2003.

Acknowledgments

279   *Danica Remy gently but:* This was five years before *dot* as in *dot-com* was voted the most
useful new word in the American Dialect Society's "Word of the Year" election for
1996, held by the Linguistic Society of America. Gayle Worland, "Coming to Terms
with 1997: Linguists Pick the Words Minted for the Year," *Washington Post*, January
12, 1998, page B1.

Index

Connect with the Season's Spirit"
(Garreau), 348n
Bees, research with, 38–39
Beil, Laura, 317n
Being Humans (Roughley, ed.), 345n
*Being and Nothingness: An Essay on
Phenomenological Ontology* (Sartre), 310
*Being There: Putting Brain, Body and World
Together Again* (Clark), 288
Bell, Kurt R., 321n
Bellah, Robert N., 309
Bellamy, Edward, 315
Bellis, Mary, 347n
Bell Labs, 40, 111–12
Beniger, James, 283
Berners-Lee, Tim, 255, 290
Berry, Wendell, 301
*Better Than Well: American Medicine Meets the
American Dream* (Elliott), 286, 345n
"Beyond the Bio-Revolution, Maintaining
Soldier Performance" (Giroir), 317n,
350n
"Beyond Computation: A Talk with
Rodney Brooks" (Brockman), 330n
Beyond Freedom and Dignity (Skinner),
153
*Beyond Genetics: Putting the Power of DNA to
Work in Your Life* (McGee), 287
*Beyond Therapy: Biotechnology and the Pursuit
of Happiness* (Kass), 234, 245, 302–3,
345n, 346n, 349n
Bhattacharya, Shaoni, 328n
Bible
end-of-the-world predictions and
societal collapse in, 149, 333n
gift economy in, 254
tower of Babel and fears of technological
advance, 149–50, 333n
See also by book, e.g. Genesis
Bidney, David, 307
Bielitzki, Joe, 32–34, 43, 320n, 321n
Billy the Kid (William Henry McCarty),
190, 340n
Web sites, 340n
Bioethics: An Anthology (Singer and Kuhse,
eds.), 306
"Biography: Francis Fukuyama," 335n
"Biography of Thomas Alva Edison"
(Beals), 330n
Biological Input/Output Systems program,
40

Biology, human, and biotechnology
anti-aging products, 11, 244, 264, 349n
bio-engineered children (better
children), 234–35
bio-engineered disease agents, 100
body part regeneration, 28, 39, 42, 43,
321n
brain damage, repairing, 29
brain damage, replacement, 38
cloning, 12, 21, 171, 174, 318n, 337n
cosmetic surgery, 264, 349n
DARPA interest in, 24, 26–43
embedded technology and, 8, 101–2
fat-fighting, 11, 12, 34, 128
fertility technology and delayed
motherhood, 247–48, 346n
in fiction, 109, 151–52
food, elimination of need, 32–34
gene doping, 5, 21, 287, 316n
"genetic personality," 31–32
in Heaven scenario, 98–99, 100, 127–29,
229–30
in Hell scenario, 155–64, 229–30, 235–36
human genome cracking, 21, 53, 100
IBM's Blue Gene and protein fold, 53
James Watson on "curing stupidity," 117
mental abilities, enhanced (e.g. child's
SAT scores), 7, 8, 11, 29, 38, 71, 73,
264–65, 318n, 349n
metabolism, changing, 32–33
perceptions, enhanced, 71, 101
physique and muscularity, enhanced, 5,
7, 56–57, 317n, 323n
reptilian brain, reaction of, 62
sleep elimination, 8, 28–29, 56, 317n,
323n
stem cell research, 95, 167, 302, 337n
U.S. government projection of impact on
society, 113
See also Genetic technology
*Biology of Belief, The: How Our Biology Biases
Our Beliefs and Perceptions* (Giovannoli
et al), 310
Bionic Man, The (TV series), 110
*Biotech Century, The: Harnessing the Gene and
Remaking the World* (Rifkin), 303
"Biotech Loses Its Innocence" (Cohen),
333n
BlackBerry, 64
Blade Runner (film), 66, 110, 168, 169,
335n

About the Author

JOEL GARREAU is a student of culture, values, and change. The author of the bestselling *Edge City: Life on the New Frontier* and *The Nine Nations of North America*, he is a reporter and editor at the *Washington Post*, as well as a member of the scenario-planning organization Global Business Network, and has served as a senior fellow at George Mason University and the University of California at Berkeley. He has appeared on such national media as *Good Morning America, Today, CBS Evening News, NBC Nightly News, ABC's World News Tonight*, and NPR's *Morning Edition* and *All Things Considered*. He lives in Broad Run, Virginia. Visit the author's website at www.garreau.com.